Fluorescence Microscopy and Fluorescent Probes

Fluorescence Microscopy and Fluorescent Probes

Edited by

Jan Slavík
Czech Academy of Sciences
Prague, Czech Republic

Springer Science+Business Media, LLC

Library of Congress Cataloging-in-Publication Data

Fluorescence microscopy and fluorescent probes / edited by Jan Slavík
p cm
"Based on the proceedings of the conference on Fluorescence Microscopy and Fluorescent Probes, held June 25-28, 1995, in Prague, Czech Republic"--T p verso.
Includes bibliographical references and index

1 Fluorescence microscopy--Congresses 2 Fluorescent probes--Congresses. I. Slavík, Jan, Ph D
QH212 F55F55 1996
578--dc20 96-43731
CIP

Based on the proceedings of the conference on Fluorescence Microscopy and Fluorescent Probes, held June 25–28, 1995, in Prague, Czech Republic

ISBN 978-1-4899-1868-0 ISBN 978-1-4899-1866-6 (eBook)
DOI 10.1007/978-1-4899-1866-6

Originally published by Plenum Press, New York in 1996
Softcover reprint of the hardcover 1st edition 1996

10 9 8 7 6 5 4 3 2 1

CONTRIBUTORS

H. Acker
Max-Planck-Institute for Molecular Physiology
Dortmund, Germany

Jolana T.P. Albrechtová
Institute of Biology II, Botany
Albert-Ludwigs-University
Freiburg im Breisgau, Germany

Marcel Ameloot
Limburgs Universitair Centrum
Universitaire Campus
Diepenbeek, Belgium

Robert E. Anderson
Neurosurgical Research Laboratory
Mayo Clinic
Rochester, MN, USA

K. Autio
VTT Biotechnology and Food Research
Espoo, Finland

Vladimir Baltic
Institute of Oncology
Department of Experimental Oncology
Sremska Kamenica, Yugoslavia

M. Bartunek
Department of Genetics and Microbiology
Faculty of Natural Sciences
Charles University
Prague, Czech Republic

G. Bellomo
Department of Medical Sciences
University of Torino
Novara, Italy

Mehdi Benchaib
Cytologie Analytique
University of Claude Berbnard
Lyon, France

Rolf Bjerkvig
University of Bergen
Department of Anatomy and Cell Biology
Bergen, Norway

Joachim Bradl
Institute of Applied Physics
University of Heidelberg
Heidelberg, Germany

Jana Brezinová
Institute of Hematology and Blood Transfusion
Prague, Czech Republic

Odd Terje Brustugun
University of Bergen
Department of Anatomy and Cell Biology
Bergen, Norway

Paul-Andre Bryon
Cytologie Analytique
University of Claude Bernard
Lyon, France

Petr Cimprich
Institute of Physiology
Czech Academy of Sciences
Prague, Czech Republic

Robert M. Clegg
Max Planck Institute for Biophysical Chemistry
Department of Molecular Biology
Göttingen, Germany

Christoph Cremer
Institute of Applied Physics
University of Heidelberg
Heidelberg, Germany

Ben Crystall
Department of Chemistry
Imperial College
London SW7 2AY, UK

Jean Davoust
Centre d'Immunologie CNRS-INSERM de Marseille-Luminy
Marseille, France

Richard Delorme
Cytologie Analytique
University of Claude Bernard
Lyon, France

Denis Demandolx
Centre d'Immunologie CNRS-INSERM de Marseille-Luminy
Marseille, France

Barbora Denksteinová
Institute of Physics
Charles University
Prague, Czech Republic

Tom Deutsch
Wellman Laboratories of Photomedicine
Department of Dermatology
Harvard Medical School
Massachusetts General Hospital
Boston, MA 02114, USA

Stein Ove Doskeland
Department of Anatomy and Cell Biology
University of Bergen
Bergen, Norway

Ralph E. Durand
B C Cancer Research Centre
Vancouver, B.C., Canada

Ian Durrant
Research and Development
Amersham International
Amersham, UK

Markus Durm
Institute of Applied Physics
University of Heidelberg
Heidelberg, Germany

Ilya V Eigenbrot
Department of Chemistry
Imperial College
London SW7 2AY, UK

R.J. Errington
Oxford University
Physiology Department
Oxford OX1 3PT, UK

M.D Fricker
Oxford University
Plant Sciences Department
Oxford OX1 3RB, UK

Dana Gášková
Institute of Physics
Charles University
Prague, Czech Republic

Hans C. Gerritsen
Debye Institute
Department of Molecular Biophysics
University of Utrecht
Byus Ballot Laboratory
Utrecht, The Netherlands

Bjorn Tore Gjertsen
Department of Anatomy and Cell Biology,
University of Bergen
Bergen, Norway

Eva Grapengiesser
Department of Medical Cell Biology
Uppsala University
Biomedicum
Uppsala, Sweden

Martin Gregor
Department of Physiology
Faculty of Science
Charles University
Prague, Czech Republic

Michael H. Gschwend
Institut für Lasertechnologien in der Medizin und Meßtechnik
Universität Ulm, Ulm, Germany

Leslie Gubba
Amersham Life Science Inc.
Pittsburgh, PA, USA

Lars Guldfeldt
Department of Dairy and Food Science
Food Microbiology,
The Royal Veterinary and Agricultural University
Frederiksberg C, Denmark

Heinz Gundlach
Carl Zeiss Jena GmbH
Division of Microscopy
Jena, Germany

Petr Hach
Institute of Histology and Embryology
1st Medical Faculty
Charles University
Prague, Czech Republic

Tayyaba Hasan
Wellman Laboratories of Photomedicine
Department of Dermatology
Harvard Medical School
Massachusetts General Hospital
Boston, MA 02114, USA

Michael Hausmann
Institute of Applied Physics
University of Heidelberg
Heidelberg, Germany

Brian Herman
Laboratories for Cell Biology
Department of Cell Biology and Anatomy
University of North Carolina at Chapel Hill
Chapel Hill, NC 27599, USA

Petr Herman
Institute of Physics
Charles University
Prague, Czech Republic

Martin Hof
Department of Physical Chemistry
Faculty of Natural Sciences
Charles University
Prague, Czech Republic

Zsolt Holló
National Institute of Haematology, Blood Transfusion and Immunology
Budapest, Hungary

Aleš Holoubek
Institute of Physics
Charles University
Prague, Czech Republic

László Homolya
National Institute of Haematology, Blood Transfusion and Immunology
Budapest, Hungary

Jan Jaap ter Horst
Limburgs Universitair Centrum
Universitaire Campus
Diepenbeek, Belgium

Rudi Hutterer
Department of Physical Chemistry
University of Würzburg
Wurzburg, Germany

Seiichi Iinuma
Wellman Laboratories of Photomedicine
Department of Dermatology
Harvard Medical School
Massachusetts General Hospital
Boston, MA 02114, USA

Elzbieta Interewicz
Department of Physiology
Faculty of Pharmacy, Medical Academy
Warsaw, Poland

Mogens Jakobsen
Department of Dairy and Food Science
Food Microbiology
The Royal Veterinary and Agricultural University
Frederiksberg C, Denmark

Blanka Janderová
Department of Genetics and Microbiology
Faculty of Natural Sciences
Charles University
Prague, Czech Republic

Bruce G. Jenks
Department of Animal Physiology
Nijmegen Institute for Neurosciences
Nijmegen, The Netherlands

Marie Jirkovská
Institute of Histology and Embryology
1st Medical Faculty
Charles University
Prague, Czech Republic

Petr Karen
Institute of Physiology
Czech Academy of Sciences
Prague, Czech Republic

Rolf K.H. Kinne
Max-Planck-Institute for Molecular Physiology
Dortmund, Germany

Ralf Kinscherf
Department of Anatomy and Cell Biology III
University of Heidelberg
Heidelberg, Germany

Christoph Köhler
Department of Anatomy and Cell Biology III
University of Heidelberg
Heidelberg, Germany

Karsten Kònig
Institut für Molekulare Biotechnologie
Jena, Germany

Werner J.H Koopman
Department of Animal Physiology
Nijmegen Institute for Neurosciences
Nijmegen, The Netherlands

Maarit Koskinen
VTT Biotechnology and Food Research
Espoo, Finland

Ivan Krekule
Institute of Physiology
Czech Academy of Sciences
Prague, Czech Republic

Lucie Kubínová
Institute of Physiology
Czech Academy of Sciences
Prague, Czech Republic

Margaret E Langmuir
Covalent Associates, Inc
Woburn, MA 01801, USA

Lotte Lammert
Department of Dairy and Food Science
The Royal Veterinary and Agricultural University
Frederiksberg C, Denmark

Karen A. LeCompte
Covalent Associates, Inc
Woburn, MA 01801, USA

Horst Ludwig
Institute of Physical Chemistry
University of Heidelberg
Heidelberg, Germany

Jan Malínský
Institute of Physics
Charles University
Prague, Czech Republic

Wolf Malkusch
Image Analysis Division
Kontron Elektronik GmbH
Eching, Germany

Vera Maravic-Stojkovic
Institute of Oncology
Department of Experimental Oncology
Sremska Kamenica, Yugoslavia

E.K. Matthews
Department of Pharmacology
University of Cambridge
Cambridge, CB2 1QJ, UK

Gunnar Mellgren
Department of Anatomy and Cell Biology
University of Bergen
Bergen, Norway

E. Merten
Max-Planck-Institute for Molecular Physiology
Dortmund, Germany

Jürgen Metz
Department of Anatomy and Cell Biology
University of Heidelberg
Heidelberg, Germany

Fredric B. Meyer
Neurosurgical Research Laboratory
Mayo Clinic
Rochester, MN, USA

Kyra Michalová
Third Medical Department
General Hospital, Charles University
Prague, Czech Republic

Vera Michalová
Third Medical Department
General Hospital, Charles University
Prague, Czech Republic

Marianna Müller
National Institute of Haematology, Blood Transfusion and Immunology
Budapest, Hungary

José-Enrique O'Connor
Departament de Bioquimica i Biologia Molecular
Facultad de Medicina
Universitat de Valencia
Valencia, Spain

Norbert Opitz
Max-Planck-Institute for Molecular Physiology
Dortmund, Germany

Izabela Orlanska
Department of Physical Chemistry and Department of Physiology
Faculty of Pharmacy, Medical Academy
Warsaw, Poland

Giuseppina Palladini
Department of Internal Medicine
IRCCS, Policlinico S.Matteo
Pavia, Italy

Daniel Palouš
Institute of Histology and Embryology
1st Medical Faculty, Charles University
Prague, Czech Republic

T Parkkonen
VTT Biotechnology and Food Research
Espoo, Finland

Peter Pavlov
Department of Anatomy and Cell Biology III
University of Heidelberg
Heidelberg, Germany

D. Phillips
Department of Chemistry
Imperial College
London SW7 2AY, UK

Jaromír Plášek
Institute of Physics
Charles University
Prague, Czech Republic

Torsten Porwol
Max-Planck-Institute for Molecular Physiology
Dortmund, Germany

Jacek Przybylski
Department of Physical Chemistry and Department of Physiology
Faculty of Pharmacy, Medical Academy
Warsaw, Poland

D. Rickwood
Biology Department
University of Essex
Colchester CO4 3SQ, UK

Bernd Rinke
Institute of Applied Physics
University of Heidelberg
Heidelberg, Germany

Eric W Roubos
Department of Animal Physiology
Nijmegen Institute for Neurosciences
Nijmegen, The Netherlands

Carlos P. Rubbi
Biology Department, University of Essex
Colchester CO4 3SQ, UK

Reinhard Sailer
Institut für Lasertechnologien in der Medizin und Meßtechnik
Universität Ulm, Ulm, Germany

Balázs Sarkadi
National Institute of Haematology, Blood Transfusion and Immunology
Budapest, Hungary

Wim J J.M Scheenen
Department of Biomedical Sciences
University of Padova
Padova, Italy

Herbert Schneckenburger
Institut für Lasertechnologien in der Medizin und Meßtechnik
Universität Ulm, Ulm, Germany

Bernhard Schneider
Institute of Applied Physics
University of Heidelberg
Heidelberg, Germany

Peter C Schneider
Max Planck Institute for Biophysical Chemistry
Department of Molecular Biology
Göttingen, Germany

Kevin T Schomacker
Wellman Laboratories of Photomedicine
Department of Dermatology
Harvard Medical School
Massachusetts General Hospital
Boston, MA 02114, USA

Henrik Siegumfeldt
Department of Dairy and Food Science
Food Microbiology
The Royal Veterinary and Agricultural University
Frederiksberg C, Denmark

Karel Sigler
Institute of Microbiology
Czech Academy of Sciences
Prague, Czech Republic

Jan Slavík
Institute of Physiology
Czech Academy of Sciences
Prague, Czech Republic

Catherine Souchier
Cytologie Analytique
University of Claude Bernard
Lyon, France

Paul Steels
Limburgs Universitair Centrum
Universitaire Campus
Diepenbeek, Belgium

Wolfgang S L. Strauß
Fachhochschule Aalen
Optoelektronik
Aalen, Germany

T M. Sundt Jr.
Neurosurgical Research Laboratory
Mayo Clinic
Rochester, MN, USA

Jan Tachezy
Department of Parasitology
Faculty of Science
Charles University
Prague, Czech Republic

Anders Tengholm
Department of Medical Cell Biology
Uppsala University
Biomedicum
Uppsala, Sweden

B.R. Terry
BioImage:S
Novo Nordisk Research Satellite
Soborg, Denmark

Hanna Tinel
Max-Planck-Institute for Molecular Physiology
Dortmund, Germany

Claus Usinger
Department of Anatomy and Cell Biology III
University of Heidelberg
Heidelberg, Germany

L. Valášek
Department of Genetics and Microbiology
Faculty of Natural Sciences
Charles University
Prague, Czech Republic

Jaroslav Vecer
Institute of Physics
Charles University
Prague, Czech Republic

Vladimir Vondrejs
Department of Genetics and Microbiology
Faculty of Natural Sciences
Charles University
Prague, Czech Republic

Edgar Wagner
Institute of Biology II, Botany
Albert-Ludwigs-University
Freiburg im Breisgau, Germany

Georges A. Wagnieres
Institute of Environmental Engineering
Swiss Federal Institute of Technology
Lausanne, Switzerland

Frank Wehner
Max-Planck-Institute for Molecular Physiology
Dortmund, Germany

Nick S White
Oxford University
Plant Sciences Department
Oxford OX1 3RB, UK

Jacek Wierzchowski
Department of Physical Chemistry and Department of Physiology
Faculty of Pharmacy, Medical Academy
Warsaw, Poland

Michael H.F. Wilkinson
Department of Medical Microbiology
University of Groningen
Groningen, The Netherlands

J. L. Wood
Oxford University
Plant Sciences Department
Oxford OX1 3RB, UK

Piotr Wroczynski
Department of Physical Chemistry and Department of Physiology
Faculty of Pharmacy, Medical Academy
Warsaw, Poland

Jun-Rui Yang
Covalent Associates, Inc
Woburn, MA 01801, USA

Zuzana Zemanová
Third Medical Department
General Faculty Hospital
Charles University
Prague, Czech Republic

PREFACE

Fluorescence microscopy images can be easily integrated into current video and computer image processing systems. People like visual observation; they like to watch a television or computer screen, and fluorescence techniques are thus becoming more and more popular. Since true *in vivo* experiments are simple to perform, samples can be directly seen and there is always the possibility of manipulating the samples during the experiments; it is an ideal technique for biology and medicine.

Images are obtained by a classical (now called wide-field) fluorescence microscope, a confocal scanning microscope, upright or inverted, with epifluorescence or transmission. Computerized image processing may improve definition, and remove glare and scattered light signal. It also makes it possible to compute ratio images (ratio imaging both in excitation and in emission) or lifetime imaging. Image analysis programs may supply a great deal of additional data of various types, starting with calculations of the number of fluorescent objects, their shapes, brightness, etc. Fluorescence microscopy data may be complemented by classical measurement in the cuvette or by flow cytometry.

Fluorescent probes are dyes that are environmentally sensitive. They can be employed as tiny molecular reporters that, encoded in the fluorescence signal, pass on detailed information about their molecular neighborhood. Their location can be determined from fluorescence microscopy images. From measurement of fluorescence intensity, quantum yield, emission spectrum, excitation spectrum, polarization (anisotropy), lifetime, time-resolved fluorescence intensity, and time-resolved fluorescence polarization, one can obtain data on environmental polarity, membrane or cytosol fluidity, membrane potential, and intracellular and intraorganellar ionic composition (pH, calcium, magnesium, sodium, potassium, chloride, iron, zinc, and heavy metals). Furthermore, fluorescent molecules can be used as specific labels to mark any type of cell structure *in vivo*. There are various fluorescent or fluorogenic enzyme substrates and fluorescent analogues (lipids, nucleosides, lectins). The catalogues list literally hundreds of fluorescent dyes; they are quite alluring.

The immense potential inherent in fluorescent techniques may be illustrated by the example of intracellular ion measurement. In the past, each measurement gave only one single value of ion concentration, while ion-sensitive fluorescent dyes yield maps of intracellular ion concentrations with spatial resolution of 200 nm, millisecond temporal resolution, possibility of 3-D reconstruction, and ion concentration changes as small as 0.01 pH or 20 nM [Ca^{2+}]. The enormous amount of new data yielded some surprising results, e.g., calcium waves moving across the cell cytoplasm, calcium spikes in individual cells, and various pH and calcium heterogeneities.

The progress in fluorescence instrumentation offers further improvements. Sensitivity can be boosted with single-photon correlation spectroscopy to the observation of single molecules. The definition limit of 150 to 200 nm can be improved by video-enhanced contrast to 20 nm; fluorescent antibody markers can be seen even at a size of 5 to 10 nm.

Recently developed scanning near-field optical microscopy allows a 50 nm definition for flat biological objects.

The future of fluorescent probes will be probably directed to a simultaneous application of several dyes. Using selective excitation, selective detection, delayed fluorescence dyes, and lifetime imaging, it is easy to monitor several physiological parameters (membrane potential, pH, calcium) simultaneously. I believe that confocal microscopy supplemented with lifetime imaging will become common during the next few years.

The following volume comprises plenary lectures and papers issuing from the conference "FLUORESCENCE MICROSCOPY AND FLUORESCENT PROBES," which took place in Prague from June 25th to 28th, 1995. The selection of well-known scientists for plenary lectures resulted in a surprisingly large number of participants, some of whom learned about the conference at the last moment. The conference presented 90 posters by more than 180 participants from 20 countries. There was an exhibition of fluorescence microscopes supplemented by technical lectures by all leading manufacturers. Plenary lectures, poster sessions, exhibitions, and technical lectures together with a rich social program, created a remarkably exciting atmosphere. This book tries to recall the memories of the conference and to invite the readers to the next Prague conference in April 1997.

Jan Slavík

CONTENTS

FLUORESCENCE MICROSCOPY AND FLUORESCENT PROBES

Fluorescence Microscopy: State of the Art 1
B. Herman

Fluorescence Lifetime-Resolved Imaging Microscopy: A General Description of Lifetime-Resolved Imaging Measurements 15
R. M. Clegg and P. C. Schneider

Confocal Fluorescence Lifetime Imaging 35
H. C. Gerritsen

Multidimensional Fluorescence Microscopy: Optical Distortions in Quantitative Imaging of Biological Specimens 47
N. S. White, R. J. Errington, M. D. Fricker, and J. L. Wood

Fluorescent Probes 57
J. Slavík

Flow Cytometry *versus* Fluorescence Microscopy 61
J. -E. O'Connor

Multichannel Fluorescence Microscopy and Digital Imaging: On the Exciting Developments in Fluorescence Microscopy 67
H. Gundlach

Fluorescence Lifetime Imaging and Spectroscopy in Photobiology and Photomedicine. 71
H. Schneckenburger, M. H. Gschwend, K. König, R. Sailer, and W. S. L. Strauß

A Versatile Time-Resolved Laser Scanning Confocal Microscope 79
I. V. Eigenbrot, B. Crystall, and D. Phillips

ION-SENSITIVE FLUORESCENT PROBES

Disappearance of Cytoplasmic Ca^{2+} Oscillations Is a Sensitive Indicator of Photodamage in Pancreatic β-Cells 85
A. Tengholm and E. Grapengiesser

Distribution of Individual Cytoplasmic pH Values in a Cell Suspension 91
P. Cimprich and J. Slavík

The Effect of Lysosomal pH on Lactoferrin-Dependent Iron Uptake in *Tritrichomonas foetus* 95
M. Gregor, J. Tachezy, and J. Slavík

On the Protein-Error of the Calcium-Sensitive Fluorescent Indicator Fura-Red 101
N. Opitz, T. Porwol, E. Merten, and H. Acker

Cytoplasmic Ion Imaging: Evidence for Intracellular Calibration Heterogeneities of Ion-Sensitive Fluoroprobes 107
N. Opitz, T. Porwol, E. Merten, and H. Acker

The Effect of Protein Binding on the Calibration Curve of the pH Indicator BCECF .. 113
J. Plášek, J. Jaap ter Horst, M. Ameloot, and P. Steels

Artifacts in Fluorescence Ratio Imaging 119
P. Cimprich and J. Slavík

Use of Fluorescent Probes and CLSM for pH-Monitoring in the Whole Plant Tissue: pH Changes in the Shoot Apex of *Chenopodium rubrum* Related to Organogenesis 125
J. T. P. Albrechtová, J. Slavík, and E. Wagner

Spatial Resolution of Cortical Cerebral Blood Flow and Brain Intracellular pH as Measured by *in Vivo* Fluorescence Imaging 133
R. E. Anderson, F. B. Meyer, and T. M. Sundt, Jr.

MEMBRANE POTENTIAL-SENSITIVE FLUORESCENT PROBES

Is a Potential-Sensitive Probe diS-c_3(3) a Nernstian Dye?: Time-Resolved Fluorescence Study with Liposomes as a Model System 139
P. Heřman, J. Večeř, and A. Holoubek

Kinetic Behavior of Potential-Sensitive Fluorescent Redistribution Probes: Modelling of the Time Course of Cell Staining 145
J. Večeř and P. Heřman

Speed of Accumulation of the Membrane Potential Indicator dis-c3(3) in Yeast Cells 151
B. Denksteinová, D. Gášková, P. Heřman, J. Večeř, K Sigler, J. Plášek, and J. Malinský

Spectral Effects of Slow Dye Binding to Cells and Their Role in Membrane Potential Measurements 157
J. Plášek and K. Sigler

Exploitation of Rhodamine B in the Killer Toxin Research 163
L. Valášek, V. Vondrejs, M. Bartúňek, and B. Janderová

FLUORESCENT PROBES FOR NUCLEIC ACIDS

"In Situ" Estimates of the Spatial Resolution for "Practical" Fluorescence Microscopy of Cell Nuclei 169
B. Rinke, J. Bradl, B. Schneider, M. Durm, M. Hausmann, H. Ludwig, and C. Cremer

Requirements for a Computer-Based System for FISH Applications 175
W. Malkusch

Fluorescent Dyes and Dye Labelled Probes for Detection of Nucleic Acid Sequences in Biological Material 179
I. Durrant and L. Gubba

Fluorescence in Situ Hybridization (FISH) in Cytogenetics of Leukemia 185
K. Michalová, Z. Zemanová, J. Březinová, and V. Michalová

Estimation of "Start" in *Saccharomyces cerevisiae* by Flow Cytometry and Fluorescent Staining of DNA and Cell Protein 191
L. Guldfeldt, H. Siegumfeldt, L. Lammert, and M. Jakobsen

Fluorescence Image Cytometry of DNA Content: A Comparative Study of Three Fluorochromes and Four Fixation Protocols 197
M. Benchaib, R. Delorme, P. -A. Bryon, and C. Souchier

FLUORESCENT LABELS, FLUORESCENT AND FLUOROGENIC SUBSTRATES

In Vivo Tissue Characterization Using Environmentally Sensitive Fluorochromes 203
G. A. Wagnières, S. Iinuma, K. T. Schomacker, T. Deutsch, and T. Hasan

Sensitive and Rapid Detection of ß-Galactosidase Expression in Intact Cells by Microinjection of Fluorescent Substrate 211
O. T. Brustugun, G. Mellgren, B. T. Gjertsen, R. Bjerkvig, and S. O. Døskeland

Fluorogenic Substrates Reveal Genetic Differences in Aldehyde-Oxidating Enzyme Patterns in Rat Tissues 217
J. Wierzchowski, P. Wroczynski, E. Interewicz, I. Orlanska, and J. Przybylski

Binding of Prothrombin Fragment 1 to Phosphatidylserine Containing Vesicles: A Solvent Relaxation Study 223
R. Hutterer and M. Hof

New Thiol Active Fluorophores for Intracellular Thiols and Glutathione Measurement 229
M. E. Langmuir, J. -R. Yang, K. A. LeCompte, and R. E. Durand

Quantification of Macrophages in the Cardiovascular System of Hypercholesterolemic Rabbits by Use of Digital Image Processing 235
J. Metz, P. Pavlov, R. Kinscherf, C. Köhler, and C. Usinger

Fluorescence Assay for Studying P-Glycoprotein Function at Single Cell Level 241
L. Homolya, M. Müller, Z. Holló, and B. Sarkadi

Alterations of Vimentin-Nucleus Interactions as an Early Phase in Cholesterol Oxide-Induced Endothelial Cell Damage 247
G. Palladini and G. Bellomo

Fluorescence Microscopy of Rye Cell Walls from Kernels to Incubated Doughs 253
M. Koskinen, T. Parkkonen, and K. Autio

Practical Approach for Immunohistochemical Staining of Muscle Biopsies 257
V. Maravić-Stojković and V. Baltić

DIGITAL IMAGE ANALYSIS

Rapid Automatic Segmentation of Fluorescent and Phase-Contrast Images of Bacteria 261
M. H. F. Wilkinson

Use of Confocal Microscopy for Absolute Measurement of Cell Volume and Total Cell Surface Area 267
C. P. Rubbi and D. Rickwood

Cell Volume Measurements Using Confocal Laser Scanning Microscopy 273
H. Tinel, F. Wehner, and R. K. H. Kinne

Subcellular Cytofluorometry in Confocal Microscopy 279
D. Demandolx and J. Davoust

Application of Confocal Microscopy to 3-D Reconstruction and Morphometrical Analysis of Capillaries 285
L. Kubínová, M. Jirkovská, P. Hach, D. Palouš, P. Karen, and I. Krekule

Retrieving Spatio Temporal Information from Confocal Data: A Study Using Melanotrope Cells of *Xenopus laevis* 291
W. J. H. Koopman, B. G. Jenks, E. W. Roubos, and W. J. J. M. Scheenen

Dynamics of Actin Measured by Fluorescence Correlation Microscopy (FCM) 297
B. R. Terry and E. K. Matthews

Index 303

FLUORESCENCE MICROSCOPY: STATE OF THE ART

Brian Herman

Laboratories for Cell Biology
Department of Cell Biology & Anatomy
CB #7090, 232 Taylor Hall
University of North Carolina at Chapel Hill
Chapel Hill, NC 27599

INTRODUCTION

Fluorescence microscopy allows the dynamic acquisition of information about the spectroscopic properties of fluorescent reporter molecules at levels of resolution to small to be seen with the naked eye. The overall goal of research and applications of fluorescence microscopy are to: 1) **observe** in greater detail, (i.e. improve spatial resolution); 2) **measure** more specifically, sensitively and accurately (i.e. improve spectroscopic information); and 3) **manipulate** more precisely (i.e. improve control) a single sample--all at the same time. Towards this end, a number of different types of fluorescence microscopes have been developed. The goal of the present communication is to review the status of the field in terms of the fundamentals of fluorescence microscopy, some of the types of fluorescence microscopes that currently exist, their applications and future developments.

FUNDAMENTALS OF FLUORESCENCE MICROSCOPY

Fluorescence

Fluorescence is the property of some atoms and molecules to absorb light of a particular wavelength and after a brief interval, termed the fluorescence lifetime, to emit light at longer wavelengths. Fluorescence is proportional to the amount of light absorbed x quantum yield. Beer's law states that for low optical densities, the amount of fluorescence is proportional to the product of the light absorbed ($I_O\varepsilon[c]x$) by the fluorophore and the quantum yield, Q:

$$F = I_O\varepsilon[c]xQ, \qquad (1)$$

where I_o is the intensity of the light beam falling on the solution, ε is the molar extinction

Fluorescence Microscopy and Fluorescent Probes
Edited by J. Slavík, Plenum Press, New York, 1996

coefficient of the fluorophore, c is the concentration of the fluorophore and x is the pathlength through the solution.

Wavelength Selection Devices

The fluorescence microscope is basically a filter fluorometer; the excitation light from a Mercury or Xenon arc lamp or a laser is isolated using a narrow band exciter filter and the emission is viewed through a cutoff or barrier filter which blocks all wavelengths below the emission band. A variety of excitation wavelength selection devices currently exist: tunable lasers, monochromators, individual vs. multiple bandpass filters, acousto-optical tunable filters (AOTF) and liquid crystal tunable filters (LCTF).

Color filtered or absorption glass (including gelatin), attenuate light solely by absorption and the amount of attenuation is dependent on the thickness of the glass. While this type of filter is long lasting and inexpensive, its usefulness is limited by high autofluorescence, low peak transmittance and poor discrimination of closely spaced wavelengths.

Thin-film coating of either a metallic nature (for fully reflective mirrors and neutral density filters) or interference coatings (for interference filters) are the most widely used type of wavelength selection devices. Advantages of thin film coating are that almost any filter type can be designed (including multiple bands filters) and that the desired wavelength, bandwidth and blocking level can be controlled with high precision. Drawbacks of this type of filter include a finite blocking wavelength range, the materials used for coating are limited in their range of transparency and these interference coatings are sensitive to the angle of incidence of the light, becoming polarizing at angles greater than 45°. Several standard exciter filter-dichromatic beam splitting mirror-barrier filter combinations are available from each microscope manufacturer. The choice of which combination to employ will depend on the fluorochrome(s) employed. Broad band filters provide more signal but less contrast, while narrow band exciters provide less signal but more contrast.

Two other newer wavelength selection devices include the acousto-optical tunable filter (AOTF) and the liquid crystal tunable filter (LCTF). In an AOTF, high frequency vibrations (50-200 MHz) from a piezo-electric device are used to generate compression waves in different types of crystal material. This leads to the generation of a diffraction grating due to the fact that the refractive index of glass increases as it is compressed. Advantages of this device include that it has no moving parts, no instability in intensity throughput, no wavelength drift, an efficiency of diffraction of the first order wave of ~85% and wavelength switching times with a temporal resolution of 2-3 msec.

LCTF's are liquid crystals whose polymer orientation is voltage sensitive. By varying the voltage in the crystal medium, one can control the degree of retardation of polarized light. Stacking such variable retarders results in an interferometer which can perform wavelength selection. LCTF's are a relatively new wavelength selection device; their advantages include the lack of mechanical parts, 10-50 msec temporal resolution and transmission of ~30-50%. However, LCTF's display polarization sensitivity, an inability to transmit light below 400 nm, and the possibility of damage at high light flux levels. Table 1 compares various properties of AOTF's and LCTF's.

Optics

By far the most commonly employed illumination system is that invented by Ploem and termed Ploem or epi--illumination. In such a setup, the exciting light is reflected on to

Table 1. Characteristics of AOTF and LCTF Tunable Filters

	AOTF	LCTF
Working range (nm)	340-4500	400-720
Switching time	< 5 msec	50 msec
Angle of acceptance	$\pm 5^{o}$	$\pm 5^{o}$
Input aperture	5 x 5 mm	18 mm
Maximum power output	$500W/cm^2$	$500mW/cm^2$
Polarization sensitivity	no	yes
Quality of diffracted beam		
out-of-band rejection	10^5x	10^4x
minimum bandwidth	30 Å	1 nm
Transmittance	80%	5-20 %

the back aperture of the objective (which acts as a condenser) by a dichromatic beam splitting mirror. Fluorescence is collected by the objective and light forming the image passes through the dichromatic beam splitting mirror to either the eyepieces or an electronic detection device. The dichromatic mirror has the property that it will reflect light of shorter wavelength (the excitation) but will transmit light of longer wavelengths (the emission).

One should employ as high transmittance objectives as possible. A higher aperture objective will increase the signal as the light concentrating power of the objective (serving as a condenser) is proportional to $(NA)^2$ and the light gathering power of the objective is also proportional to $(NA)^2$. Thus, at least up to everything but the highest NA objectives, brightness varies as $(NA)^4$. However, brightness goes as $1/(Mag)^2$- therefore brightness $\propto (NA)^4/(Mag)^2$. In addition, the more the intermediate image produced by objective is magnified by the eyepieces, the dimmer will the final image be. (The brightness of the image will be inversely proportional to the square of the eyepiece magnification). Therefore, as low a power eyepiece as is consistent with convenient observation should be employed. The wavelength of light which will be transmitted through the objective as well as the application for which it is being used, also need to be considered when deciding on an appropriate objective. For example, fluorite lenses can be used at > 350 nm, while quartz lenses can be used at <350 nm. Planapo lenses are excellent for photography but multi elements will diminish transmittence. Interestingly, the light throughput of modern objectives is only 15-50%. The selected objectives should also have negligible autofluorescence, no solarization (color changes due to prolonged irradiation) and picked to optimize brightness in the image.

A number of objectives exist for fluorescence microscopy. A partial list includes: 1) ultra-fluors (200-700 nm transmission); 2) Plan-Achromats (flat field, color corrected, useable at near UV wavelengths); 3) Plan-Neofluars (flat field corrected with high transmission near UV); 4) Fluars which have a wide spectral range; 5) Multi-immersion objectives, which can be used with or without a coverslip in a variety of different medium; 6) Quartz objectives for use in the UV; 7) Epiplan Achromats and Neofluars for use in reflected light ; and 8) newly developed high numerical aperture water immersion objectives (40x-60x, N.A. 1.2). These latter lenses are becoming increasingly popular as they can be used with medium bathing cells and tissues allowing fluorescence imaging of real time cellular physiology and provide excellent images.

Detectors

Many of the specimens biologists examine contain low levels of fluorescence either due to low probe density in the sample or the need for minimization of excitation intensity to reduce fading or phototoxicity. The conventional wisdom about excitation is to minimize phototoxicity, while getting enough signal. A signal of 10^{-6} f.c. or 10 μlux (about 10% of what dark adapted eye can see), results in 4,000,000 photons/cm^2/sec or only < 1 photon/pixel/video frame. Thus, the use of the most sensitive fluorescence detection schemes is of the utmost importance.

A number of very sensitive detectors are available for fluorescence microscopy, and can be divided into photon detectors (photomultiplier tubes) or image detectors (cameras). Photomultipliers are the most sensitive detection devices, capable of single photon detection, but suffer from the fact that they are unable to provide information from anything other than a single spot. In terms of image detectors, ultra-low-light-level video-or solid state cameras have been developed to provide highly sensitive detection of fluorescence with rapid temporal and high spatial resolution. A variety of these detectors exist, from single stage silicon intensified targets (SIT), two stage SITs (ISIT), microchannel plate (MCP) intensifiers, charged coupled device (CCD) detectors (cooled or non-cooled, slow scan) and gated intensifiers where the voltage to the detector can be turned on & off on the order of nanoseconds. These latter imaging devices are used primarily for fluorescence lifetime imaging microscopy (see below). Each of these imaging detectors differs in terms of their sensitivity, linearity, wavelength sensitivity, and temporal and spatial resolution; selection of an appropriate detector will be usage dependent.

CURRENT TYPES AND APPLICATIONS OF FLUORESCENCE MICROSCOPES

Digitized Video Microscopy (DVM)

DVM consists of capturing the fluorescent image produced by a standard fluorescence microscope using an ultra sensitive spatial camera, dividing the image up into discrete picture elements (pixels) and converting the intensity of fluorescence in each pixel to a number. These detectors allow the quantitative analysis of variations in two dimensional intensity distributions in time. The output from the camera is fed into the image processing computer which divides the image into a 512 X 512 pixel array. The fluorescence intensity falling on each pixel is digitized to one of 256 gray levels within a frame acquisition time (30 ms).

Digital imaging is the process of converting visual information into numeric form, which allows the quantitative study of image data. Because this can be done for every point in the image at the same time, such a system gives the investigator both temporal and spatial photometric information, with a maximum sampling rate of 30 Hz. This rapid acquisition allows extremely fast image processing operations to be performed. For example, images can be averaged as fast as they are acquired producing an improvement in the signal to noise ratio of the final image which is proportional to the square root of the number of frames averaged. This is an extremely important tool to improve detection of weakly fluorescent specimens which can be barely visible or invisible to the dark adapted eye. Background sources of fluorescence from an unstained specimen can be stored and subtracted from the actual image to provide a more accurate representation of the distribution of the fluorescent label. Shading and geometric distortion can be removed and the resultant image "gray stretched" to improve contrast. Finally, of course digital frame storage allows for several standard ways of image display: pseudo color, isointensity contour map and isomorphoric (rubber sheet map).

DVM techniques have found great utility in measuring the distribution and mobility of fluorescently labeled components on and within single living cells. To undertake such studies, one needs the ability to maximize signal, keeping temporal resolution as high as possible and yet not damage the cell. There are a number of ways one can maximize signal and signal to noise in DVM. These include the use of probes with the highest quantum yields coupled with the use of detectors with the lowest dark counts and highest quantum efficiency and protection against photobleaching.

One of the most popular applications of the use of DVM is in the measurement of ion concentrations in the cytoplasm and organelles of single intact living cells. As an example, Figure 1 illustrates a macrophage loaded with the Ca^{2+}-sensitive dye Fura-2. After a basal image of the resting Ca^{2+} level was obtained, platelet activating factor (PAF) was locally released at the leading edge of the cell (shown in the top left hand corner of the images) and the Ca^{2+} monitored in real time every 10 sec for a period of 3 min. A number of findings immediately emerge from such an exercise and underscore the utility of DVM. Addition of PAF causes a transient increase in Ca^{2+} which differs in concentration in a spatially and temporally defined manner. Thus, DVM allows real time observation of cellular physiology with high temporal and spatial resolution in single intact living cells.

Multiparameter Digitized Video Microscopy (MDVM)

MDVM is used to monitor a variety of cellular parameters in single living cells with quantitative temporal and spatial resolution. The very low light levels involved in this system minimize damage to the cells, and so permit the monitoring of cellular functions without disrupting the normal physiology. Cultured cells may be loaded with multiple probes whose fluorescence are specific for environmental parameters and whose excitation and/or emission are non overlapping. By selecting excitation and emission filters under computer control, the MDVM system samples repeatedly over time the parameter-specific fluorescence of each of the probes. Thus, several different parameters can be followed simultaneously over time in single living cells. Spatial patterns of these parameters within cells are also determined through the imaging capabilities inherent in the MDVM technique.

The recent availability of photolabile compounds, which upon photolysis with ultraviolet (UV) light release putative second messengers (2), provides the ability to study the role and regulation of agonist-induced alterations in cellular second messengers involved in cellular physiology. The principle behind the use of caged compounds is based on the ability of UV irradiation to photolytically alter the chemical structure of the caged compound such that the binding and/or affinity of the overall structure for the "caged" component of the complex is reduced, allowing its release from the caging structure. Photolytic conversion generally occurs with the redistribution of electronic charges in the compound, thereby reducing its affinity for the molecule it was harboring and allowing its release from the caging structure. The conversion is often accompanied by the cleavage and release of the 'chemically caged' moiety from the compound. Once the caged material has been introduced into the cytoplasm of a cell (via microinjection), it may be released from the cage with a pulse of UV light and subsequent changes in cellular Ca^{2+} can be monitored **in the same cell** using Ca^{2+}-sensitive fluors (e.g. Fluo-3, Calcium Green, Calcium Crimson, etc.) that do not absorb at the wavelength of light used for photolysis. A number of currently available caged compounds exist which can modulate intracellular Ca^{2+} levels, including IP_3, a non-hydrolyzable analogue of IP_3 ($GPIP_2$), Ca^{2+} itself (NITR-7, NITR-5 or Nitrophenyl-EGTA), and Ca^{2+}-chelators (diazo-2 and caged-BAPTA).

Fluorescence Recovery After Photobleaching (FRAP)/Video-FRAP

This technique allows the translational mobility of fluorescently labeled molecules to be monitored in very small (2-5 μm) regions of a single cell's surface or cytomatrix (3). The basic concept of a spot photobleaching measurement is as follows. The emission (F_i) produced by laser excitation of a fluorescence label on or in a region of a membrane, surface or thin volume is measured. Next, an intense, short photobleaching pulse of light is directed to the same region of the specimen, rapidly destroying a substantial amount of the fluorescence in that region. After this photobleaching pulse, the end of which is defined as time zero, the recovery of fluorescence within the photobleached region is measured as a function of the time with a beam coincident with the photobleaching beam but of the same intensity as that used to measure F_i. This fluorescence intensity ranges from F_o at time zero to $F\infty$ at the time when the recovery is judged complete. The kinetics of recovery can be empirically characterized by the time ($\tau_{1/2}$) required to reach 50% of complete recovery. The mobile fraction [$mf \equiv (F\infty - F_o) / (F_i - F_o)$] characterizes the extent to which the initial fluorescence F_i is regained. Fluorophores immobile on the time scale of the measurement are indicated by an mf of less than 100%. If spontaneous recovery of fluorescence does not occur and the probe beam does not itself induce photobleaching, the recovery kinetics can be related to the rate of lateral transport of neighboring unbleached fluorophores into the previously bleached region in a straightforward manner. For example, in the case of diffusion-limited fluorescence recovery in two dimensions, the diffusion coefficient is directly proportional to the square of the laser spot radius in the specimen plane and inversely proportional to the half-time for recovery, $\tau_{1/2}$.

In a typical FRAP experiment, a photomultiplier measures the total brightness levels of the bleached region in the sample. In video-FRAP, the photomultiplier tube is replaced by a low light level detector so that the FRAP data is acquired just as if the investigator were

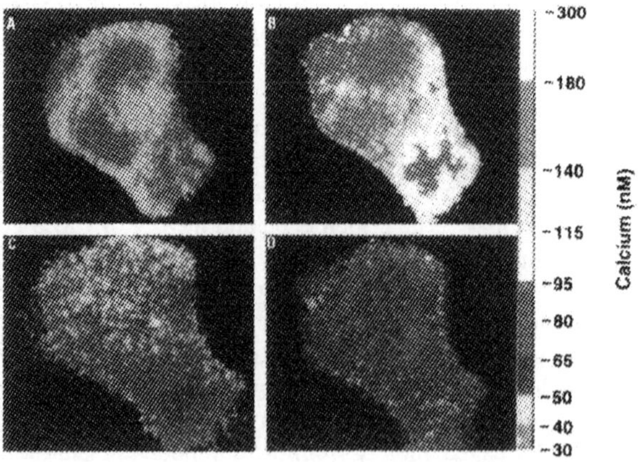

Figure 1. Alteration in Cytosolic Ca^{2+} in macrophages stimulated with platelet activating factor (PAF). Murine peritoneal macrophages were isolated and loaded with Fura-2 AM (1 μM) for 20 min at 23°C. Following washing, cells were placed on the stage of a DVM and Ca^{2+} measured as previously described (1), prior to and after addition of 1 nM PAF. A, before PAF addition; B, C and D, 2, 15, and 30 sec after PAF addition.

viewing the experiment through the microscope. Thus, video-FRAP offers much more information in the form of a two-dimensional array of brightness, allowing analysis of such data for evidence of anisotropic diffusion and flow.

Total Internal Reflectance Fluorescence (TIRF) Microscopy

The Fresnel formula describing the reflection of a light beam incident on interfaces between two media having different refraction indices, predicts that a light beam propagating in a medium with a higher refractive index is totally reflected at the boundary between a medium with a lower refractive index if the angle of incidence equals or exceeds a critical angle (for quartz/water: 65.4°). However, a so-called evanescent wave enters the second medium to a depth of roughly 0.1 μM and is capable of exciting only those fluorophores which are located at the interface or within this small depth. This special type of illumination has been used for the measurement of adsorption-desorption kinetics and surface diffusion rates of proteins at the quartz aqueous buffer interface (4). It has also been used to study only molecules in the ventral (substrate attached) membrane of tissue culture cells (5).

Confocal Microscopy

Conventional microscopes create images with a depth-of-field at high power of 2-3 μ m. Since the resolving power of optical microscopes is about 0.2 μm, superimposition of detail within this plane of focus obscures structural detail that would otherwise be resolved. In addition, for specimens thicker than this depth of field, light form out of focus planes creates diffuse halos around objects of study. These halos are especially prominent in fluorescence microscopy. By contrast, confocal microscopes create optical sections which are ~0.5 μm thick. Importantly, confocal microscopy rejects light from out-of-focus planes, and minimizes superimposition of detail producing images which are remarkably detailed. Confocal microscopy can basically be thought of as a CAT scanner for cells. Recently, UV laser scanning confocal microscopes have become available that expand the range of confocal applications to include UV-excited fluorophores. Increasingly, confocal microscopy has become an essential analytical tool to study the structure and physiology of living cells (6).

The temporal resolution of a laser scanning confocal microscope is dictated by the speed at which it can scan a focused spot of light across the specimen and in doing so, generate an image with suitable signal-to-noise. Alternatively, if higher temporal resolution is required and two dimensional spatial information can be sacrificed, a single line can be repetitively scanned across the field at intervals between 10-100 msec. Should higher temporal resolution be required, the line scanning component of the confocal could be deactivated and repetitive measurement of the same spot could be made repeatedly.

Automated Fluorescence Image Cytometry (AFIC)

The purpose of an automated image acquisition system is to perform the desired experiment, and gather and store data for later retrieval or analysis. The AFIC consists of an epi-fluorescence microscope, low-light-level camera and computer (7). The images from the microscope are captured by a CCD camera, digitized and stored in computer memory. For precise quantification, the images are corrected for background (by subtracting an image not containing stained objects) and variations in the efficiency of the system over the imaging area (shading). Shading is assessed by measuring a standard fluorescent object at different positions over the imaging area. Cell nuclei are automatically detected by recording images of them, correcting the images for background and shading and segmenting the images into regions corresponding to stained nuclei and (unstained) background. In addition, the AFIC has a computer-controlled stage, which automatically records the location of the images relative to a fixed reference point to a precision of 1 μm. This feature allows the user to remove the slide from the stage (for further staining etc.) and then re-image the same scenes as before. Also, the focus drive of the microscope is under computer control in order acquire images at different focal depths.

The ability of AFIC to automatically scan and identify individual nuclei (cells) on a slide is based on the fact that when stained with the fluorescent DNA dyes Hoechst or DAPI, the nuclei are very bright compared to the background allowing highly reliable identification of nucleus on a slide using image segmentation techniques (8). These techniques use the fact that nuclei: (1) have higher intensities than the background in the images; (2) are approximately circular, and (3) have a limited range of sizes. Following filtering of the raw image, the stained nuclei in the original and filtered images are detected by calculating adaptive, gradient-weighted threshold intensities that are between the intensities of the nuclei and background. Next, the algorithm finds the optimum edges of the nuclei by assuming such edges correspond to pixels with a locally maximum slope.

Once the individual nuclei have been detected, numerous properties of the individual nuclei can be directly quantified. These properties routinely include: their location in the image; area; perimeter; total fluorescence intensity from the DNA stain, and variation of the DNA staining within the nucleus. The same properties measured from other fluorescent labels in the same nuclei can be quantified, by mapping the regions corresponding to each nucleus over the images of each label. The output from the algorithms is an image showing the regions automatically defined as being nuclei. Additional algorithms are available for reporting the results as histograms (e.g. DNA ploidy distributions) and scattergrams for comparing the distribution of one property with respect to another.

AFIC was combined with a fluorescence in situ hybridization (FISH) technique to screen cervical specimens in an investigation of the prevalence of HPV-16 in women with varying grades of cervical dysplasia/carcinoma (9). The sensitivity and specificity of the FISH procedure for detecting the presence of HPV-16 genome was compared to PCR and Southern blot analysis. 29% of the specimens tested were positive for HPV-16 by PCR while 14% of the specimens tested were positive for HPV-16 by the FISH procedure. When the sensitivity and specificity of the FISH procedure was compared to that of PCR in terms of HPV detection and clinical disease status, it was found that the specificity of FISH and PCR with respect to predicting CIN 2-3 and carcinoma in situ was 90% and 75% respectively, while the sensitivity of FISH and PCR with respect to predicting CIN 2-3 and carcinoma in situ was 21% and 36% respectively. Because the FISH procedure preserves cellular morphology, has a higher specificity yet comparable sensitivity to PCR, use of the FISH procedure for detection of high risk HPV may serve as a useful adjunct to cytological screening for detection of high grade cervical disease.

Fluorescence Resonance Energy Transfer Microscopy

This technique allows the study of cell component interactions at the molecular (10-100 Angstroms) level (10). RET can be used to probe cell-substrate interactions, growth factor receptor dimerization and the interaction of electron transport chain components. For RET to occur, the donor must be fluorescent and of sufficiently long lifetime. The energy transfer to the acceptor is nonradiative (is not mediated by a photon) and requires the distance between the chromophores to be relatively close (usually not exceeding 50-100 Angstroms). RET varies most importantly as the inverse of the sixth power of the distance separating the chromophores. The dependence of the energy transfer efficiency on the donor-acceptor separation provides the basis for the utility of this phenomenon in the study of cell component interactions.

The rate of energy transfer (K_T) or the energy transfer efficiency (E_T) are both related to the lifetime of the donor in the presence or absence of the acceptor:

$$K_T = (1/\tau_D)\,(Ro^6/R) \qquad (2)$$

when Ro is the critical Forster distance, the donor-acceptor separation for which the transfer rate equals the donor de-excitation rate in the absence of acceptor, τ_D the lifetime of the donor in the absence of the acceptor and R the distance separating the donor and acceptor molecules. R_o can be calculated from:

$$R_o = \left(JK^2\ Q_o\ n^{-4}\right)^{1/6} K \text{ where} \qquad (3)$$

J is the overlap integral, a measure of overlap of donor emission and acceptor absorption spectra, K^2= orientation factor; K = 2/3 if the orientation of the donor and acceptor are random, Q= donor quantum yield (no acceptor present) and n = index of refraction of medium separating donor and acceptor. E_T is related to R by:

$$R = Ro\,(1/E_T\text{-}1)^6 \qquad (4)$$

$$\text{where } ET = (\tau_{DA}/\tau_D) \qquad (5)$$

where τ_{DA} is the lifetime of the donor in the presence of the acceptor. Thus, by measuring the lifetime of the donor with and without acceptor, one can calculate the distance between donor and acceptor.

Because energy transfer occurs over distances of 10 to 100 Å, a FRET signal corresponding to a particular field (or pixels) within a microscope image provides additional information beyond the microscopic limit of resolution down to the molecular scale. The ability to measure interactions and distances of molecules provides the principal and unique benefits of FRET for microscopic imaging (11). FRET imaging is particularly useful in examining *temporal* and *spatial* changes in the distribution of fluorescently conjugated biological molecules in living cells. Recently, the studies of epidermal growth factor (EGF) receptors have been carried out using time-resolved FRET imaging (12). By using frequency-domain lifetime imaging microscopy and FRET, EGF receptor clustering during signal transduction was monitored and a stereochemical model for the tyrosine kinase of the EGF receptor has been investigated. Time-resolved FRET imaging has also been applied to study the extent of membrane fusion of individual endosomes in single cells (13). Using time-domain FLIM and FRET, the extent of fusion, and the number of fused and unfused endosomes were clearly visualized and quantitated.

Fluorescence Anisotropy Microscopy

Localized alterations in lipid order are thought to modulate the activity of plasma membrane proteins (e.g. receptors, transport enzymes) (14). Thus, the ability to measure plasma membrane lipid order with high temporal and spatial resolution would be of great value in understanding the effects of lipid organization on plasma membrane protein activity. Fluorophores preferentially absorb photons whose electron vectors are aligned parallel to the absorption transition dipole moments of the fluorophore. Using polarized light, it is possible to selectively excite those fluorophore molecules whose absorption transition dipoles are parallel to the electric vector of the excitation. The transition dipole moments for absorption and emission have a fixed orientation within each fluorophore, and the angle between those dipole moments determines the maximum (limiting) measured anisotropy. Rotational motion (that is further displacement of the emission dipole from its starting position) which occurs during the lifetime of the excited state, will further lower the observed emission anisotropy. Polarized decay curves ($I_{\parallel}(t)$ and $I_{\perp}(t)$) can be collected with a polarizer placed in the excitation light path and an analyzer placed in the emission light path oriented either parallel or perpendicular to the excitation polarizer. The time-dependent emission anisotropy decay curves r(t) can then be calculated from:

$$r(t) = (I_{\parallel}(t) - I_{\perp}(t)) / (I_{\parallel}(t) + 2I_{\perp}(t)) \qquad (6)$$

Microscopic fluorescence polarization measurements on single cells present certain problems not associated with fluorometric measurements (15). The major difficulties include the depolarizing effects of the microscope optics, the orientational dependence of the observed fluorescence polarization, and the lower fluorescence signals obtainable from individual cells. The depolarization effects of the microscope optics arises from the necessity of using objective lenses with high apertures for maximal collection of fluorescence, rather than a narrow slit of effectively zero aperture. The orientational dependence of the observed fluorescence polarization is due to the fact that the surface of all cells is three dimensional, rather than a two dimensional bilayer. Thus, surface curvature, in addition to the rotational mobility of the fluorophore during its excited state lifetime, will determine the final observed depolarization. Corrections for the effects of surface curvature requires knowledge of the probe excited state orientation distribution in order to determine orientation-dependent fluorescence polarization (16). Lastly, since microscopic measurements are performed on single living cells, it is imperative to keep the probe concentration as low as possible in order to minimize alterations in membrane function.

In studies examining hepatocyte hypoxic injury, the hypoxic hepatocyte and its plasma membrane blebs were modeled as the simple case of a spherical membrane surface labeled with a rod-shaped probe of known orientation in the membrane (TMA-DPH) (17). Using this approach it was possible to describe the plasma membrane lipid order profile around the bleb perimeter with a set of four fluorescence images. A different cell geometry, or a probe with different absorption/emission dipole orientations or diffusion properties, would require the appropriate modifications to the equations used in order to determine the functional dependence of the polarization ratio on lipid order and membrane surface location. Image acquisition at other polarizer orientations might also be necessary.

A set of polarization measurements consisted of four images; fluorescence polarization ratios ($F_{\parallel}/F_{\perp}$) were computed on a pixel-by-pixel basis from digitized image pairs after background subtraction and image alignment. From the ratio images the corresponding cone angles were calculated. Hypoxic hepatocytes form plasma membrane blebs; these blebs displayed lipid order parameters of S=0.91-0.98 [$S=1/2\cos q_{max}(1+\cos q_{max})$]. For comparison, using lipid vesicles roughly the size of rat hepatocytes, cone angles for TMA-DPH of 32±4° (S=0.78±0.05) and 17±2° (S=0.935±0.015) in fluid and gel-phase lipid vesicles respectively were found (17). The data indicated that the fluorophore resided in a rigid membrane environment after hypoxic injury in rat hepatocytes.

Fluorescence Lifetime Imaging Microscopy

Steady-state fluorescence microscopy is limited in its ability to quantitate and study

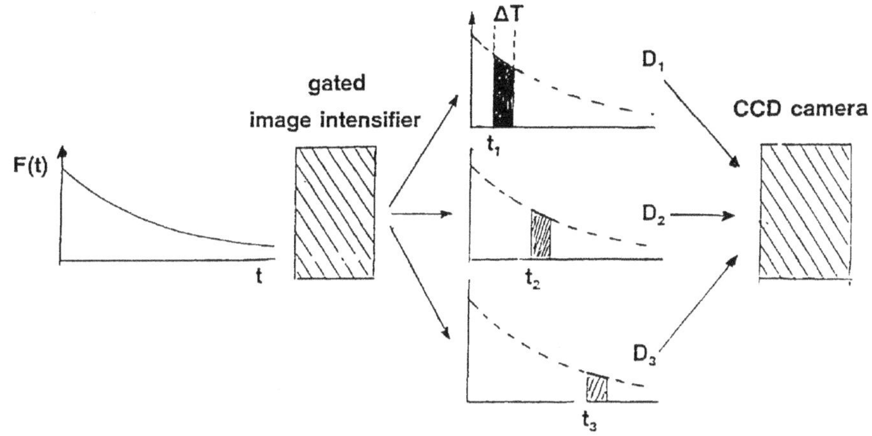

Figure 2. Functional description of rapid lifetime determination for time-resolved fluorescence lifetime imaging using a gated multichannel plate image intensifier.

rapid dynamic events (i.e. on the order of nano- to milliseconds) in cells. Measurements of time-dependent fluorescence emission anisotropy and time resolved spectra of fluorescent plasma membrane protein or lipid probes can provide quantitative dynamic information about plasma membrane lipid fluidity, topography and interactions amongst membrane components. Use of these techniques requires the ability to measure the fluorescent lifetime of the probe of interest in a spatially defined manner at high temporal resolution in intact single living cells. The development of FLIM is an extremely important advance, as it allows for the first time, the combination of the sensitivity of fluorescence lifetime to environmental parameters to be monitored in a spatial manner in single living cells (18). As such FLIM provides a sensitive technique capable of obtaining data regarding the dynamics and heterogeneous nature of plasma membrane components which can not be obtained by using the methods mentioned above which are based on time averaging of the fluorescent signal.

In order to make time- and spatially-resolved lifetime measurements at the same time, a time gated microchannel plate (MCP) image intensifier coupled to a CCD camera is used (Figure 2). The gated MCP is turned on for a very brief interval (i.e. 2 nsec) at some time interval (t_1) after the exciting pulse and the emitted intensity from the gated MCP is acquired on the CCD (which is continually left on). This process is repeated a large number of times (i.e. 10,000) at this time t_1 with the emitted intensity from the gated MCP continually being accumulated on the CCD. The CCD is then read out, the time of the gate window with respect to the excitation pulse is temporally shifted (t_2) and the whole process is repeated. Time-resolved fluorescence images can be directly detected in short time and stored in a computer.

Recently, FLIM has been applied to various biological studies. One of these has used FLIM to monitor the extent of membrane fusion of individual endosomes in single cells (13). By using fluorescence resonance energy transfer techniques and FLIM, the extent of fusion, and the number of fused and unfused endosomes was clearly visualized. In another study, FLIM has been used to monitor levels of cytoplasmic and nuclear Ca^{2+} in cells (18). These studies were carried out because Ca^{2+} was being monitored with the Ca^{2+}-sensitive dye Fluo-3. Fluo-3 is not a ratiometric dye and quantitative calibration of Ca^{2+} is not possible due to the problems of pathlength and accessible volume. The FLIM approach allows quantitative measurements of intracellular Ca^{2+}, using non-ratiometric visible wavelength Ca^{2+}-sensitive indicators, such as Fluo-3 and Calcium Crimson. Importantly, measurement of Ca^{2+} using FLIM is independent of probe concentration and photobleaching. In addition, FLIM also provides the opportunity to study the dynamics of the environment surrounding the Ca^{2+}-sensing probes; two or three-dimensional images of the fluorescent lifetime of Ca^{2+}-sensitive fluors and their rotational mobility (using time resolved emission anisotropy) can be obtained during response of single living cells to stimulation. This will allow the quantitative determination of cellular Ca^{2+} levels as well as document whether any change in probe characteristics (i.e. binding to cytoplasmic or nuclear components) occurs during the measurements. In the future, FLIM could also be adapted to the confocal microscope, combining the quantitative nature of FLIM with the optical sectioning power of confocal microscopy.

CONCLUSIONS

Form this brief review, it should be apparent that there are a number of different types of fluorescence microscopic techniques which are available to the scientist to observe, measure and manipulate biological specimens. This review is by no means complete as other types of fluorescence microscopes in various stages of development/application include laser trapping microscopy, nearfield scanning fluorescence optical microscopy (NSOM), intravital fluorescence microscopy, fourier spectroscopy and spectral dispersion microscopy and nanovid and atomic force microscopes. These newer types of microscopes, in conjunction with those that are already in existence, provide scientists with an unparalleled array of ways to monitor biological specimens. Of course, many of these different types of microscopes currently , and will in the future, find applications outside of the biological sciences. Similarly, microscopes developed for non-biological purposes could be applied to biological sciences, thus further increasing the already substantial versatility of the fluorescent microscope.

ACKNOWLEDGEMENTS

Much of the work described in this review would not be possible without the substantial contributions of many of my colleagues. While the list is long, particularly important contributions were made by David Albertini, Jim DiGuiseppi, Pamela Diliberto, Salvatore Fernandez, Kathryn Florine-Casteel, Gerry Gordon, Charles R. Hackenbrock, Ken Jacobson, Dennis Koppel, John J. Lemasters, Steve Lockett, Ammasi Periasamy, Deborah K. Smith, Xue Feng Wang, and Pawel Wodnicki. Support from the NIH, NSF, ACS, AHA, Gustavus and Louise Pfeiffer Foundation and the North Carolina Biotechnology Center are also gratefully acknowledged.

REFERENCES

1) Diliberto, P.A., Gordon, J., and Herman, B. Regional and Mechanistic Differences in PDGF-Isoform-Induced Alteration in Cytosolic Free Calcium in Porcine Vascular Smooth Muscle Cells. *J. Biol. Chem.* 266:12612 (1991).

2) Kao, J.P.Y., Harootunian, A.T., and Tsien, R.Y. Photochemically generated cytosolic calcium pulses and their detection by Fluo-3. *J. Biol. Chem.* 264:8179 (1989).

3) Gordon, G.W., Chazotte, B., Wang, X.F., and Herman, B. Two Diffusional Analysis of Fluorescence Recovery After Photobleaching Data. *Biophys. J* 68:766 (1995).

4) Tamm, L.K and Bartoldus, I. Antibody binding to lipid to model lipid membranes: the large ligand effect. *Biochem.* 27: 7453 (1988).

5) Lanni, F., Waggoner, A.S., and Taylor, D.L. Structural organization of interphase 3T3 fibroblasts studied by total internal fluorescence microscopy. *J. Cell Biol.* 100: 1091 (1985).

6) Chacon. E., Reece, J.M., Nieminen, A.-L., Zahrebelski, G., Herman, B., and Lemasters, J.J. Distribution of Electrical Potential, pH, Free Ca^{2+} and Volume Inside Cultured Rabbit Cardiac Myocytes During Chemical Hypoxia: A Multiparameter Digitized Confocal Microscopic Study. *Biophys. J.* 66:942 (1994).

7) Periasamy, A. and Herman, B. Computerized Microscopic Vision in Cell Biology. *J. Comp. Assit. Micro.* 6: 1 (1994).

8) Lockett, S.J., Jacobson, K., and Herman, B. Quantification Accuracy of an Automated, Fluorescence-Based Image Cytometer. *Anal Quat. Cytol. Histol.* 14: 187 (1994).

9) Siadat-Pajouh, M., Periasamy, A., Ayscue, A., Moscicki, A.B., Palefsky, J.B., Walton, L., DeMars, L.R., Power, J.D., Herman, B., and Lockett, S.J. Detection of Human Papilloma Virus Type 16/18 DNA in Cervicovaginal Cells by Fluorescence Based In Situ Hybridization and Automated Image Cytometry. *Cytometry* 15: 245 (1994).

10) Herman, B. Resonance Energy Transfer Microscopy. *in* Fluorescence Microscopy of Living Cells in Culture. Methods in Cell Biology 30 (II) eds. Taylor, D.-L. and Wang, Y.-L. Academic Press, San Diego, CA. (1989).

11) Wang, X.F., Lemasters, J.J., and Herman, B. Plasma Membrane Architecture During Hypoxic Injury in Rat Hepatocytes Measured by Fluorescence Quenching and Resonance Energy Transfer. *J. BioImag.* 1: 30 (1994.).

12) Gadella, T.W., J.J., Jovin, T.M. and Clegg, R.M. Fluorescence lifetime imaging microscopy (FLIM): spatial resolution of microstructures on the nanosecond time scale. Biophys. Chem. 48, 221 (1994).

13) Oida, T., Sato, Y., and Kusumi, A. Fluorescence lifetime imaging microscopy (flimscopy): methodology development and application to studies of endosome fusion in single cells. *Biophy. J.* 64: 676 (1993).

14) Alvia, R. C., Curtain, C. C., and Gordon, L. M. *Lipid Domain and the Relationship to Membrane Function.* Alan R. Liss Press, New York, (1988).

15) Axelrod, D. Fluorescence polarization microscopy. *in* Fluorescence Microscopy of Living Cells in Culture. Methods in Cell Biology 30 eds. Taylor, D.-L. and Wang, Y.-L. Academic Press, San Diego, CA. (1989).

16) Florine-Casteel, K, Phospholipid order in gel- and fluid-phase cell size liposomes measured by digitized video fluorescence polarization microscopy. *Biophys. J.* 57: 1199 (1990).

17) Florine-Casteel, K., Lemasters, J.J., and Herman, B. Lipid Order in Hepatocyte Plasma Membrane Blebs During ATP Depletion Measured by Digitized Video Fluorescence Polarization Microscopy. *FASEB J.* 5: 2078, (1991).

18) Periasamy, A., Wang, X.F., Wodnicki, P., Gordon, G., Kwon, S., Diliberto, P.A., and Herman, B. High Speed Fluorescence Microscopy: Lifetime Imaging in the Biomedical Sciences. *J. Micro. Soc. Am.* 1: 13, (1995).

FLUORESCENCE LIFETIME-RESOLVED IMAGING MICROSCOPY: A GENERAL DESCRIPTION OF LIFETIME-RESOLVED IMAGING MEASUREMENTS

Robert M. Clegg and Peter C. Schneider

Max Planck Institute for Biophysical Chemistry
Department of Molecular Biology
Am Fassberg 11
D-37077 Göttingen
Germany

INTRODUCTION

Fluorescence lifetime-resolved imaging microscopy — FLIM — is a relatively new fluorescence imaging technique by which the temporal attributes of luminescence emission can be measured directly at each location of a microscope image. The mean lifetime of the emission can be determined at every pixel of a digital image with a high spatial and temporal resolution. Different fluorescence components with differing decay times can be enhanced or suppressed throughout a fluorescence image. The measurements can be carried out at every pixel simultaneously. In this presentation a account of the experiment is given in terms of a comprehensive integrated theoretical framework.

LIMITATIONS OF STATIC STEADY-STATE IMAGING AND ADVANTAGES OF TEMPORAL RESOLUTION

The most common application of fluorescence microscopy is to determine the spatial localization of a fluorescence probe, or of fluorescently labeled macromolecules, in a microscopic specimen. More recently there has been increased interest to derive more detailed information concerning the physical and chemical environment of fluorescent probes, as this provides additional information about the immediate molecular surroundings of the probes.

Often the information derived from a simple static fluorescence image is sufficient to answer the question at hand; however, there are many complications that can limit the reliability of static measurements. For instance: 1) The luminescence emission (from intrinsic as well as from extrinsic probes) from biological samples viewed in a fluorescence microscope can be complex and often has contributions from several emitting components. 2) The molecular environment of fluorophores in a biological specimen is generally considerably more heterogeneous than that in cuvette experiments which usually employ relatively highly purified components. Such heterogeneous surroundings can cause major difficulties in the interpretation of fluorescence signals because the spectroscopic characterization of many fluorescent dyes are often affected dramatically by their molecular environment (polarity, hydrophobicity, ionic strength, pH, etc.). Depending on the location of even a singular probe in a biological sample, its emission spectrum can exhibit large shifts and its quantum yield can vary widely. 3) Even if the emission of a fluorescent probe is not significantly affected by its environment, different fluorescent species are often co-localized

Fluorescence Microscopy and Fluorescent Probes
Edited by J. Slavík, Plenum Press, New York, 1996

in a biological specimen. The discrimination of separate emitting species within an image is traditionally achieved by utilizing the dispersion characteristics of their static excitation and emission spectra; however, this is sometimes insufficient for a reliable discrimination. 4) The distribution of many fluorophores in a biological specimen is not homogeneous, and it is the actual concentration distribution of certain molecular species that is often the major concern of the experimenter. However, due to dynamic excited-state processes, such as dynamic quenching and energy transfer, it can be difficult to estimate the true concentration of a fluorophore, and therefore its true distribution. 5) The wavelength of background emission often considerably overlaps the emission spectrum of the fluorescent species of interest. The fluorescence background can hinder the identification of a particular component and the accurate determination of its concentration. These artifacts can contribute major difficulties for detecting either the presence of, or the relative concentration of, a single fluorescent species, especially with samples as complex as living cellular systems, even using extrinsic probes with well characterized spectroscopies. Even without these difficulties, the averaging process intrinsic to static measurements resulting from integrating over long times conceals a wealth of useful information available from time-resolved measurements.

Resolving directly the temporal decay of luminescence in the fluorescence microscope

Until recently, static methods provided the only option available for imaging fluorescence microscopy. If the mechanism of the excited state decay or of the molecular motion is known in detail, then even parameters measured under non-temporally-resolved (i.e. static) conditions can yield information on the rates of deactivation through specific pathways. Examples of this are static determinations of the quantum yield of luminescence and the emission anisotropy, and the slow cumulative rate of some selected deactivation process that continually removes selected molecules from the total molecular ensemble (such as the slow rate of photobleaching (Jovin and Arndt-Jovin, 1989)). However, only seldom does one know sufficient photophysical and photochemical details to exclude inaccuracies and substantial errors when interpreting time-averaged signals in terms of the underlying dynamic parameters of a kinetic phenomenon. The mechanism of luminescence phenomena can be elucidated most reliably by measuring directly the kinetics of decay from the excited state. A comprehensive, quantitative understanding of the different deactivation pathways operable for fluorophores at different positions in an extended object can be obtained only by temporally resolving the fluorescence decay; this in turn provides valuable information concerning the environment of the molecular probe. In addition, if the decay of different fluorescence components can be temporally resolved, the temporal characteristics can be used to improve the discrimination between different fluorophores, to increase the contrast of an image and to suppress or enhance particular lifetime components.

FLIM complements static fluorescence microscopy

FLIM has been developed to improve the proficiency of selectivity and discrimination of fluorescence imaging measurements in a microscope environment. Fluorescence lifetime-resolved imaging techniques make use of additional information directly related to the dynamic aspects of the fluorescence process. The conditions and requirements of many present-day static fluorescence microscopy experiments place extreme demands on the sensitivity and overall performance of the optical and electronic instrumentation, and the same applies to lifetime-resolved imaging measurements. Because optical imaging technology has developed rapidly and the sensitivity of the measurement, the spatial resolution, the dynamic range, and the convenience and the speed of image acquisition, analysis and display have improved considerably in the last years, the FLIM technique has been developing rapidly.

Several technically reliable instruments have been reported for measuring mean luminescence lifetimes in an image and for discriminating different fluorescence components according to their fluorescence lifetimes in the time range from nanoseconds (Clegg et al., 1992; Clegg et al., 1994; Clegg et al., 1990; Gadella Jr. et al., 1994; Gadella Jr. et al., 1993; Lakowicz and Berndt, 1991; Morgan et al., 1992; Piston et al., 1992; So et al., 1994; Wang et al., 1991) to many seconds (Marriott et al., 1991). Because lifetime-resolved imaging can make unique and novel contributions to a wide spectrum of problems it is desirable to make

FLIM readily available to many investigators. With this in mind we have been concerned in developing a FLIM technique which is not only technically sound but which is also adaptable to many situations and can be operated effectively by users with diverse and dissimilar interests. One aspect of an effective application of FLIM is concerned with the instrumentation and with a "user friendly" operation. For instance: 1) the measurement should be routinely and easily performed by non-technical personnel, 2) the measurement should be carried out rapidly so that experiments with live cells, which continually change their state, are possible, 3) the method should be convenient enough for routine medical and laboratory diagnostics. The realization of these technical objectives is continually being enhanced. Another aspect required for effectively applying FLIM is a good basic understanding of the theoretical description of the experimental methodology. This leads to a more effective implementation of the technique, more proficient interpretations of the results, and a better understanding of possible pitfalls and artifacts of the measurements. This communication is concerned with the latter topic.

This chapter presents a comprehensive theoretical treatment of fluorescence lifetime measurements with sufficient generality for describing the many-sided aspects of frequency domain FLIM measurements. To effect optimal use of FLIM, it is expedient to have a general and unified theoretical treatment of lifetime-resolved fluorescence measurements that can describe the complex circumstances that arise when imaging biological samples. Features of the lifetime-resolved imaging in a complex biological milieu are discussed in the context of the general theoretical presentation. Although these topics are not especially difficult and have been dealt with previously in the literature[1], a consistent and comprehensive presentation of all aspects, especially in relation to imaging, does not exist. In the context of such a general and unified theoretical description the possible artifacts and difficulties of the time-resolved imaging can be better perceived, and the advantages of FLIM can be better appreciated. An understanding of these general concepts and signal definitions is required for any but the most rudimentary analysis of lifetime-resolved imaging data.

GENERAL DESCRIPTION OF TIME-RESOLVED LUMINESCENCE SIGNALS

Temporally-resolved spectroscopic measurements can provide detailed accounts of the state of a luminescent molecule and its immediate environment. The excited state of a molecule becomes deactivated through several pathways, and lifetime-resolved measurements enable more direct procurement of information concerning these different deactivation processes than does a static measurement. The temporal profile of fluorescence decay from a population of excited molecules (even in a homogeneous solution of identical molecules) is governed by several underlying dynamic processes of deactivation. The kinetic equations describing the rate of deactivation involve a competition between many different pathways of deactivation. The measured lifetime of an excited state of a molecule is defined as the inverse of the total rate of deactivation — and the total rate of deactivation is the sum of all rates over every pathway available to the excited molecule to pass away from the excited state in question. An analysis of the decay profile of fluorescence emission can provide valuable information concerning fundamental physical and chemical processes that occur while the molecules are in the excited state, e.g. quenching processes, energy transfer, the rate of intersystem crossing from the excited singlet to the triplet state, excited-state reactions, excited state complexes, diffusion processes, molecular rotation, etc.. Because a static fluorescence signal arises from the integral over times much longer than the fluorescence lifetimes, invaluable information concerning the fluorescent molecule and its immediate molecular environment is lost by integrating the dynamic process of emission to produce a time invariant static signal. For this reason, there has always been great interest in developing and using time-resolved techniques in fluorescence; however, until recently, only static fluorescence signals have been available for imaging fluorescence emission in microscopy.

---

[1]We will not attempt here to give a complete list of references on this topic. For non-imaging fluorescence measurements the literature is extensive; a few general references with many references to primary publications are (Cundall and Dale, 1983; Jameson and Reinhart, 1989). A general discussion of frequency domain measurements can be found in (Jameson et al., 1984; Lakowicz and Maliwal, 1985; Spencer and Weber, 1969).

General description of the time course of the fluorescence decay for the different methods used for measuring fluorescence lifetimes

All methods for determining the rate of luminescence decay involve ultimately the change or variation in some molecular or physical parameter within the duration of the excited state, even those methods that are measured in a static steady-state manner. In both static and time-resolved steady-state measurements the data can be averaged over long times; however, in static steady-state measurements the integration process of the data acquisition is carried out in such a way that direct information concerning the lifetimes is not available. The methods for measuring the rate of luminescence decay in a fluorescence microscope are similar in many respects to those used for measuring the luminescence decay in a homogeneous solution. The two most widespread methods for measuring the decay of luminescence are: 1) direct time-resolved data acquisition following a pulse of excitation light and 2) phase and modulation techniques operating in the frequency domain with repetitive (sometimes sinusoidally) modulated excitation light.

In order to facilitate the exposition, some comments concerning general aspects of time-resolved measurements are given before presenting those topics relevant for the experimental realizations. Because the treatments of the general case are scattered throughout the literature, and have not been presented in the context of lifetime imaging, the derivations of the general equations are given in enough detail so that the reader is not required to search the literature. This allows a comprehensive and uniform description of the experimental realization, and facilitates the discussion of the analyses of the data and the applications.

Throughout the derivations we point out attributes of the equations that express characteristics of the fluorescence signal that are important for imaging work.

For simplicity, in this paper we use the convenient exponential complex number notation for the time varying quantities. The real part of any expression can be extracted at the end of a calculation by applying the equality, $\cos(\theta) = [\exp(j\theta) + \exp(-j\theta)]/2$. As a reminder, Euler's relation between the Cartesian and polar notation for complex numbers is $a + jb = \sqrt{a^2 + b^2} \cdot \exp\left(j \cdot \tan^{-1}(b/a)\right)$.

In general, the luminescence decay from a sample following a single short excitation event (i.e. on the time scale of the excited state lifetime) is not intense enough to record the decay with sufficient accuracy; therefore, the excitation event is repeated many times and the emission event is averaged over many repetitions to increase the signal-to-noise level. This is true of most lifetime-resolved luminescence measurements (in both the time and the frequency domain). The various methods for measuring fluorescence lifetimes are actually quite similar; besides the obvious differences in the experimental method of recording and analyzing the signal, their main differences lie in the pattern or form of the repetitive modulation of the excitation light. All methods (repetitive or not) can be expressed as a convolution of the response of the fluorescence system to a delta function excitation pulse with a function representing the actual excitation.

$$F(t) = Q \cdot \int_0^t E(t') \cdot F_\delta(t - t') \cdot dt' \qquad (1)$$

Q is a multiplicative term so that $F(t)$ represents the actual fluorescence signal; it accounts for the quantum yields of the fluorophores, instrumentation effects, the total power of the excitation light, etc.. The letters E and F refer to the excitation perturbation and the fluorescence response. The symbol $F_\delta(t - t')$ refers to the fluorescence response at time t to a delta function excitation pulse at time t' (see equation 3, below). In the general experimental case the transient terms resulting from the sudden application of the excitation light have subsided, so that we need only consider the steady-state (but temporally-resolved) response of the system to the repetitive excitation modulation.

The excitation light is usually a repetitive modulation that can be represented by a Fourier series (in complex notation, $j=(-1)^{1/2}$):

$$E(t) = \sum_{n=-\infty}^{\infty} E_n \cdot e^{j\omega_n t} = \sum_{n=-\infty}^{\infty} E_n \cdot e^{j\left[2\pi nt/T\right]} \tag{2}$$

where T is the period of the repetitive excitation, E_n is the amplitude of the nth frequency component, $\omega_n = 2\pi n/T = n \cdot \omega_T$ (where $\omega_T = 2\pi/T$) and

$$E_n = \frac{1}{T}\int_0^T E(t) \cdot e^{-j\omega_n t} dt \tag{2a}$$

It should be noted that the E_ns are complex numbers, and as such have a phase and magnitude; i.e. $E_n = |E_n| \cdot e^{j\phi_n^E}$, where $|E_n|$ is the magnitude and ϕ_n^E is the phase of E_n. The phase of the nth frequency component depends on the shape of the repetitive pulse. However, the quantity $E(t)$ is a real quantity. Because $E(t)$ is real, $E_{-n} = E_n^*$.

The fluorescence response to a single delta function excitation pulse, δ (in effect this is a pulse with a width that is short on the scale of the fluorescence lifetime), generally consists of several exponentially decaying components.

$$F_\delta(t) = \underset{\substack{fluorescence \\ components}}{\sum_{s=1}^{s=S}} a_s \cdot \exp(-t/\tau_s) \tag{3}$$

where a_s is the relative amplitude of the sth component of the fluorescence decay, and τ_s is the corresponding lifetime. Ultimately, if we could make perfect measurements the fluorescence components denoted by the subscript "s", correspond either to different molecular species, or to different lifetime components of a particular fluorophore. For example, a single fluorophore could exist in different environments that induce disparate lifetimes, or the probe could be attached (covalently or non-covalently) to different molecular species. If there is a continuous distribution of fluorescence lifetimes, the sum over "s" should be replaced by an integral. However, it must be kept in mind that practically the sum (or integral) in Eq. 3 is defined operationally by the fitting procedure. That is, the experimentalist may not be able to distinguish the emission from the separate fluorescing molecular species (this is often the case, especially when dealing with an image of a complex biological sample). Then Eq. 3 represents an approximation to the data (or rather a projection of the data onto an exponential functional space) and the different components do not correspond strictly to singular physical molecular species. Usually in the fitting procedure the sum of the amplitudes $\sum_{s=1}^{s=S} a_s$ is normalized to unity, and the actual measured amplitude can be accounted for by a simple multiplicative factor (see Eq. 6 below). It is important to remember in the discussion below, that the number and identity of fluorescence components probably vary over the image. This may be due to the actual variance of the fluorescence species, but it may also be due to the approximate nature of the fitting procedure, leading to a discreet spectrum (one or more) of lifetimes that do not correspond in a direct manner to the actual molecular species.

Assuming that the fluorescence components of Eq. 3 correspond to actual fluorescing molecular species, the relative values of the a_s components are proportional to the relative number of photons absorbed by each fluorescing species, "s". Therefore, if the different components "s" with separate fluorescence decay times of τ_s originate from different states of a single molecular species, and if the absorbance coefficient of all molecules of this chemical species are the same (this is often the case, at least approximately), the a_s values are independent of the relative quantum yields of the individual fluorescing states of this particular fluorescent molecular species. The relative quantum yields are then controlled only by the relative decay times. In this case the a_s values are a measure of the relative number of molecules being excited; however, the static fluorescence intensity may not be interpreted so easily. We emphasize this simple point because as we will see below, the static fluorescence signal usually measured in the microscope does not enjoy this simple direct interpretation of the intensity in terms of the number of excited molecules without making the often unwarranted assumption that all molecules of any particular molecular species are in identical surroundings, and therefore have identical decay times.

TIME-RESOLVED IMAGING FOR THE SPECIAL CASES OF PURE SINUSOIDALLY MODULATED EXCITATION LIGHT

Because it is the simplest — and most commonly presented — case, we consider first an example where the excitation light $E_i(t)$ is a simple sinusoidal excitation, with a single sinusoidal frequency component $\omega_1 = 2\pi/T$. The subscript "i" identifies the parameters for the ith pixel of the image.

$$E_i(t) = E_{0,i} + E_{1,i} \cdot 2 \cdot \cos(\omega_1 t) = E_{0,i} + E_{1,i} \cdot \left(e^{j\omega_1 t} + e^{-j\omega_1 t}\right) \tag{4}$$

Note in Eq. 4 we have arbitrarily set the phase of the cosine term to zero. Considering that the measured fluorescence signal can have many exponentially decaying components (see Eq. 3), the convolution integral is,

$$F_i(t) \propto \int_0^t \left\{ \left(E_{0,i} + \frac{E_{1,i}}{2} \cdot \left(\exp(j\omega_1 t') + \exp(-j\omega_1 t')\right) \right) \cdot \sum_{s=1}^{s=S_i} a_{s,i} \cdot \exp\left(-(t-t')/\tau_{s,i}\right) \right\} dt' \tag{5}$$

Carrying out the integration, the convolution equation describing the fluorescence for a purely sinusoidally modulated excitation light becomes (at sufficiently long times so that the transient terms become insignificant),

$$F_i(t) = Q_i \cdot \left\{ \begin{array}{l} \left[\sum_{s=1}^{s=S_i} E_{0,i} \cdot a_{s,i} \cdot \tau_{s,i} \right] \\ + E_{1,i} \cdot \left\{ \begin{array}{l} \left[\exp(j \cdot \omega_1 \cdot t) \cdot \sum_{s=1}^{s=S_i} \left\{ \dfrac{a_{s,i} \cdot \tau_{s,i}}{1 + j \cdot \omega_1 \cdot \tau_{s,i}} \right\} \right] \\ + \left[\exp(-j \cdot \omega_1 \cdot t) \cdot \sum_{s=1}^{s=S_i} \left\{ \dfrac{a_{s,i} \cdot \tau_{s,i}}{1 - j \cdot \omega_1 \cdot \tau_{s,i}} \right\} \right] \end{array} \right\} \end{array} \right\} \tag{6}$$

Q_i is the same term as Q in Eq. 1 that accounts for the quantum yields of the fluorophores, instrumentation effects, the total power of the excitation light, etc. at every pixel. Note that the two expressions in square brackets in the last time dependent term are complex conjugates of each other.

By transforming the complex numbers in the summation over "s" into polar notation, Eq. 6 can easily be shown to be,

$$F_i(t) = Q_i \cdot \left\{ \begin{array}{l} \left[\sum_{s=1}^{s=S_i} E_{0,i} \cdot a_{s,i} \cdot \tau_{s,i} \right] \\ + E_{1,i} \cdot \left\{ \sum_{s=1}^{s=S_i} \frac{a_{s,i} \cdot \tau_{s,i}}{\left(1 + \left(\omega_1 \cdot \tau_{s,i}\right)^2\right)^{1/2}} \cdot \left\{ \begin{array}{l} \exp\left(j \cdot \left[\omega_1 t + \phi_{s,i}\right]\right) + \\ \exp\left(-j \cdot \left[\omega_1 t + \phi_{s,i}\right]\right) \end{array} \right\} \right\} \end{array} \right\} \quad (6a)$$

where we have defined the phases, $\phi_{s,i}$, as $\phi_{s,i} = \tan^{-1}\left(\omega_1 \tau_{s,i}\right)$.

In real quantities Eq. 6b becomes,

$$F(t) = Q_i \cdot \left\{ \begin{array}{l} \left[E_{0,i} \cdot \sum_{s=1}^{s=S_i} a_{s,i} \cdot \tau_{s,i} \right] \\ + \left[2 \cdot E_{1,i} \cdot \sum_{s=1}^{s=S_i} \frac{a_{s,i} \cdot \tau_{s,i}}{\left(1 + \left(\omega_1 \cdot \tau_{s,i}\right)^2\right)^{1/2}} \cdot \left\{\cos\left(\omega_1 t + \phi_{s,i}\right)\right\} \right] \end{array} \right\} \quad (6b)$$

The signal represented by equation 6b is a constant term (the average time independent fluorescence) plus a weighted sum of cosines with the same frequency ω_1 , but different phases, $\phi_{s,i}$. The weighting factors are functions of the frequency, ω_1 , and the lifetime of each fluorescence component, $\tau_{s,i}$, as well as of the amplitudes of the individual fluorescence components, $a_{s,i}$.

Eq. 6b demonstrates explicitly a useful property of signals varying sinusoidally with identical frequencies — but different phases — that is very important for image work. The emission from every point of the image is a linear combination of time-varying sinusoidal fluorescence signals with different phases. Because the sinusoidal components differ only in their phase, the resultant overall time-varying signal that is measured can be expressed as a single sinusoidal component with the frequency ω_1 and an overall phase and amplitude (e.g., see Eq. 14). However, it is possible to employ phase sensitive methods for recording images that can suppress selectively and exclusively only those sinusoidally varying signals with phases corresponding to one particular fluorescence component (in this case, the phase detection is said to be orthogonal to this selected fluorescence component). An image can then be constructed from the residual time-varying fluorescence consisting of contributions from only those fluorescence components with phases different from this one selected phase; thereby, a single particular fluorescence component can be completely suppressed at all points of the imaged sample; this can be achieved rather easily at every pixel of an image simultaneously (Clegg et al., 1994; Gadella Jr. et al., 1993).

TIME-RESOLVED FLUORESCENCE IMAGING FOR A GENERAL REPETITIVE MODULATION OF THE EXCITATION LIGHT

For a general repetitive modulation of the excitation light (i.e. more than one frequency component of the excitation light) and with multiple fluorescence components, we simply sum over all terms of the general Fourier expansion of $E(t)$ (see Eqs. 2 and 2a). As with the simpler case above the Q_i, $a_{s,i}$, S_i and $\tau_{s,i}$, all depend on the position in the sample and the pixel, i, in the image; in general the $E_{n,i}$s (i.e. the frequency components of the excitation intensity) are also dependent on the location in the image.

After integration of Eq. 1 using the Fourier expansion of a generalized excitation modulation of Eq. 2, and assuming long times (i.e. $t \to \infty$ so the transient terms have subsided) the convolution equation for the fluorescence signal becomes (where, as previously, "i" refers to the ith pixel element),

$$F_i(t) = Q_i \cdot \left\{ \left[\sum_{s=1}^{s=S_i} E_{0,i} \cdot a_{s,i} \cdot \tau_{s,i} \right] + \sum_{\substack{n=-\infty \\ n \neq 0}}^{\infty} \left[\exp(j \cdot \omega_n \cdot t) \cdot E_{n,i} \cdot \sum_{s=1}^{s=S_i} \left\{ \frac{a_{s,i} \cdot \tau_{s,i}}{1 + j \cdot \omega_n \cdot \tau_{s,i}} \right\} \right] \right\} \tag{7}$$

The reader is reminded that the $E_{n,i}$s are complex numbers. The second sum over "s" in the second square brackets of Eq. 7 refers to the fluorescence decay of an ensemble of fluorescence components at the location in the sample corresponding to pixel "i" of the recorded digital image (as in Eq. 6); each term of this sum (over "s") can be written in polar notation as a single overall amplitude and phase as shown in Eq. 8,

$$\sum_{s=1}^{s=S_i} \left\{ \frac{a_{s,i} \cdot \tau_{s,i}}{1 + j \cdot \omega_n \cdot \tau_{s,i}} \right\} = \sum_{s=1}^{s=S_i} \left\{ \frac{a_{s,i} \cdot \tau_{s,i}}{\left(1 + \left(\omega_n \tau_{s,i}\right)^2\right)^{1/2}} \cdot \exp\left(-j \cdot \phi_{s,n,i}\right) \right\} \tag{8}$$

where the phase delay of each fluorescence component depends on the frequency component ω_n as well as the decay time $\tau_{s,i}$ according to $\phi_{s,n,i} = \tan^{-1}\left(\omega_n \tau_{s,i}\right)$.

Similar to Eq. 6b and using Eq. 8, Eq. 7 can be written in real form as:

$$F_i(t) = Q_i \cdot \left\{ \begin{aligned} & \left[\sum_{s=1}^{s=S_i} E_{0,i} \cdot a_{s,i} \cdot \tau_{s,i} \right] \\ & + 2 \cdot \sum_{n=1}^{n=\infty} \left[\left|E_{n,i}\right| \cdot \sum_{s=1}^{s=S_i} \frac{a_{s,i} \cdot \tau_{s,i}}{\left(1 + \left(\omega_n \cdot \tau_{s,i}\right)^2\right)^{1/2}} \cdot \left\{ \cos\left(\omega_n t + \phi^E_{n,i} - \phi_{s,n,i}\right) \right\} \right] \end{aligned} \right\} \tag{9}$$

where $\phi_{s,n,i}$ is a phase, defined to be $\phi_{s,n,i} = \tan^{-1}\left(\omega_n \tau_{s,i}\right)$. In Eq. 9 we have used the expression $E_{n,i} = \left|E_{n,i}\right| \cdot e^{j\phi^E_{n,i}}$.

The discussion following the special case represented by Eq. 6b also applies to the general case of Eq. 9. The contribution of the separate fluorescence components to the fluorescence signal are indicated explicitly in Eq. 9. We have derived Eq. 9 and written it out in detail in this fashion for later reference, and in order to point out the following characteristics concerning the static and the temporally-resolved steady-state fluorescence signals.

1) The time-independent fluorescence measured in a normal static fluorescence microscope (corresponding to the first term in square brackets of Eq. 9) depends only on the

average static time-integrated light intensity at that pixel, $E_{0,i}$, (see the 2nd paragraph following Eq. 17 for a discussion of this point). It does not depend on the time-varying frequency components of the excitation light. When the static fluorescence signal is expanded in this way it becomes clear why the static fluorescence is often not a good measure of the relative concentrations of the fluorophores at different positions of an image. The static fluorescence can vary for several reasons: (a) the intensity of the excitation light may not be distributed evenly (i.e. $E_{0,i}$ may not be the same for all i), (b) the number (e.g. the concentration) of molecules with any distinct lifetime may vary throughout the sample (i.e. the $a_{s,i}$ can vary), and (c) the lifetimes of the different components can vary (i.e., there may be variations in the nonradiative rates of deactivation (such as quenching and energy transfer), which change the value of $\tau_{s,i}$ from point to point of the image). Only by resolving the lifetimes can we reliably explicate variations in the static fluorescence intensity throughout an image.

2) In the frequency domain, the amplitudes are weighted by $\tau_{s,i}$ in addition to the amplitudes of the exponential decay, $a_{s,i}$. Due to this, the components with very rapid decay times can become significantly de-emphasized (compared to a direct measurement of the components of the exponentially decaying signal); this is true for both the static fluorescence signal, and for the amplitudes corresponding to the different steady-state, temporally-resolved frequency components.

3) This last point becomes important when comparing the fluorescence signals originating from different locations in an image where excited-state processes — that can reduce the lifetimes considerably, such as dynamic quenching, or energy transfer — can occur. In this case, both the static and the temporally-resolved fluorescence signals can become considerably reduced at certain localities of a sample, although the $a_{s,i}$s (which are related directly to the concentrations of the fluorescing species) may not be affected. This means that the average fluorescence intensity should not be used uncritically as a measure of the concentrations of the fluorescing molecular species in the image. An important motivation for measuring fluorescence lifetimes at every location of an image is to be able to check whether the fluorescence intensities are affected by processes that can reduce the $\tau_{s,i}$s; and if so, to correct the intensity images to obtain quantitative estimates of local concentrations.

4) The fluorescence components with slower decay times can be suppressed in the time-varying part of the signal (the second square brackets of Eq. 9) by choosing the fundamental frequency, ω_T, high enough so that the slower components are significantly suppressed relative to the faster components. Because the modulation depth of the time-dependent fluorescence at every pixel is easily determined and can be stored as a separate image, this is an easy and rapid way to produce an image showing the distribution of the faster fluorescence components without making a full analysis to determine the separate lifetimes (which would require making image lifetime measurements at several frequencies and then fitting the frequency dispersion at every pixel). Another more sophisticated method of suppressing/enhancing particular fluorescence components with separate well-defined lifetimes (faster or slower) will be discussed below.

5) Each frequency component contributing to the total fluorescence in Eq. 9 has an associated phase value, $\phi_{s,n,i}$, that is dependent on $\tau_{s,i}$ and ω_n, (see Eq. 8), but the phases of the different lifetime components are independent of each other. This phase can be used to suppress or enhance selectively particular lifetime components of the overall fluorescence, in spite of the fact that the overall fluorescence has single experimentally determined phase and modulation values (see Eqs. 11 and 12). This represents one of the most useful applications of FLIM.

THE OVERALL MEAN MEASURED PHASE AND MODULATION OF THE FLUORESCENCE AT EVERY PIXEL

At each ω_n the overall measured phase and modulation can be expressed easily in terms of the individual phases and modulation depths of all the lifetime components.

Experimentally, only a single overall amplitude and a single overall phase can be determined at each pixel of an image for each frequency component. The lifetime components in Eq. 8 can be written in terms of the experimentally determined overall amplitude $F_{t,\omega_n,i}$ and phase $\Phi_{F,\omega_n,i}$, for each ω_n frequency component as

$$\sum_{s=1}^{s=S_i}\left\{\frac{a_{s,i}\cdot\tau_{s,i}}{\left(1+\left(\omega_n\tau_{s,i}\right)^2\right)^{1/2}}\right\}\cdot\exp\left(-j\cdot\phi_{s,n,i}\right)=F_{t,\omega_n,i}\cdot\exp\left(-j\cdot\Phi_{F,\omega_n,i}\right) \quad (10)$$

where,

$$F_{t,\omega_n,i}=\left(\left[\sum_{s=1}^{S_i}\frac{\alpha_{s,i}}{1+\left(\omega_n\tau_{s,i}\right)^2}\right]^2+\left[\sum_{s=1}^{S_i}\frac{\alpha_{s,i}\cdot\omega_n\tau_{s,i}}{1+\left(\omega_n\tau_{s,i}\right)^2}\right]^2\right)^{1/2} \quad (11)$$

and

$$\Phi_{F,\omega_n,i}=\tan^{-1}\left\{\sum_{s}^{S_i}\frac{\alpha_{s,i}\cdot\omega_n\tau_{s,i}}{1+\left(\omega_n\tau_{s,i}\right)^2}\Bigg/\sum_{s}^{S_i}\frac{\alpha_{s,i}}{1+\left(\omega_n\tau_{s,i}\right)^2}\right\} \quad (12)$$

Here we have defined the low frequency ($\omega_n \rightarrow 0$) amplitude of the sth component in the frequency domain to be $\alpha_{s,i}=a_{s,i}\cdot\tau_{s,i}$. In addition, if we define $F_{0,i}$ to be the static part of the fluorescence,

$$F_{0,i}=E_{0,i}\cdot\sum_{s=1}^{s=S_i}\alpha_{s,i} \quad (13)$$

then Eq. 9 can be written as simply as,

$$\begin{aligned}F_i(t)&=Q_i\cdot\left\{F_{0,i}+\sum_{\substack{n=-\infty\\n\neq 0}}^{\infty}\left[\left|E_{n,i}\right|\cdot F_{t,\omega_n,i}\cdot\exp\left[j\cdot\left(\omega_n\cdot t+\phi_{n,i}^E-\Phi_{F,\omega_n,i}\right)\right]\right]\right\}\\&=Q_i\cdot\left\{F_{0,i}+2\cdot\sum_{n=1}^{\infty}\left[\left|E_{n,i}\right|\cdot F_{t,\omega_n,i}\cdot\cos\left(\left(\omega_n\cdot t+\phi_{n,i}^E-\Phi_{F,\omega_n,i}\right)\right)\right]\right\}\end{aligned} \quad (14)$$

Eq. 14 is the generalized expression of the *measured* temporally-resolved fluorescence signal at the ith pixel of a digital image. The derivation has been given in detail to demonstrate clearly the contribution of the separate fluorescence lifetime components for each ω_n frequency component in Eq. 9 to the measured signal of Eq. 14. Comparison of Eqs. 9 and 14 shows explicitly the connection between the underlying separate contributions of the different fluorescing components with different phases and modulations to the overall phase and modulation at each frequency that is actually measured in an experiment. It will be shown below how to determine the overall phase of the nth frequency component, $\Phi_{F,\omega_n,i}$, and the amplitude of the nth frequency component, $Q_i \cdot E_{n,i} \cdot F_{t,\omega_n,i}$, using heterodyne or homodyne frequency techniques. These are the fundamental experimental quantities that can be determined at every pixel. In general, at the present time only the phase and modulation of the fundamental frequency component ($\omega_1 = \omega_T$) is usually determined; however, if the data is acquired at multiple frequencies with sufficient phase density (that is a sufficient number of phase points - see below), similar experimental methods can be used to determine several frequency components simultaneously. Such a frequency dispersion can be analyzed with the above expressions to render estimates of the underlying separate multiple fluorescence components.

Analyzing the amplitudes of the separate frequency components requires normalizing to the depth of modulation of the excitation light

Eq. 14 is written in terms of absolute fluorescence intensity terms; however, fluorescence intensities must be compared to standards, because the amplitudes of fluorescence measurement are not independently defined quantities. To analyze the amplitudes of these dynamic fluorescence signals they must be normalized in order to become independent of the actual instrumentation effects and the excitation level. The following ratio, $M_{F,n,i}$, of the amplitude of the nth component to the static fluorescence component is defined as the modulation depth of the nth component (see Eq. 14).

$$M_{F,n,i} = \left(Q_i \cdot E_{n,i} \cdot F_{t,\omega_n,i}\right) \Big/ \left(Q_i \cdot F_{0,i}\right) = \left(E_{n,i} \cdot F_{t,\omega_n,i}\right) \Big/ F_{0,i} \tag{15}$$

In $M_{F,n,i}$ the instrumentation factors, Q_i , cancel, so that the signals become normalized for instrumentation effects at every pixel (optics, laser power, electronic amplification, etc.). To determine the modulation depth of the fluorescence, the modulation depth of the excitation light (which is not 100%) must be taken into account. The modulation depth of the excitation light, $M_{E,n,i}$, can be measured directly by simply reflecting the excitation light through the same optical path in the microscope as the fluorescence. For a light scattering, or light reflection, signal, we have $M_{E,n,i} = \lim_{\tau_{s,i} \to 0} M_{F,n,i}$. The ratio $M_{E,n,i}$ for the excitation light can be written as

$$M_{E,n,i} = E_{n,i} \Big/ E_{0,i} \tag{16}$$

Therefore, by taking the ratio of the fluorescence modulation in Eq. 15 (see Eq. 13 for $F_{0,i}$) to the excitation modulation of Eq. 16 it is possible to correct for the depth of modulation of the different frequency components of the excitation light,

$$M_{n,i} = \frac{M_{F,n,i}}{M_{E,n,i}} = \left(\left[\sum_{s=1}^{s=S_i} \frac{\left(\alpha_{s,i} \Big/ \sum_{s=1}^{s=S_i} \alpha_{s,i} \right)}{1 + \left(\omega_n \tau_{s,i}\right)^2} \right]^2 + \left[\sum_{s=1}^{S_i} \frac{\left(\alpha_{s,i} \Big/ \sum_{s=1}^{s=S_i} \alpha_{s,i} \right) \cdot \omega_n \tau_{s,i}}{1 + \left(\omega_n \tau_{s,i}\right)^2} \right]^2 \right)^{\frac{1}{2}} \tag{17}$$

$M_{n,i}$ is the true measured modulation of the fluorescence. If a frequency dispersion is acquired (multiple ω_ns), this expression can be used to analyze the amplitude data for estimating the separate decay times and their corresponding frequency-domain relative amplitudes $\left(\alpha_{s,i}/\sum_{s=1}^{s=S_i}\alpha_{s,i}\right)^{1/2}$.

COMMENTS ON THE DERIVATION OF THE TIME-RESOLVED SIGNAL

This completes the theoretical section describing the temporal characteristics of the fluorescence signal at every pixel of an image. We have shown how to represent the fluorescence of a sample that is excited by a generalized repetitive modulated light source, and we have developed expressions to show explicitly how the overall phase and modulation of the different frequency components of the excitation light are related to the different lifetime components of the fluorescence decay. Similar detailed expressions have been given in the literature to describe non-imaging phase-and-modulation experiments in solutions contained in cuvettes. In general, only a single frequency component of the excitation light is assumed. We have derived the equations in detail so that the role of the fluorescence components with different lifetimes in the frequency response of the fluorescence is clearly presented. It is especially important for image work that the composition of the fluorescence signal be described clearly and be well defined. The methods for detecting the phase-and-modulation of fluorescence signals in images does not in general allow the application of narrow band-pass frequency methods as they are applied in single-channel measurements; thus, it is imperative for a correct estimation of the errors to take the higher frequency components of the measured signal into account if they are present. Between 10^3 to 10^6 separate fluorescence lifetime measurements can be made in a single image. To assist in the interpretation of FLIM data, to guard against possible artifacts of such a complex data set and to utilize the lifetime-resolved data for determining more reliably photophysical parameters that can inform us of the probe environment, it is helpful to have a unified and detailed account of the various processes contributing to the measured fluorescence signal. This is necessary for estimating the errors, as well as interpreting the data. The treatment is easily expanded to include other effects, such as those discussed in the next paragraph.

An important omission has been made in deriving the equations above. It has been assumed that the fluorescence recorded in each pixel of the image arises only from molecules excited by the excitation light impinging on each corresponding individual location. Because many samples are optically complex, there may be contributions to the fluorescence recorded by an individual pixel that arise from neighboring (or farther away) locations of the sample. These spurious contributions can originate either from light scattering or refraction of the excitation beam striking other locations of the sample, or from emission processes in neighboring locations. In addition, if the sample is thick, fluorescence from out-of-focus layers (that can also be derived effectively from neighboring locations of the two-dimensional sample) can also significantly contribute to the measured fluorescence of a pixel. These factors can complicate the matter considerably, and a good understanding of the frequency spectrum (i.e. the ω_n components) and the lifetime spectrum of the composite fluorescence signal is crucial for a reasonable interpretation of such effects. For thick samples (such as biological tissue) the multiple light scattering of the excitation and emission photons can also cause major changes in the phases and amplitudes of a lifetime-resolved signal. This is an active and very large field of interest that cannot be discussed here.

DETECTING THE TIME-RESOLVED FLUORESCENCE AT EACH PIXEL OF THE IMAGE WITH HOMO/HETERODYNE TECHNIQUES

The notation for the fluorescence signal in Eq. 7 can be simplified as,

$$F_i(t) = Q_i \cdot \left\{ \left[\sum_{s=1}^{s=S_i} E_{0,i} \cdot \alpha_{s,i} \right] + \sum_{\substack{n=-\infty \\ n \neq 0}}^{\infty} \left[E_{n,i} \cdot \sum_{s=1}^{s=S_i} \left\{ A_{n,s,i} \cdot \exp\left(j\left[\omega_n \cdot t - \phi_{s,n,i} \right] \right) \right\} \right] \right\} \quad (18)$$

where we have defined $A_{n,s,i} = a_{s,i} \cdot \tau_{s,i} \Big/ \left(1+\left(\omega_n \tau_{s,i}\right)^2\right)^{1/2}$.

The amplification (i. e. the gain) at every location of an image intensifier, $G_i(t)$, can be varied at high frequencies in a time-dependent manner. As a result the signal impinging at every pixel of the digital imaging camera is modulated with a repetitive modulation of the gain by the following form,

$$G_i(t) = G_{0,i} + \sum_{\substack{m=-\infty \\ m\neq 0}}^{m=\infty} G_{m,i} \cdot \exp\left(j\omega_m^G t\right) = G_{0,i} + \sum_{m=-\infty}^{m=\infty} \left|G_{m,i}\right| \cdot \exp\left(j\phi_{m,i}^G\right) \cdot \exp\left(j\omega_m^G t\right) \quad (19)$$

where we have shown explicitly that the Fourier component of the gain, $G_{m,i}$, is a complex entity. $\phi_{m,i}^G$ is the phase of the nth frequency component of the gain modulation at the ith pixel. Note that we have included the possibility that the gain modulation depends on the location within the image. The frequencies of the Fourier expansion of $G_i(t)$ do not have to be the same as those for $F_i(t)$. Even when m=n, ω_m^G is not necessarily equal to ω_n.

Homodyne detection techniques are those where the fundamental frequency of the image intensifier modulation is the same as for the fluorescence. In *heterodyne* detection the frequency components of the fluorescence and the amplification are different, but very close (see Eqs. 21 and 22). In both techniques, only the very low frequency components of the resultant signal is retained. Both detection methods are applied in the frequency domain; the time dependent result of this high frequency amplification can be described analytically by multiplying $F_i(t)$ (Eq. 18) by $G_i(t)$, (Eq. 19), which results in Eq. 20.

$$\begin{aligned}
&F_i(t) \cdot G_i(t) = \\
&\left\{Q_i \cdot \left[\sum_{s=1}^{s=S_i} E_{0,i} \cdot \alpha_{s,i}\right] \cdot G_{0,i}\right\} \\
&+\left\{Q_i \cdot \left[\sum_{s=1}^{s=S_i} E_{0,i} \cdot \alpha_{s,i}\right] \cdot \sum_{\substack{m=-\infty \\ m\neq 0}}^{m=\infty} \left|G_{m,i}\right| \cdot \exp\left(j\phi_{m,i}^G\right) \cdot \exp\left(j\omega_m^G t\right)\right\} \\
&+\left\{G_{0,i} \cdot Q_i \cdot \sum_{\substack{n=-\infty \\ n\neq 0}}^{\infty} \left[E_{n,i} \cdot \sum_{s=1}^{s=S_i} \left\{A_{n,s,i} \cdot \exp\left(j\left[\omega_n \cdot t - \phi_{s,n,i}\right]\right)\right\}\right]\right\} \\
&+\left\{\begin{aligned} &Q_i \cdot \left\{\sum_{\substack{n=-\infty \\ n\neq 0}}^{\infty} \left[E_{n,i} \cdot \sum_{s=1}^{s=S_i} \left\{A_{n,s,i} \cdot \exp\left(j\left[\omega_n \cdot t - \phi_{s,n,i}\right]\right)\right\}\right]\right\} \\ &\cdot \left\{\sum_{\substack{m=-\infty \\ m\neq 0}}^{m=\infty} \left|G_{m,i}\right| \cdot \exp\left(j\phi_{m,i}^G\right) \cdot \exp\left(j\omega_m^G t\right)\right\}\end{aligned}\right\}
\end{aligned} \quad (20)$$

The second and third terms in {} brackets on the right hand side of Eq. 20 always vary sinusoidally at very high frequencies (both ω_m^G and ω_n are between 1 and 400 MHz). Because both the homo- and heterodyne methods involve a low frequency filtering (certainly below 1 KHz) both these terms will be filtered out completely. The first term on the right is derived from the static components of $G_i(t)$ and $F_i(t)$, and will survive all low frequency filtering. The last term is the cross multiplication of the time-dependent sums of $G_i(t)$ and $F_i(t)$, which results in a sum of time-dependent terms varying with the sums and differences of the frequencies ω_n and ω_m^G. All terms of the sums for which there is a high frequency component (in the MHz region) will go to zero due to the low frequency filtering. However, the terms where m=-n, and where $\omega_n^G = -\omega_n$ (homodyne case), or where $\omega_n^G = -\omega_n + \Delta\omega_n$ and $\Delta\omega_n/\omega_n \ll 1$ (heterodyne case), will survive the low frequency (LF) filtering. We rewrite the Eq. 20 retaining only the terms that contribute following the low frequency filtering,

$$\{F_i(t)\cdot G_i(t)\}_{LF} =$$
$$\left\{Q_i\cdot\left[\sum_{s=1}^{s=S_i} E_{0,i}\cdot\alpha_{s,i}\right]\cdot G_{0,i}\right\}$$
$$+\left\{Q_i\cdot\sum_{n=1}^{\infty}\left[\begin{array}{l}|E_{n,i}|\cdot|G_{n,i}|\cdot\exp\left(j\left(\omega_n-\omega_n^G\right)t\right)\\ \cdot\sum_{s=1}^{s=S_i}\left\{A_{n,s,i}\cdot\exp\left(j\left[\phi_{n,i}^E-\phi_{s,n,i}-\phi_{n,i}^G\right]\right)\right\}\end{array}\right]\right\}$$
$$+\left\{Q_i\cdot\sum_{n=1}^{\infty}\left[\begin{array}{l}|E_{n,i}|\cdot|G_{n,i}|\cdot\exp\left(-j\left(\omega_n-\omega_n^G\right)t\right)\\ \cdot\sum_{s=1}^{s=S_i}\left\{A_{n,s,i}\cdot\exp\left(-j\left[\phi_{n,i}^E-\phi_{s,n,i}-\phi_{n,i}^G\right]\right)\right\}\end{array}\right]\right\} \quad (21)$$

Here we have taken advantage of the fact that $\phi_{s,n,i} = -\phi_{s,-n,i}$, $\phi_{n,i}^E = -\phi_{-n,i}^E$, $\phi_{n,i}^G = -\phi_{-n,i}^G$, $A_{n,s,i} = A_{-n,s,i}$ and we have written the frequency component of the excitation light as a polar complex quantity, $E_{n,i} = |E_{n,i}|\cdot\exp\left(\phi_{n,i}^E\right)$. Eq. 21 is easily seen to be equal to,

$$\{F_i(t)\cdot G_i(t)\}_{LF} =$$
$$\left\{Q_i\cdot\left[\sum_{s=1}^{s=S_i} E_{0,i}\cdot\alpha_{s,i}\right]\cdot G_{0,i}\right\}$$
$$+\left\{2\cdot Q_i\cdot\sum_{n=1}^{\infty}\left[\begin{array}{l}|E_{n,i}|\cdot|G_{n,i}|\\ \cdot\sum_{s=1}^{s=S_i}\left\{A_{n,s,i}\cdot\cos\left(\Delta\omega_n t+\left[\phi_{n,i}^E-\phi_{s,n,i}-\phi_{n,i}^G\right]\right)\right\}\end{array}\right]\right\} \quad (22)$$

where in Eq. 22 we have used the identity $\omega_n - \omega_n^G = \Delta\omega_n$. Eq. 22 is the general equation representing the measured signal at every pixel "i"; homodyne detection corresponds to $\Delta\omega_n = 0$; heterodyne detection corresponds to $\Delta\omega_n \neq 0$.

Eqs. 22 defines the terms important for the experimental realization of FLIM in the frequency domain.

THE EXPERIMENTAL REALIZATION OF FLIM

In this section a brief description of the construction and operation of our frequency domain FLIM apparatus will be given, and the performance characteristics of some of the key components will be pointed out. For details and applications the reader is referred to the literature (Clegg et al., 1992; Clegg et al., 1994; Gadella Jr. et al., 1994; Gadella Jr. and Jovin, 1995; Gadella Jr. et al., 1993).

The intensity of the light source (usually a cw Ar^+-laser) is passed through an optical fiber into the side port of the microscope, through the objective onto the sample (epi-illumination). The illuminating light is repetitively modulated (up to several hundred MHz) with a standing wave acoustic-optical modulator (AOM); this is a convenient and inexpensive method to modulate laser light at these high frequencies. As illustrated in Eq. 2, any shape of modulation can be used. The light emitted from the sample is directed through the normal optical path of a fluorescence microscope, including the spectral filtering components, and is focused onto the cathode of a MCP image intensifier. The intensified image at the output phosphor of the intensifier is focused onto a CCD camera for digital recording, and the data is transferred to a computer for further analysis (see below).

The gain of the intensifier ($G_i(t)$ of Eq. 19) is modulated at a high frequency, ω_m^G (see Eq. 19); the frequency is either the same as that of the excitation light (homodyne method), or at a frequency very close (within a few tens to hundreds of Hz) of the excitation modulation frequency (heterodyne method) — see Eqs. 21 and 22. The MCP intensifier can be modulated by varying the acceleration voltage of the intensifier cathode, or by varying the voltage across the MCP that controls the gain. The outcome of both is a repetitive HF gain characteristic $G_{n,i}(t)$, that can be expressed as a Fourier expansion as in Eq. 22. Varying the voltage across the MCP modulates the amplification of every channel separately, and the spatial contrast and resolution of the image is retained during the modulation (Clegg et al., 1992; Clegg et al., 1994; Gadella Jr. et al., 1994; Gadella Jr. et al., 1993). Earlier systems of cathode-modulated intensifiers exhibited image defocusing and iris effects. Defocusing results if during a period of the modulation waveform the voltage is too low and the secondary electrons are not sufficiently accelerated to avoid lateral spreading before hitting the front face of the MCP. Iris effects are due to non-uniformity of the phase of the high frequency gain voltage over the surface of the cathode. However, recently instruments have become available that avoid these problems by employing present state-of-the-art MCP intensifiers with thicker coatings that reduce the cathode impedance and thereby lead to a higher frequency response. We employ both methods of intensifier modulation.

The low frequency filtering (Eq. 21, 22) is attained through the long-time (usually >100 ms) integration operation on the surface of the CCD camera. After integrating, the image is digitized and transferred to a computer. In the homodyne case, where $\Delta\omega_n = 0$, the camera shutter is simply opened and the signal averaged until a sufficient signal-to-noise ratio is achieved. A series of steady-state images are recorded at different $\phi_{n,i}^E - \phi_{n,i}^G$ values (see Eq. 22). This is done by varying either $\phi_{n,i}^E$ or $\phi_{n,i}^G$ separately. Several images are recorded at equally spaced phase points over a total phase shift of 2π. These images form a series of luminescence intensity readings equally spaced in phase at every pixel of the fluorescence image. The phase and modulation values of the fundamental frequency component of this series of points at each pixel are compared to the phase and modulation values of the excitation light (either scattered or reflected) from the microscope sample. In this way the fluorescence lifetime at the image location i can be evaluated from the phase

$\phi^{E}_{n,i} - \phi_{s,n,i} - \phi^{G}_{n,i}$ and the modulation[2] values of the hetero- or homodyne signals (Eq. 22) at each pixel in a completely analogous manner as explained for the direct HF fluorescence signal (Eqs. 14 - 17). The phase and modulation parameters of the HF fluorescence signals (Eq. 14) are retained in the hetero- and homodyne signals (Eq. 22). In the expressions for the hetero- and homodyne signals there is an additional phase and an additional factor in each term due to the HF gain modulation; however, these cancel out when calculating the phase and modulation values relative to those of the excitation light.

In the heterodyne case, where $\Delta\omega_n \neq 0$, $\phi^{F}_{n,i} - \phi^{E}_{n,i} = const$ but the total relative phase is constantly changing at the frequency $\Delta\omega_n$ (period $T = \Delta\omega_n/2\pi$). Here the phase detection is carried out in a boxcar fashion. The "on state" of the cathode of the image intensifier is gated synchronously with a relatively narrow time window ($\Delta t << T$) at the $\Delta\omega_n$ heterodyne frequency, the gate delay is shifted by a small amount (e.g. $T/10$), and another steady-state image is collected on the CCD camera. After a series of images has been acquired, a reference measurement of the reflected light has to be taken in the same fashion. Then data at every pixel is processed in the same way as in the homodyne case.

AN ALGORITHM FOR FITTING THE DATA FOR THE PHASES AND MODULATION AT EVERY PIXEL

The most efficient algorithm for fitting data with equally spaced points is a digital Fourier transform. The discrete Fourier transform of such a series of data points at each pixel determines directly the sine and cosine components, and thereby the phase and modulation of the signal, at every pixel. These components are identical to the best least squares fit to the data using the harmonic functions (Brigham, 1974; Hamming, 1973). Due to the orthogonality of the basis functions, the sine and cosine contributions for each frequency component of the Fourier analysis (multiple frequencies of the fundamental repetition frequency, T^{-1}) are determined individually, and independently. We will only treat the case where the excitation modulation is approximately a sine wave, and in this case only the fundamental frequency component of the signal is of interest An extension to the higher frequency components is straightforward. The data set is a collection of K equally spaced time points $D_{k,i}$ for each pixel (the subscripts k,i refers to the kth time (phase) point for the ith pixel). Each $D_{k,i}$ is multiplied by the sine or cosine function at the appropriate phase θ_k (, the phase of that data point) within the period,

$$F_{\sin,i} = \Sigma_{k=1}^{K} \sin(\theta_k) \cdot D_{k,i} \tag{23}$$

$$F_{\cos,i} = \Sigma_{k=1}^{K} \cos(\theta_k) \cdot D_{k,i} \tag{24}$$

$$F_{0,i} = \left(\frac{1}{K}\right) \cdot \Sigma_{k=1}^{K} D_{k,i} \tag{25}$$

where $F_{\sin,i}$ and $F_{\cos,i}$ are the sine and cosine Fourier components at pixel i, and θ_i is the phase of data point $D_{k,i}$ in the repetitive period, assuming that the first data point has $\theta_k = 0$. $F_{0,i}$ is the phase invariant background of Eq. 13 (at that frequency) at each pixel; in the hetero- homodyne signal (Eq. 22) $F_{0,i}$ is actually equal to $Q_i \cdot \left[\Sigma_{s=1}^{s=S_t} E_{0,i} \cdot \alpha_{s,i}\right] \cdot G_{0,i}$. The phase of the measured signal at each pixel is $\Phi_{F,\omega,i} = \tan^{-1}(F_{\sin,i}/F_{\cos,i})$. The dynamic

[2]The modulation of the hetero- or homodyne signal (Eq. 22) is defined relative to the depth of modulation of the excitation light in exactly the same way as described for the high frequency fluorescence signals in Eqs. 15-17.

amplitude at each pixel is $F_{\omega,i} = \left(F_{\sin,i}^2 + F_{\cos,i}^2\right)^{1/2}$ The fractional modulation depth at every pixel is $M_{F,\omega,i} = F_{\omega,i}/F_{0,i}$; compare this to Eq. 15. The actual value at each pixel is addressed only 3xK times; for M pixels this corresponds to 3xKxM data addressing operations.

The phase and modulation of the reflected excitation light are usually used as reference values for the corresponding values of the sample. If the phase characteristics of the intensifier are the same over the entire surface of the intensifier (no iris effect), and if the impinging excitation light has the same phase and modulation at every illuminated location of the image, the average of these values over the excitation light image can be used rather than determining these values at every pixel. This simplifies the measurement process and is the case for our apparatus. The phase and modulation of the control can be determined very accurately and rapidly in this way. It is also possible to use the emission of a known fluorescence sample (usually a homogeneous sample) for setting the reference values for the sample phase and modulation. In this case the values of these parameters for the sample can be determined relative to those of the fluorescence standard, and because the phase shift and the modulation of the standard fluorescence is known, the actual values for the sample can be determined. This is a simple extension of Eq. 22 and corresponds to the practice in single-channel experiments.

EXTRACTING THE FLUORESCENCE LIFETIMES FROM PHASE AND MODULATIONS VALUES

The fluorescence values at every pixel can be treated in the same way as single-channel cuvette measurements (see references in footnote 1) to determine the fluorescence lifetimes. Eq. 14 expresses the fluorescence signal in terms of the overall *measured* singular phase and amplitude values of the separate frequency components; Eq. 9 represents the same signal expanded in terms of the separate fluorescence components. By making phase and modulation determinations at single frequencies, only mean values of lifetimes can be determined (Clegg et al., 1994; Gadella Jr. et al., 1994; Gadella Jr. et al., 1993; Lakowicz, 1983; Spencer and Weber, 1969). The mean lifetime determined from the phase measurements of each frequency component is $\left\langle \tau_{\Phi_{F,\omega_n,i}} \right\rangle = \tan\left(\Phi_{F,\omega_n,i}\right)/\omega_n$ (see Eq. 14); the mean lifetime determined from the modulation value of each frequency component can be shown to be $\left\langle \tau_{M_{n,i}} \right\rangle = \sqrt{M_{n,i}^{-2} - 1}/\omega_n$. (see Eq. 17). Because of inconveniences in carrying out the imaging measurements at many frequencies, and because of the enormous volume of data, lifetime-resolved imaging experiments have been limited to determining these mean τ values. The *individual* luminescence lifetimes of the separate fluorescence species can however be determined (at least estimated) by measuring the phase and modulation values at several frequencies, and then fitting the frequency dispersion data to Eqs. 12 and 17 (Gratton and Limkeman, 1983; Jameson et al., 1984) (the number of fluorescence components is usually determined from such a fitting procedure). Recently we have improved our FLIM instrumentation so that experiments at multiple frequencies are now a practical option; this will improve considerably the photophysical information that can be determined from FLIM experiments.

SUMMARY

Fluorescence lifetime-resolved imaging in the microscope is now a practical technique that can be implemented in many laboratories having access to electronic imaging equipment. The technique is still relatively novel, but it has the potential of being applied for investigating a wide variety of scientific problems. By modulating the gain of an image intensifier at high frequencies it is possible to make phase-resolved nanosecond measurements simultaneously at every pixel of an image. The image data acquired with a CCD camera can be analyzed rapidly to give fluorescence lifetime images, where every pixel value corresponds to the mean fluorescence lifetime at that position of the image. We have referred the reader to the literature for detailed descriptions and for applications of FLIM instrumentation.

The above presentation has given a compact, integral and comprehensive account of the theoretical framework of the experiment, with an emphasis on frequency domain measurements. Throughout the theoretical description we have discussed several features of the lifetime measurements that specifically pertain to imaging measurements. This exposition was deemed useful to provide potential new users of FLIM with an integral, unified account of the theoretical description, so that it is not necessary at the beginning to search for expositions scattered throughout the literature (most of which refers to single-channel cuvette experiments). The high resolution and quality customary for traditional steady-state imaging measurements can be retained in temporally-resolved images. Phase suppression and enhancement of selected fluorescence components with known lifetimes emphasize particular molecular species in an image. The technique is suitable for routine applications. Still, we expect rapid further improvements especially concerning the speed of data acquisition, analysis, and display of the computed images.

ACKNOWLEDGMENTS

We heartily acknowledge the on-going collaborations with T.W.J. Gadella Jr. and T.M. Jovin.

REFERENCES

Brigham, E. O. (1974). The Fast Fourier Transform. New Jersey. Prentice-Hall

Clegg, R. M., Feddersen, B., Gratton, E. and Jovin, T. M. (1992). Time resolved imaging fluorescence microscopy. *Proc.* SPIE *1604*, 448-460.

Clegg, R. M., Gadella, T. W. J. and Jovin, T. M. (1994). Lifetime-resolved fluorescence inaging. *Proc.* SPIE *2137*, 105-118.

Clegg, R. M., Marriott, G., Feddersen, B. A., Gratton, E. and Jovin, T. M. (1990). Sensitive and rapid determinations of fluorescence lifetimes in the frequency domain in a light microscope. Biophys. J., 34th Annual Meeting of the Biophysical Society *57*, 375a.

Cundall, R. B. and Dale, R. B. (1983). Time-Resolved Fluorescence Spectroscopy in Biochemistry and Biology. New York. Plenum Press

Gadella Jr., T. W. J., Clegg, R. M. and Jovin, T. M. (1994). Fluorescence lifetime imaging microscopy: pixel-by-pixel analysis of phase-modulation data. Bioimaging *2*, 139-159.

Gadella Jr., T. W. J. and Jovin, T. M. (1995). Oligomerization of Epidermal Growth Factor on A431 cells studied by time-resolved fluorescence imaging microscopy. A stereochemical model for tyrosine kinase receptor activation. J. Cell Biol. *129*, 1543-1558.

Gadella Jr., T. W. J., Jovin, T. M. and Clegg, R. M. (1993). Fluorescence lifetime imaging microscopy (FLIM): Spatial resolution of microstructures on the nanosecond time scale. Biophys. Chem. *48*, 221-239.

Gratton, E. and Limkeman, M. (1983). A continuously variable frequency cross-correlation phase fluorometer with picosecond resolution. Biophys. J. *44*, 315-324.

Hamming, R. W. (1973). Numerical Methods for Scientists and Engineers. New York. Dover Publications, Inc.

Jameson, D. M., Gratton, E. and Hall, R. D. (1984). The measurement and analysis of heterogenous emissions by multifrequency phase and modulation fluorometry. Applied Spectroscopy Reviews *20*, 55-106.

Jameson, D. M. and Reinhart, G. D. (1989). Fluorescent Biomolecules: Methodologies and Applications. New York. Plenum Press

Jovin, T. M. and Arndt-Jovin, D. J. (1989). FRET microscopy: digital imaging of fluorescence resonance energy transfer. Application in Cell Biology. In Cell structure and function by microspectrofluorometry, ed. Kohen, E. and Hirschberg, J. G., Academic Press, New York. Vol. pp. 99-117

Lakowicz, J. R. (1983). Principles of Fluorescence Spectroscopy. New York. Plenum Press

Lakowicz, J. R. and Berndt, K. W. (1991). Lifetime-selective fluorescence imaging using an rf phase-sensitive camera. Rev. Sci. Instrum. *62*, 1727-1734.
Lakowicz, J. R. and Maliwal, B. P. (1985). Construction and performance of a variable-frequency phase-modulation fluorometer. Biophys. Chem. *21*, 61-78.
Marriott, G., Clegg, R. M., Arndt-Jovin, D. J. and Jovin, T. M. (1991). Time-resolved imaging microscopy. Phosphorescence and delayed fluorescence imaging. Biophys. J. *60*, 1374-1387.
Morgan, C. G., Mitchell, A. C. and Murray, J. G. (1992). In situ fluorescence analysis using nanosecond decay time imaging. Trends Anal. Chem. *11*, 32.
Piston, D. W., Sandison, D. R. and Webb, W. W. (1992). Time-resolved fluorescence imaging and background rejection by two-photon excitation in laser scanning microscopy. *Proc.* SPIE *1604*, 379-389.
So, P. T. C., French, T. and Gratton, E. (1994). A frequency domain time-resolved microscope using a fast-scan CCD camera. *Proc.* SPIE *2137*, 83-92.
Spencer, R. D. and Weber, G. (1969). Measurements of subnanosecond fluorescence lifetime with a cross-correlation phase fluorometer. Ann. Acad. Sci. *158*, 361-376.
Wang, X. F., Uchida, T., Coleman, D. M. and Minami, S. (1991). A two dimensional fluorescence lifetime imaging system using a gated image intensifier. Applied Spectroscopy *45*, 360-366.

CONFOCAL FLUORESCENCE LIFETIME IMAGING

Hans C. Gerritsen

Debye Institute, Department of Molecular Biophysics,
University of Utrecht, Buys Ballot Laboratory, P.O. Box 80.000,
3508 TA Utrecht, The Netherlands. E-mail: h.c.gerritsen@fys.ruu.nl

INTRODUCTION

The optical microscope, arguably the most mature of all laboratory instruments, is currently enjoying a period of intense development that is opening up many new application areas. These developments have been made possible by the use of new optical arrangements, sophisticated light sources and detectors and by the application of novel fluorescence-based contrast forms. The combination of fluorescence with microscopy yields detailed and selective information at the microscopic level. For instance, different structural and functional entities in biological cells can be stained selectively with (extrinsic) fluorescent probes so that they can be easily identified in the images. Numerous fluorescent probe molecules are now available which enable the selective imaging of the electric potential of membranes, DNA, spatial variations of free ion concentrations, and specific proteins within single (living) biological cells.

In fluorescence microscopy the red-shift of the fluorescence emission with respect to the excitation wavelength is utilized to discriminate the fluorescence signal from a background of scattered excitation light. Recently several groups started working on an alternative spectroscopic imaging method whereby the fluorescence intensity decay is used as a contrast mechanism[1-8]. In its simplest form the decay is a single exponential process:

$$I(t) = I_0 \, e^{-t/\tau} \qquad (1)$$

Here I_0 is the fluorescence intensity at time t=0 and τ the fluorescence lifetime, which is usually on the order of several ns. The fluorescence lifetime τ determines the shape of the decay curve and is therefore independent of factors such as the excitation intensity, absorption effects and fading due to photobleaching. The differences in the fluorescence decay behaviour can often be associated with changes in the chemical environment of the fluorophore such as pH, ion concentration and binding to macromolecules. This makes fluorescence lifetime imaging an attractive tool for quantitative imaging. In addition, it can be used to discriminate the emission of specific fluorophores against an autofluorescence background. Moreover, fluorescence lifetime contrast based imaging provides a discrimination between molecules with overlapping fluorescence emission bands, but with different fluorescence decay times, in multi-labelling experiments.

This paper starts with the description of a confocal fluorescence lifetime imaging setup utilizing time-gated detection of fluorescence. This is followed by a number of examples of (confocal) fluorescence lifetime imaging.

METHOD

The technical difficulty in fluorescence lifetime imaging lies in the fact that in general the lifetime of the probes is of the order of several nanoseconds. Therefore fast detection schemes have to be used. At present two different approaches to fluorescence lifetime imaging are being used, phase fluorimetry and time-gating. In the first the lifetime is derived from the phase shift between the excitation light and the fluorescence emission light, see the chapter by R. Clegg, in the second the lifetime is related to the ratio of the intensities collected in two time-windows. Here, we will limit ourselves to the implementation of time-gated lifetime imaging in an existing confocal microscope[6].

TIME-GATED LIFETIME IMAGING

The principle of the time-gated detection technique used in our microscope is depicted schematically in Figure 1.

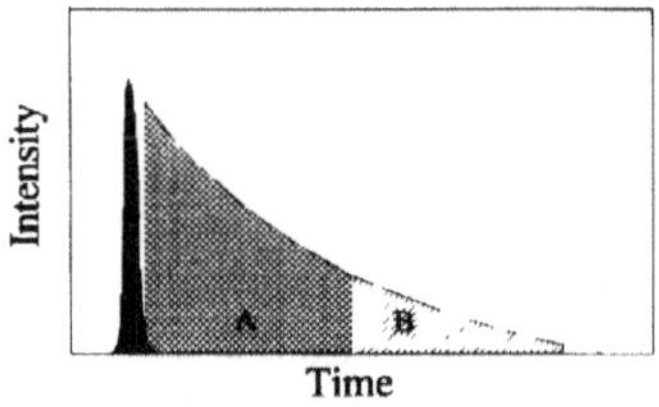

Figure 1. The principle of time gated detection.

The essence of the method is the capture of the fluorescence emission excited by a light pulse within two time windows. The ratio of the integrated intensities provides a good measure for the fluorescence decay time provided the width of the windows and their separation in time are chosen judiciously. In the case of a simple monoexponential fluorescence intensity decay the fluorescence lifetime is given by[10]

$$\tau = \frac{\Delta t}{\ln(I_A / I_B)} \tag{2}$$

where Δt is the time-offset between two windows, A and B, of equal widths and I_A and I_B are the corresponding integrated fluorescence intensities. In the monoexponential case the optimum gate width amounts to 2.5 τ[11]. In the case of a multi-exponential fluorescence decay only an effective fluorescence decay time can be determined.

Most of the work described here was performed using a conventional confocal laser scanning microscope (CLSM) equipped with a 488 nm cw Argon-ion laser[12]. In order to realize the pulsed excitation required for the time-gating method, a fast optical chopper is placed between the cw laser and the microscope[6]. In this way the sample is excited with nanosecond light pulses at rates of up to 25 MHz. The fluorescence emission is detected in two time-gated windows whose widths as well as offsets can be adjusted independently. Usually, the first window starts immediately after the light pulse and the second starts at the closure of the first (see Figure 1). Both gates open sequentially after each light pulse such that at the optimum gate widths (2.5 τ) virtually all fluorescence is collected after each excitation pulse. The acquisition of the fluorescence in both windows after each light pulse makes the method comparatively insensitive to fading caused by photo bleaching.

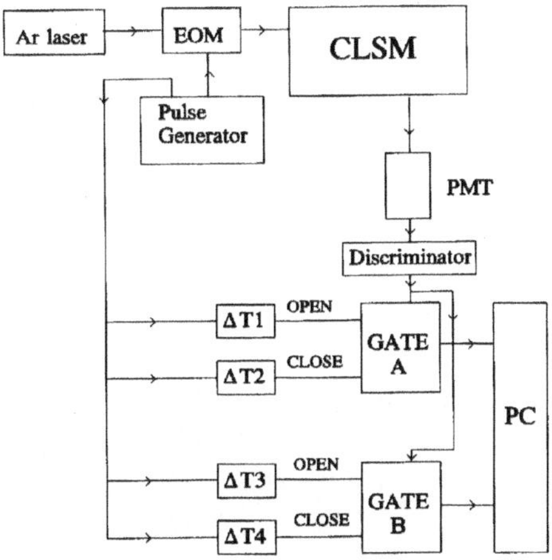

Figure 2. The schematic diagram of the confocal fluorescence lifetime microscope.

The microscope and the detection circuit are depicted schematically in Figure 2. Fluorescence photons are detected by means of a fast photo multiplier tube (Hamamatsu R1894) with a rise-time of 0.8 ns. The output pulses of the PMT are fed into a 1 GHz bandwidth preamplifier which is followed by a discriminator and two digital gates. The time offset of the windows with respect to the excitation pulse and their widths are controlled by passive delay lines and the output of these gates is connected to two fast 16-bit counters. Most of the electronics utilizes Emitter Coupled Logic (ECL) electronics with sub nanosecond rise times. Count rates of 20 MHz have been realized using this system without noticeable pile-up effects. Several hundred to one thousand laser excitation pulses are required per pixel in order to accumulate sufficient counts. After this integration period the contents of the two counters are fed into a 16-bit digital frame-grabber interfaced to a 486 personal computer. The system is capable of producing fluorescence lifetime images at a rate of 20 - 100 μs per pixel. After the acquisition of the two time-gated images, the result is displayed on the computer screen. Usually the ratio of the contents of the two windows, the window-ratio, is displayed after scaling to grey values in the range 0-255. Conversion to the parameter of interest such as pH, oxygen concentration, ion concentration, lifetime etc. can be realized by calibrating the microscope response (window-ratio). In practice this is done by recording the window-ratio using a series of samples of known composition.

EXAMPLES OF FLUORESCENCE LIFETIME IMAGING

EXPERIMENTAL

We now present several examples of fluorescence lifetime imaging to demonstrate the potential of the technique. In most of the experiments, a Zeiss multi-immersion 40x/0.9 objective in the water immersion mode was used. The lifetime images were recorded with two contiguous windows with an equal gate width of about 2.5 ns and an excitation pulse width of approximately 2 ns, except for the oxygen images. These will be discussed in detail below. In all the imaging experiments the first window started immediately after the excitation pulse. The fluorescence emission in the lifetime experiments was selected by means of a dichroic beamsplitter (510 nm) in combination with a 520 nm long pass filter. In general the selection of the emission filters is not critical since most of the excitation light is separated in time with respect to the time windows. The images were recorded at an intensity of only 2 μW as measured before the microscope objective. Despite this low power level the acquisition time of the images (512x512 pixels) was just 10 to 60 seconds.

MULTI-LABELLING

Lifetime images were obtained from the algae Gymnodinium nagasakience, which was stained with an antibody FITC conjugate against proteins in its outer membrane. In addition to the FITC probe, the algae contains autofluorescing chloroplasts. The fluorescence lifetime of the antibody amounts to 1.1 ns and the average lifetime of the chlorophyll is about 0.7 ns[6,13].

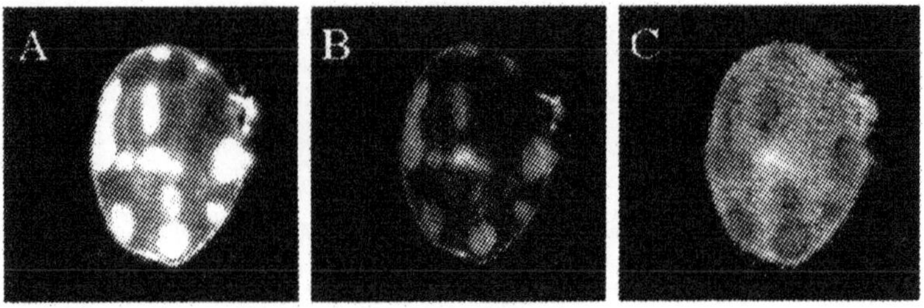

Figure 3. The fluorescence intensity images corresponding to window A (a) and window B (b) of the FITC antibody stained algae Gymnodinium nagasakience. c) The ratio image of the intensities recorded in windows A and B (I_B/I_A). Dark and light areas correspond to short and long lifetimes respectively.

The images corresponding to windows A and B are shown in Figure 3a and b respectively. As expected the image corresponding to the later time window contains a lower intensity because of the fluorescence intensity decay. Both images contain the signal from the antibody-FITC and the chloroplasts. The sum of the two images is an ordinary intensity image.

In Figure 3c the ratio image of the intensities recorded in windows A and B is shown (I_B/I_A). Here, the grey value is a measure for the fluorescence lifetime. Dark and light areas correspond to short and long lifetimes respectively, see eq. 1. Thus, the chloroplasts show up as dark spots due to the comparatively short fluorescence lifetime, while the lighter areas e.g. around the membrane are due to the longer fluorescence lifetime of FITC. The fluorescence lifetime image can be used to carry out a simple segmentation procedure. To this end two binary images are generated from the fluorescence lifetime image, corresponding to lifetimes larger and shorter than 0.9 ns. After multiplying the intensity image (I_A+I_B) with the binary images, two images are created containing intensities with decay times shorter than 0.9 ns (chlorophyll, Figure 4a) and longer than 0.9 ns (FITC, Figure 4b).

For comparison two conventional fluorescence intensity images are shown in Figure 5a (chlorophyll) and 5b (FITC). These figures were recorded using a 580 nm longpass filter and a FITC bandpass filter respectively. The morphological resemblance between the chlorophyll images (4a and 5a) as well as the FITC images (4b and 5b) is striking. The lifetime contrast images even seem to show a better discrimination between chlorophyll and FITC. However, this may be attributed to the non ideal properties of the spectral filters employed here.

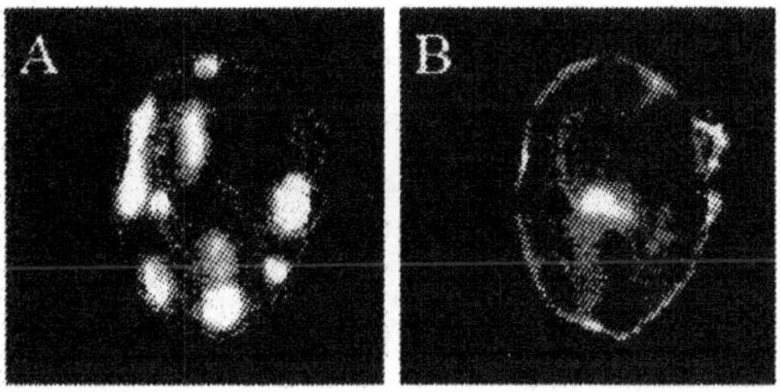

Figure 4. The chlorophyll a) and FITC b) fluorescence intensity images resulting from a lifetime based segmentation procedure.

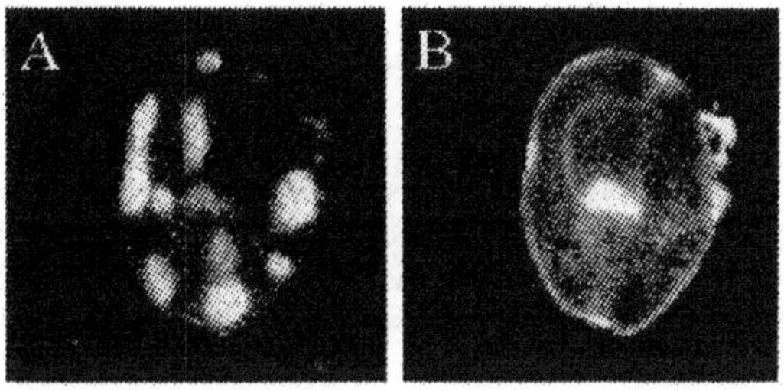

Figure 5. The chlorophyll a) and FITC b) fluorescence intensity images recorded using a longpass and a bandpass (spectral) filter respectively.

Quantitative ion-concentration imaging

Ion concentrations such as pH, Ca^{2+}, Na^{+} are of great importance in biological system and can conveniently be studied by employing fluorescent probes[14-15]. These probes are characterized by a high selectivity for the ions and a marked change in their photophysical properties upon binding. Several recent studies have revealed that the differences in the fluorescent lifetime of the free probe and the probe-ion complex can be used for the quantitative imaging of ion concentrations[16-19]. Two examples obtained in our laboratory are shown below. A full description of these experiments can be found in ref. 18 and 19.

Calcium imaging using CalciumGreen CalciumGreen is a well known calcium indicator which exhibits a large increase in the fluorescence quantum yield and thus in fluorescence intensity upon binding to calcium ions[20]. This property and the fact that the probe can be excited with visible (blue) light, makes it attractive for the imaging of free calcium ions. The main problems with the use of this and other single wavelength probes arises from inhomogeneous partitioning of the probe within or between cells, in addition to probe leakage and photobleaching. All these factors affect the fluorescence intensity and therefore hamper the quantitative interpretation of fluorescence images. However, CalciumGreen shows large differences between the fluorescence lifetimes of the free probe and the probe-calcium complex. This opens up the possibility to use the probe as a fluorescence lifetime probe.

We investigated the lifetime behaviour of CalciumGreen at different calcium concentrations. The microscope response was calibrated using CalciumGreen dissolved in a series of buffers of known ion concentrations. The window-ratio for CalciumGreen as a function of the calcium concentration in the calibration buffers is shown in Figure 6. The S-shaped calibration curve indicates that CalciumGreen can be used for fluorescence lifetime imaging in the Ca^{2+} range from about pCa 6.5 (300 nM) to 8.5 (3 nM). This makes the probe particularly suitable for lifetime imaging in the low end of the physiological free Ca^{2+} concentration range.

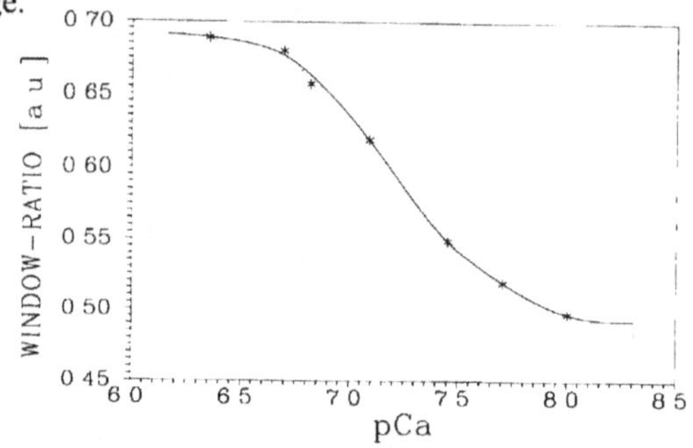

Figure 6. The window-ratio for CalciumGreen as a function of the pCa.

The shape of the calibration curve is consistent with the existence of two distinct states of the probe, one bound to Ca^{2+} and one free, each state with its own fluorescence lifetime. The validity of this description has been confirmed by standard time-correlated single photon counting measurements[18]. These indeed yielded a mono-exponential decay for calcium free and calcium saturated buffers with decay times of 0.46 ns and 3.53 respectively. Intermediate calcium concentrations on the other hand resulted in a biexponential decay with components corresponding to the bound and free states, but with amplitudes depending on the calcium concentration.

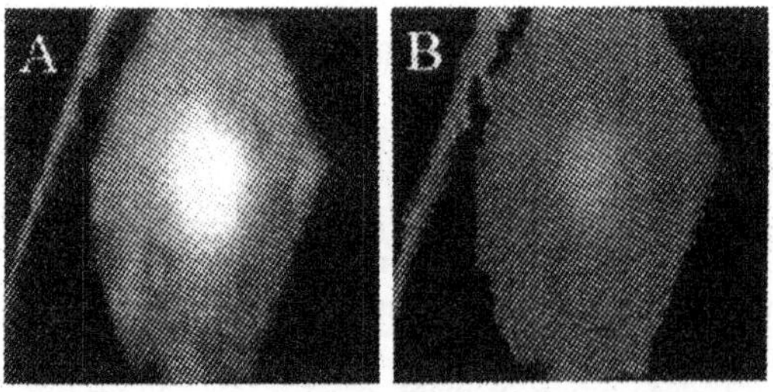

Figure 7. a) The fluorescence intensity image of a CalciumGreen stained rat myocyte b) The window-ratio (fluorescence lifetime) image of the myocyte. The grey-value is a measure for the Ca^{2+} concentration.

The fluorescence intensity image of a CalciumGreen stained cardiac rat myocyte is shown in Figure 7a. The fluorescence intensity in the nucleus is significantly higher than that in the cytoplasm of the cell. This strongly suggests a higher Ca^{2+} concentration in the nucleus of the myocyte. Figure 7b shows the window-ratio image of the same cell. Here, the grey-value is a direct measure of the window-ratio. From the calibration curve and the grey-values of the image a concentration of 20 nM ± 8% in the nucleus and 18 nM ± 18% in the cytoplasm can be found. Ratio imaging experiments using the ratio probe Fura-2[21] reported in literature indeed confirm the presence of equal Ca^{2+} concentrations of this low value in the nucleus and cytoplasm of this type of myocyte[17]. The difference between the intensity image and the window-ratio image can be explained by the preferential partitioning of CalciumGreen into the nucleus.

It should be noted that the calibration in this experiment is based on probes dissolved in calcium buffers. The calibration may not be transferable to cells since the chemical environment inside the cells is different from that in the buffers.

pH imaging using c.SNAFL-1 In order to compare the results of fluorescence lifetime imaging with those of conventional emission ratio imaging a study was undertaken using the pH probe c.SNAFL-1[22]. C.SNAFL-1 shows a 60 nm shift in the position of the emission band in the pH range 7-9[23]. In addition to the shift of the emission spectrum the probe also shows notable changes in the fluorescence intensity decay behaviour in this pH range.

For the emission ratio experiments the microscope described in the previous section was used with cw excitation and without time-gated detection. The emission was detected with a bandpass filter at 540±20 nm and with a 580 nm long pass filter. The ratio of the intensities recorded at the two different emission wavelength bands was used as a measure for the pH.

The relation between the microscope response (window-ratio and emission ratio) and the pH was again determined experimentally. First of all the microscope response was calibrated using a series of c.SNAFL-1 containing buffers over the pH range from 5 to 10.5. A second calibration of the pH response was performed on intact Chinese Hamster Ovary (CHO) cells incubated with c.SNAFL-1 diacetate using the nigericin/high potassium method[22,24]. In the latter calibration method the cell membrane is permeable to H^+ ions such that the intra cellular pH can be controlled by varying the extra cellular pH.

The calibration curve of the window-ratio versus the pH in buffer is shown in Figure 8a. From the figure the useful range for quantitative pH determinations is estimated to extend from about pH 7.0 to pH 9.5. A calibration curve of the window-ratio, recorded for cells, is shown in the same figure. The curves overlap almost perfectly, indicating that the calibration of the window-ratio versus pH is hardly sensitive to the presence of cellular components. The maximum deviation between the two curves amounts to 0.2 pH units at pH 8.0.

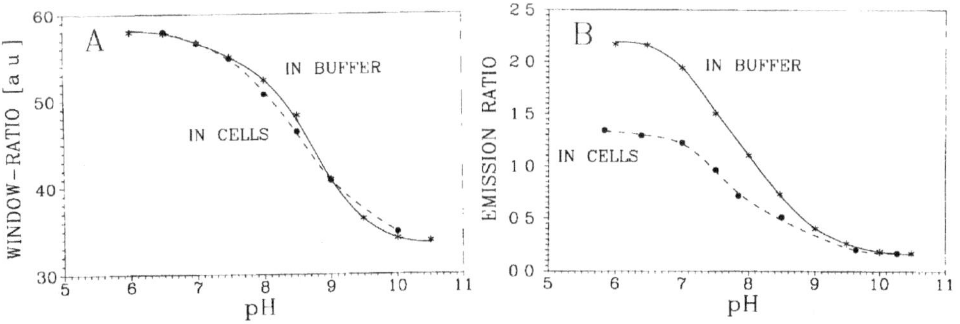

Figure 8. a) The calibration curve of the window-ratio versus the pH in buffer and cells. b) The same but now for the emission ratio.

The calibration curves of the emission ratio versus pH on the same buffers and cells are shown in Figure 8b. Now significant deviations are found between the two calibration methods. The emission ratios recorded in cells are systematically lower than those recorded in buffer. A 1 pH unit lower plateau is observed at the low end of the pH range.

These results clearly demonstrate that the fluorescence emission ratio of c.SNAFL-1 is sensitive to the detailed chemical environment and therefore requires a calibration procedure on cells. On the other hand the probe can be used as a fluorescence lifetime probe with only a simple calibration on buffers.

Living intact CHO cell stained with c.SNAFL-1 were imaged with our microscope. The confocal window-ratio images show a constant pH of 7.3 throughout the whole cell, as is also found with emission ratio imaging. The pH value agrees well with that of previous emission ratio studies using an *in vivo* calibration procedure[25,26].

Oxygen imaging

Oxygen sensing in single cells (J774 macrophages) has been realized with fluorescence lifetime contrast. To this end the dynamic quenching of the oxygen sensitive probe ruthenium tris(2,2'-dipyridyl) dichloride hydrate (RTDP[27]) was used. Dynamic quenching is a process in which the excited state is depopulated by collisions with quencher molecules. The introduction of an additional decay channel causes a reduction of the fluorescence lifetime. This opens up the possibility to use fluorescence lifetime imaging for the quantitative imaging of quenching molecules such as oxygen. For a dynamic quenching process the quencher concentration can be calculated from the fluorescence lifetime using the Stern-Volmer equation:

$$\frac{\tau_0}{\tau} = 1 + K_{SV}\,[O_2] \tag{3}$$

with τ_0 and τ the unquenched and quenched lifetimes respectively, K_{SV} the quenching coefficient and $[O_2]$ the oxygen concentration. K_{SV} depends on the temperature, viscosity of the surrounding medium, collision radius and quenching efficiency of the fluorescent molecule[28]. This parameter can be determined experimentally by measuring the ratio τ_0/τ as a function of the oxygen-concentration.

The first step in the experiments concerned the calibration of the microscope response by measuring the window-ratio of RTDP in buffers containing different oxygen-concentrations. The oxygen-concentration varied from <0.02 μM to 300 μM, which includes the range of concentrations under physiological conditions (90-260 μM).

Fluorescence lifetime images were acquired on setting the leading edge of window B about 300 ns after the leading edge of window A. Both windows had an identical width of 300 ns. Fluorescence lifetime images of RTDP were acquired using an excitation frequency of 555 kHz. The lower duty-cycle results in a longer acquisition time. Nevertheless it was possible to record a 256x256 image in about 100 seconds in single cells. The images were recorded using an Olympus oil immersion objective 100x/NA 1.25.

The window-ratio in the calibration buffers was found to decrease on increasing the oxygen-concentration in agreement with expectations, see Figure 9. The calibration curve was used in subsequent experiments to directly convert window-ratios into oxygen-concentrations. The validity of the Stern-Volmer equation was verified by converting the window-ratio into the fluorescence lifetime using equation 2. An excellent linear fit was found over the entire O_2-concentration range (0 to 300 μM): $\tau_0/\tau = 0.998 + 2.73\ 10^{-3}\ [O_2]$, with R=0.999. This observation proves that the lifetime reduction is indeed induced by dynamic quenching and that the quenching coefficient amounts to 2.73 mM^{-1} .

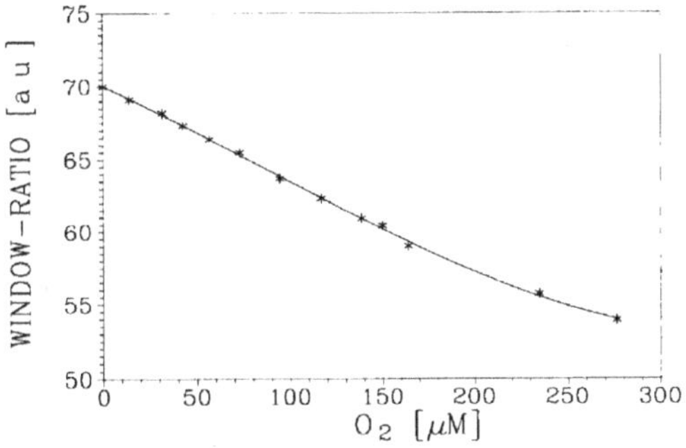

Figure 9. The RTDP window-ratio as a function of the oxygen concentration.

Importantly, no changes in the window-ratio within the experimental errors were found upon varying the pH from 7 to 8 and adding 150 mM NaCl. The influence of cellular contents on the fluorescence behaviour of RTDP was investigated using imaging experiments on J774 macrophages in buffers at two oxygen-concentrations: < 0.02 μM and 250 μM. Identical window-ratios are found inside the cells and in the surrounding buffers for each oxygen-concentration. On making the generally accepted assumption that the intra and extra cellular oxygen-concentration are the same, this shows that the window-ratios are insensitive to the presence of cellular material.

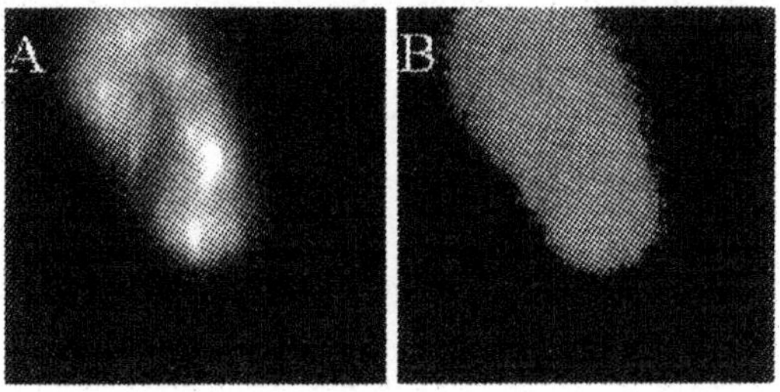

Figure 10. In a) the fluorescence intensity image of a RTDP stained macrophage is shown and in b) the corresponding window-ratio image

We used the information obtained above to carry out quantitative O_2-imaging experiments on the J774 macrophages. The fluorescence intensity image (figure 10a) shows a non-uniform intensity across the cell, while the window-ratio image (figure 10b) shows no variations across the cell. The latter rules out oxygen concentration variations inside the cell. The heterogeneities in the fluorescence intensity image can therefore be attributed to a differential partition of the probe into cell organelles in the cytoplasm.

The lifetime contrast setup may also be used to monitor the oxygen-consumption of macrophages in an air-tight sample stage. After flushing fresh buffer through the sample stage the window-ratio was measured every four seconds at the same spot in a cell. A typical variation of $[O_2]$ with time is shown in Figure 11. Initially the oxygen-concentration amounts to 250 μM, which is identical to the oxygen-concentration in the surrounding buffer. The $[O_2]$ rapidly decreases due to oxygen consumption by the macrophages and reaches a minimum of 170 μM after about 4 minutes. On reaching the minimum oxygen-concentration, the sample compartment was flushed again with fresh buffer for 3 minutes and the experiment was repeated. This operation replenishes the oxygen-concentration to 250 μM again. Interestingly, on repeating the experiment, a minimal oxygen-concentration of

125 μM is reached, though the depletion rate remains constant at about 5 μM/min. No explanation has been found yet for this behaviour.

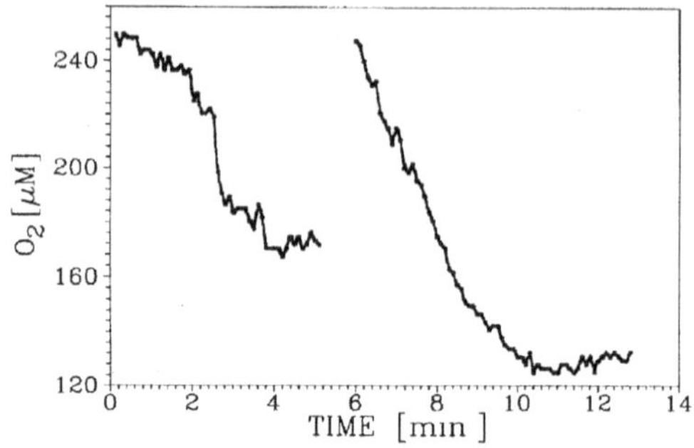

Figure 11. The variation of the oxygen concentration with time after shutting of the oxygen supply to the sample container.

DISCUSSION

Fluorescence lifetime imaging has been implemented by several groups in widefield[1-5] as well as in confocal[6,8,9] microscopes. The confocal arrangement improves the contrast of the fluorescence lifetime images compared to widefield imaging systems. An additional advantage of the confocal approach is that only a single channel detection is required, which simplifies the design of the fluorescence lifetime detection system significantly. Moreover, no correction for inhomogeneous detector response is necessary. We opted for a system utilizing pulsed excitation in combination with two time-gated photon counters which open sequentially after each light pulse. This combination offers a number of important advantages:

- High collection efficiency; at the optimum gate widths the fluorescence is collected during a total period of 5 τ after each excitation pulse.
- High sensitivity; only 225 detected photons need to be acquired for lifetime sensing with an accuracy of 10%[11,29]. This may be somewhat worse in our case because of the comparatively broad excitation pulse. Nevertheless, fluorescence lifetime images could be recorded with only 2 μW of excitation power in several tens of seconds.
- Insensitivity to excitation intensity fluctuations; even strong intensity fluctuations are canceled by the simultaneous use of two gates.
- Low intrinsic detection noise level; in photon counting the dark count rate is in general several hundred counts/s while the signal level is at many million counts/s.
- Insensitivity to photo bleaching effects; besides a reduction of the statistics fading of the fluorescence intensity due to photo bleaching does not show up in the window-ratio. The signal in both time windows fades at equal rates. However, photobleaching may introduce photo-products whose fluorescence emission bands overlap that of the probe and moreover exhibit a different fluorescence lifetime.
- Insensitivity to scattered excitation light; scattered excitation light falls outside the detection time-windows, so simple emission filters suffice.
- Sub nanosecond lifetime resolution; despite the broad excitation pulse employed here, a 0.7 ns lifetime could be discriminated from a 1.1 ns lifetime.

Fluorescence lifetime imaging is an attractive alternative to ratio imaging for the quantitative determination of ion concentration distributions in cells. It was shown here that the intensity Ca^{2+} probe CalciumGreen and the emission ratio pH probe carboxy SNAFL-1 can be used for quantitative fluorescence lifetime imaging in the physiological ion concentration range. The results on c.SNAFL-1 reveal only minor differences between the calibration on buffer and the *in vivo* calibration method. Therefore a simple calibration procedure on buffers suffices when c.SNAFL-1 is used as a fluorescence lifetime probe. In contrast the emission ratio of c.SNAFL-1 strongly depends on the method of calibration, such that *in vivo* calibrations are required for quantitative imaging studies on cells.

The fluorescence lifetime imaging technique is not restricted to the two ion indicators used here. The presence for two distinct fluorescence lifetimes for the bound and free probe appears to be a common feature of ion indicators. In addition it was demonstrated that the continuous lifetime change introduced by selective dynamic quenching can be employed for the quantitative imaging of quencher molecules such as oxygen. A limitation of the current lifetime imaging implementations is the comparatively long image acquisition time of 10 s and longer. This makes the method unsuitable for the imaging of fast Calcium transients. However, it is expected that this restriction will be lifted in the near future.

ACKNOWLEDGEMENTS

This research is supported by the Technology Foundation (STW), grant number UNS00.2225, under the auspices of the Netherlands Organization for Scientific Research (NWO). This paper would not have been possible without the excellent collaboration with Arie Draaijer and Renata Sanders. Finally, I want to thank Yehudi Levine for his inspiring comments.

REFERENCES

1. C.G. Morgan, A.C. Mitchel and J.G. Murray, Nanosecond time-resolved fluorescence microscopy: principles and practice, Trans. R. Microsc. Soc. 1:463 (1990).
2. J.R. Lakowicz and K.W. Berndt, Lifetime-selective fluorescence imaging using an rf phase-sensitive camera, Rev. Sci. Instrum. 62(7):1727 (1991).
3. T. Ni and L.A. Melton, Fluorescence lifetime imaging: An approach for fuel equivalence ratio imaging, Appl. Spectr. 45(6):938 (1991).
4. X.F. Wang, T. Uchida, D.M. Coleman and S. Minami, A two-dimensional fluorescence lifetime imaging system using a gated image intensifier, Appl. Spectr. 45(3):360 (1991).
5. G. Marriot, R.M. Clegg, D.J. Arndt-Jovin and T.M. Jovin, Time resolved imaging microscopy. Bioph. J., 60(6):1374 (1991).
6. E.P. Buurman, R. Sanders, A. Draaijer, H.C. Gerritsen, J.J.F. van Veen, P.M. Houpt and Y.K. Levine, Fluorescence lifetime imaging using a confocal laser scanning microscope, Scanning 14:155 (1992).
7. T.W.J. Gadella, T.M. Jovin and R.M. Clegg, Fluorescence lifetime imaging microscopy (FLIM) - Spatial resolutions of microstructures on the nanosecond time scale. Biophysical Chemistry 48(2):221 (1993).
8. D.W. Piston, D.R. Sandison and W.W. Webb, Time-resolved fluorescence imaging and background rejection by two-photon excitation in laser scanning microscopy, SPIE 1640:379 (1992).
9. P.T.C. So, T. French, W.M. Yu, K.M. Berland, C.Y. Dong and E. Gratton, Time-resolved fluorescence microscopy using two-photon excitation, Bioimaging 3:49 (1995).
10. R.J. Woods, S. Scypinski, L.J. Cline Love and H.A. Ashworth, Transient digitizer for the determination of microsecond luminescence lifetimes, Anal. Chem. 56:1395 (1984).
11. R.M. Ballew and J.N. Demas, An error analysis of the rapid lifetime determination method for the evaluation of single exponential decays, Anal. Chem. 61:30 (1989).

12. A. Draaijer and P.M. Houpt, A standard video-rate confocal laser scanning reflection and fluorescence microscope, Scanning 10:139 (1988).
13. W. Haehnel, A.R. Holzwarth and J. Wendler, Picosecond fluorescence kinetics and energy transfer in the antenna chlorophylls of green algae, Photochem. and Photobiol. 37:435 (1983).
14. T.J. Rink, R.Y. Tsien, and T. Pozzan, Cytoplasmic pH and free Mg^{2+} in lymphocytes, J. Cell Biol. 95:189 (1982).
15. R.Y. Tsien, and M. Poenie, Fluorescence ratio imaging: a new window into intracellular ionic signaling, Trends Biochem. Sc. 11:450 (1986).
16. J.R. Lakowicz, H. Szmacinski, K. Nowaczyk, and J.L. Johnson, Fluorescence lifetime imaging of Ca^{2+} using visible wavelength excitation and emission, SPIE 1640:390 (1992).
17. J.R. Lakowicz, H. Szmacinski, and K. Nowaczyk, Fluorescence lifetime imaging of calcium using Quin-2, Cell Calcium 13:131 (1992).
18. R. Sanders, H.C. Gerritsen, A. Draaijer, P.M. Houpt and Y.K. Levine, Fluorescence lifetime imaging of free calcium in single cells, Bioimaging 2:131 (1994).
19. R. Sanders, A. Draaijer, H.C. Gerritsen, P.M. Houpt and Y.K. Levine, Quantitative pH imaging in cells using confocal fluorescence lifetime imaging microscopy, Anal. Biochem. 227:302 (1995).
20. M. Eberhard and P. Erne, Calcium binding to fluorescent calcium indicators: calcium green, calcium orange and calcium crimson, Biochem. Bioph. Res. Comm. 180:209 (1991).
21. J.R. Berlin, and M. Konishi, Ca^{2+} transients in cardica myocytes measured with high and low affinity Ca^{2+} indicators, Biophys. J. 65:1632 (1993).
22. T.J. Rink, R.Y. Tsien, and T. Pozzan, Cytoplasmic pH and free Mg^{2+} in lymphocytes, J. Cell Biol. 95:189 (1982).
23. J.E. Whitaker, R.P. Haugland and F.G. Prendergast, Spectral and photophysical studies of benzo[c]xanthene dyes: dual emission pH sensors, Anal. Biochem. 194:330 (1991).
24. J.A. Thomas, R.N. Buchsbaum, A. Zimniak and E. Racker, Intracellular pH measurements in Ehrlich ascites tumor cells utilizing spectroscopic probes generated in situ, Biochemistry 18:2210 (1979).
25. J.A. Cook and M.H. Fox, Effects of chronic pH 6.6 on growth, intracellular pH, and response to 42°C hyperthermia of chinese hamster ovary cells, Cancer Res. 48:2417 (1988).
26. R. Gonzalez-Mendez, G.M. Hahn, N.G. Wade-Jardetzky and O. Jardetzky, Magn. Res. Med. 6:373 (1988).
27. M.E. Lippitsch, J. Pusterhofer, M.J.P. Leiner and O.S. Wolfbeis, Fibre-optic oxygen sensor with the fluorescence decay time as the information carrier, Anal. Chim. Acta. 205:1 (1988).
28. J.R. Lakowicz, J. Biochem. Biophys. Methods 2:90 (1980).
29. H.C. Gerritsen, R. Sanders, A. Draaijer and Y.K. Levine, The photon economy of fluorescence lifetime imaging, Scanning 18:55 (1996).

MULTIDIMENSIONAL FLUORESCENCE MICROSCOPY: OPTICAL DISTORTIONS IN QUANTITATIVE IMAGING OF BIOLOGICAL SPECIMENS

N S White[1], R J. Errington[2], M.D Fricker[1] and J L Wood[1]

Oxford University
[1]Plant Sciences Department,
South Parks Rd, Oxford OX1 3RB, UK
[2]Physiology Department,
Parks Rd, Oxford. OX1 3PT, UK

INTRODUCTION

Photometric and morphometric measurement of biological microscopy specimens often poses problems due to their optical properties and those of the microscope (Fricker and White, 1992) 3-D Visualisation (White, 1995) and quantification (Sandison *et al*, 1995, Taylor and Wang, 1989b) of fluorescent probes in living tissue is possible by computer assisted confocal microscopy (e g Brackenhoff *et al*, 1979, Shotton, 1989, Wilson, 1990, Inoué 1995). The specimen is both the object under investigation and also a key component of this integrated imaging system

FLUORESCENCE MICROSCOPES

Fluorescence microscopes are designed for two optical processes (i) illuminating the object field and (ii) imaging the illuminated field Each has an intensity response The partial-response for one point of the field is given by multiplication of the component point responses Summing responses as the detection is moved in steps over the entire illuminated field (a convolution) gives the overall performance for a fluorescence microscope (Wilson and Sheppard, 1984). This performance can be modified by reducing the field and thus the range of the convolution (Lukosz, 1966).

Wide-field (conventional) illumination is not directly focused but is spread over the field aperture and then projected into the sample Fluorescent molecules throughout the sample are evenly illuminated. A camera, observers eye etc., collects in-focus and out-of-focus light to a wide-field image The conventional microscope response is a 3D detection Airy disk convolved with a flat illumination response, thus condenser aberrations are less important than those of imaging optics

Confocal illumination is focused to a point which is scanned through the specimen Many confocal implementations are commercially available (Pawley, 1995), this discussion is restricted to single-point laser scanning confocal microscopes The confocal detector is also focused within the specimen but the field is reduced to a single point, which is scanned to form an image Detection and illumination are thus approximately equivalent and co-aligned, giving z-discrimination and optical sectioning. The overall confocal response is now a 'convolution' over a point field, which is just the product of the illumination and detection responses (Wilson and Sheppard, 1984) Illumination and detection are equally important, aberrations of co-alignment (confocality) influencing confocal performance (Entwistle and Noble 1994, Fricker and White, 1992)

Two-photon techniques (Denk *et al*, 1990) use the combined energy of two photons to excite fluorescent molecules The two-photon point illumination response is approximately the equivalent confocal excitation squared (Jacobsen *et al,* 1994) Wide-field detection (i.e with a condenser) should ideally be used The overall two-photon microscope response is thus the convolution of squared illumination with a flat detection response. This mimics an incoherent confocal reflection instrument (Gu and Sheppard, 1995), with a performance determined by the illumination

OPTICAL DISTORTIONS IN 3-D FLUORESCENCE MICROSCOPY

Microscopes may distort the optical field (spatial error) and/or signal intensity (photometric error) Absorption, path length, refractive index, polarisation, etc., are widely exploited for image contrast (Inoué, 1986; Lacey, 1989) but fluorescence relies on specific probes and optical filters (Taylor and Wang, 1989a) Interactions with the specimen (fig. 1) and/or optics that give interference, amplitude/phase contrast etc , also degrade fluorescence

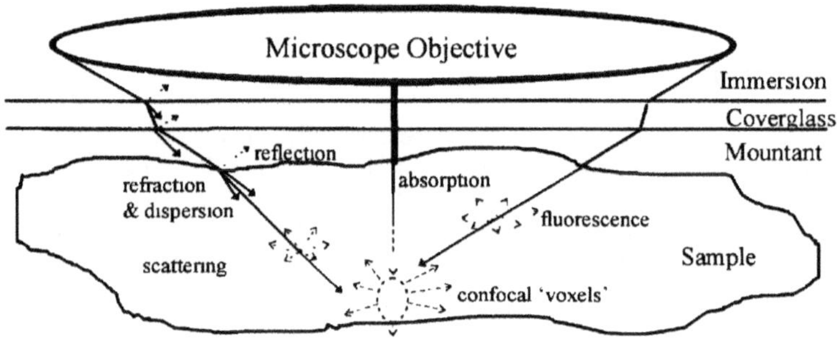

Figure 1. Absorption, refractive index and dispersion depend on the specimen and wavelength(s) and determine the 'confocal voxel' imaged by CLSM. Absorption by fluorophores does not significantly reduce illumination but scattering losses may be important (e g. with un-cleared woody tissue, dense cartilage, etc.).

Spatial Distortions

Computer assisted 3-D fluorescence microscopes produce a digital array (Carlsson and Liljeborg, 1989). The corresponding specimen volume depends on several spatial distortions

Bending of light by an optical boundary (refraction) results from the refractive index gradient (see Born and Wolf, 1991). Light rays from a point source radiate in a spherical wavefront, but a plane boundary will introduce distortion, preventing refocussing to a point image This is spherical aberration which blurs (i) the illumination focus within the specimen (important for confocal and two-photon) and (ii) fluorescence rays at the detector (all microscopes) Optical boundaries at the sample e.g., biological material (Cogswell and Larkin, 1995), mountant, coverglass and immersion medium (fig 2), increase spherical errors and deviate the axial focus The image is thus blurred and spatially distorted (Carlsson, 1991, Visser *et al,* 1992, Hell *et al,* 1993)

Refractive index often varies with wavelength (Born and Wolf, 1991) Thus colours are refracted through different angles (dispersed) Chromatic aberration (fig 2) may (i) focus wavelengths at different points in the specimen (confocal and two photon) or (ii) disperse multiple fluorescence emissions at the detector (conventional and confocal) or (iii) illuminate a sample volume that is spatially separated from the detected volume (confocal only). All objectives exhibit some chromatic focus error (Fricker and White, 1992) and lateral chromatic (magnification) error Dispersion in the sample, mountant and immersion media will increase chromatic errors.

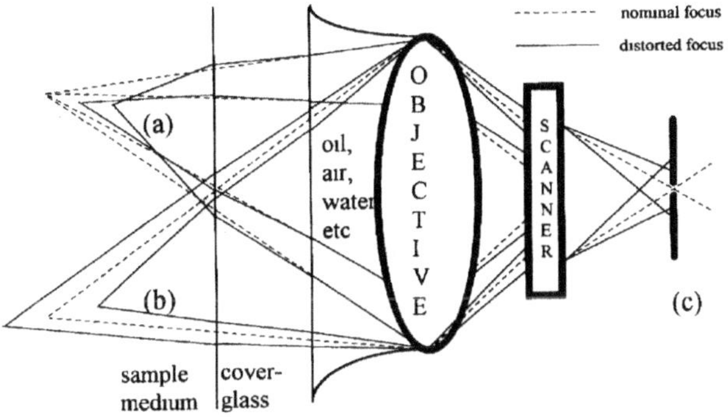

Figure 2. Refractive boundaries lead to differential focusing and blurring of point illumination and fluorescence (a). Differential focus of wavelengths (b), by chromatic aberration, affects axial and in-plane focus Both distortions produce attenuation by blurring and differential focus at the confocal detector (c).

Photometric Errors

All non-absorption degradations arise from spatial distortions in illumination or detection (conventional, confocal and two-photon), or mis-registration (confocal only) Biological specimens are often largely transparent (Visser *et al,* 1991) but local absorption may occur, e.g plant chlorophyll and other pigments absorb (Moss and Loomis, 1952; Wooley, 1971) particularly around 440±25nm and 675±25nm (near some laser lines) Diffuse scattering (e g. in woody tissue, cartilage etc) also contributes to attenuation, which in confocal and two-photon imaging also arises by focus error from refractive structures (cell and vascular walls, cuticle, organelles and membranes etc)

In conventional microscopy monochromatic blurring increases when focusing into biological specimens, reducing peak values, contrast and resolution but with no appreciable attenuation Blurring from chromatic distortion reduces contrast and resolution with multiple fluorescent probes, or probes with a wide emission range (Inoué and Oldenbourg, 1993) There is no attenuation in the absence of absorption In confocal microscopes, fluorescence away from the overlapping excitation and emission probes is severely attenuated. Combined optical sectioning and monochromatic blurring therefore result in axial, and some lateral, attenuation (Hell *et al,* 1993) All confocal spatial distortions gives rise to attenuation. A unique property of confocal microscopy is that chromatic mis-alignment of illumination and emission will blur and, by z-sectioning, attenuate the signal (Sandison *et al,* 1995) Illumination distortions degrade two-photon performance (Jacobsen *et al,* 1994), monochromatic blurring decreasing the probability of excitation This gives a unique detection-independent attenuation The wavelength is longer than for single-photon excitation, so two-photon (with wide-field detection) is less sensitive to spherical errors than confocal, but there is a slight chromatic based attenuation by lengthening of the laser pulses (Denk *et al,* 1995).

Critical optical components must be optimised (Inoué and Oldenbourg, 1993), including the choice of objective (Keller, 1995) Immersion and mountant may be matched by a water or glycerol lens but live specimens cannot be matched to the mountant (fixed material can be 'cleared' in glycerol, DPX etc.) The correct coverglass (refractive index = 1.525, thickness = 170μm) should be used unless the objective is designated NCG (no coverglass) or has a coverglass correction collar. High quality objectives optimise resolution and contrast in conventional microscopy, but chromatic corrections can be relaxed for single channel fluorescence Lenses with high transmission fluorite elements (e g for UV) are often acceptable Imaging at the field edge and photomicrography demand 'plan' (flat field) objectives In-focus images of multiple fluorophores over the entire field require plan apochromatic lenses (corrected for three or four wavelengths). Confocality is extremely sensitive to spatial distortions. The resulting attenuation is minimised by plan apochromats, which are (i) incompatible with highest transmission, (ii) corrected over a restricted of their nominal 3-D field of view range (axial and in-plane) and (iii) usually not available for different immersion media. Lenses for two-photon epi-illumination cannot be ideal for both excitation and detection, illumination needs a plan lens but chromatic correction is unnecessary for single wavelength excitation. Wide-field detection needs only high transmission, but epi-illumination precludes using a simple collector

The 3-D Sampled Volume

A 3-D fluorescence image is related to the corresponding fluorochrome distribution (White, 1995) via a transfer function (fig.3) with three components (i) a spatially varying 3-D point spread function or PSF, (see also Agard and Sedat, 1983; Shaw, 1995; Tao and Nicholson, 1995) (ii) a 3-D attenuation function and (iii) a distorted 3-D sampling lattice

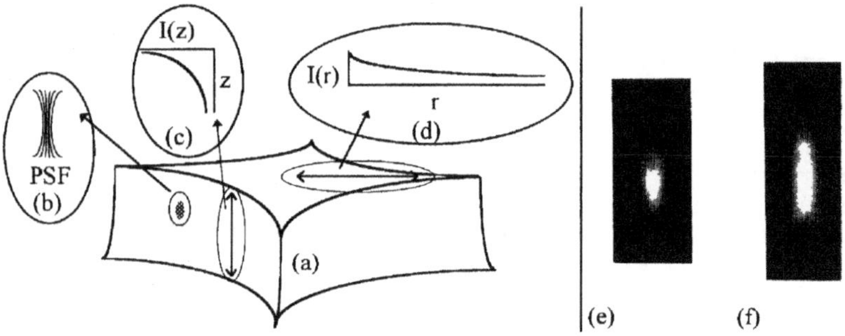

Figure 3. The 3-D optical field (a) is distorted and the PSF (b) varies through the sample, depending on corrections and sample properties In confocal and two-photon fluorescence (a) combined with (b) give axial (c) and lateral (d) attenuation. Measured PSFs (e) above and (f) below leaf epidermis the deep PSF is more asymmetric and axially blurred FWHM (x,y) (e) = 0 2 μm (f) = 0.25μm z· (e) = 0 6μm (f) = 1.2μm

We model monochromatic focus error (fig.4) by tracing all rays passing through the lens This is more reliable than just using (i) extreme rays (Visser *et al*, 1992), which over estimate errors, (ii) near-axis rays (Carlson, 1991) which under-estimate confocal and two-photon errors, or (iii) a single average ray (Gahm and Witte, 1986) which overestimates, ignoring total internal reflection Our method is simpler than PSF derivations (e g Hell *et al*, 1993).

The consequences of monochromatic and chromatic blurring of illumination and detection is illustrated diagramatically in fig.5, based on digitally modelled PSF's (White *et al*, 1995)

Measuring Distortions in Test Samples

To independently test our ray tracing model, we measured aqueous medium/ glass profiles in confocal reflection and fluorescence for a 'sandwich' of fluorescent buffer between coverglass and slide (Hell *et al*, 1993) This exactly duplicated the geometry of our model Measured distances

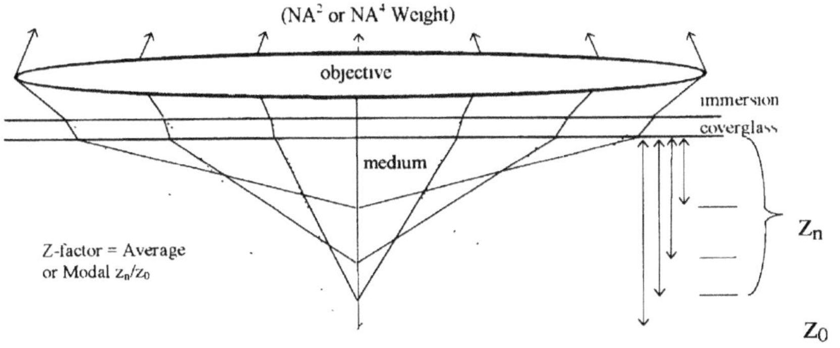

Figure 4. Axial distortion from specimen/mountant refractive boundaries modelled by ray tracing using Snell's Law (Born and Wolf, 1991). Rays at increasing angles focus at displaced positions (z_n) The ratio of displaced to nominal focus (z_n/z_0) for each ray (Carlsson, 1991) was weighted by its contribution to the total efficiency (Gahm and Witte, 1986) with factors = NA^2 (conventional) or NA^4 (confocal/two-photon) to produce average or modal corrections Total internal reflection may limit the maximum aperture to less than the nominal NA. when the immersion refractive index is higher than the mountant (Hell *et al,* 1993)

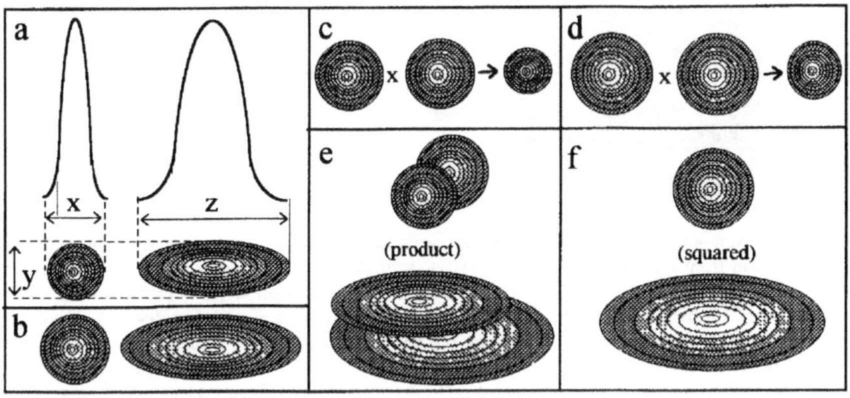

Figure 5. 3D 'Gaussian PSFs' crudely model the image intensity from a 'point' object with blue fluorescence excitation and blue/green/red emissions (see White *et al*, 1995) The unaberrated response (a) shows (x,y) and z-resolution anisotropy in all cases Increasing the Gaussian radius (b) simulates monochromatic blurring, which is reduced by confocal (c) and two-photon (d) product responses, at the expense of attenuation Chromatic error is simulated by a wavelength dependent magnification (x,y) or shift (z) which gives misalignment of illumination/detection in confocal (e), resulting in attenuation. The two-photon response (f) has no chromatic shift, but resolution is limited by the red illumination

between the two inner glass /medium boundaries were compared to the 'correct' distance measured with a water immersion lens (White *et al*, 1995) Focus error in glass/medium reflection profiles and our NA^4-weighted modal z-factors from ray tracing agree well; in confocal reflection, maximum intensity indicates the position of an edge Our measured focus errors for 1.4 NA oil/water (0 83) agree with our NA^4-weighted model factor (0 84) and the NA-independent z-factor (0 82)of Hell *et al*, (1993) These workers do not distinguish between confocal and conventional focus error

We used a second model of fluorescent beads (fig 6), to assess spatial distortions from non-planar structures Our focus error from beads (0.69) agrees with the average, but not the more reliable modal, ray model Refraction by the bead blurs the image axially with an added focusing effect There is a small error using a 50% threshold to find the bead edge (van Vliet, 1993, Verbeek and van Vliet, 1994) Guilak (1993), also calibrated with beads, but their confocal z-factor (0 87) was closer to our conventional model with planar boundaries than our estimates from beads. This may be partly explained by their negatively stained beads and 25%/75% threshold, or their 'partially confocal' slit detection (see Pawley, 1995), that would be closer to the conventional case

In situ calibration

There is no general correction of axial attenuation for all biological specimens, even for a particular confocal system (Pawley, 1995) or objective (Keller, 1995) Z-focus error is slightly dependent on the NA and attenuation follows from increased z-discrimination at high NA We have used a pragmatic approach to calibrate specimen induced fluorescence attenuation using *in situ* sea responses in a standard botanical preparation Abaxial epidermal strips were peeled from *Commelina communis* leaves (Weyers and Travis, 1981) For *in situ* PSF measurements we used

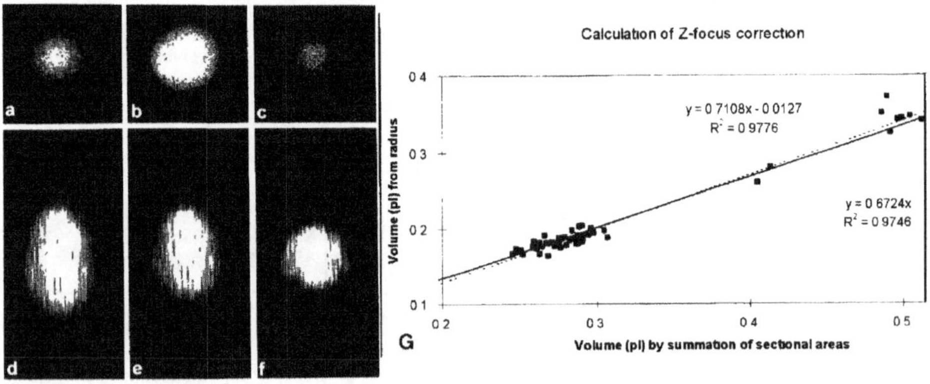

Figure 6. Confocal optical sections of a 7 μm fluorescent bead (a-c) (x,z) vertical sections uncorrected (d), after z-correction from the ray model (assuming flat boundaries) (e), and after empirical z-correction (f) accounting for focusing and spherical aberration by the bead Correlation (g) between volume estimates from the central confocal section (x-axis) and by summing areas in all sections (y-axis), using a 50% intensity threshold to detect the edge of the bead, provides a slope giving the z-correction· y = 0 672x, R^2 = 0 975 (zero intercept - solid) or y = 0 718x - 0 013, R^2 = 0 978 (non-zero intercept - dotted) This correction gave volumes by confocal microscopy consistent with Coulter counter calibrations (see also Guilak, 1993)

unfixed material in buffer *In situ* sea response attenuation profiles were measured in epidermal strips that were fixed and made permeable through a graded (25 - 100%) ethanol series Strips were taken via 50 % ethanol to buffer or glycerol containing 53μM fluorescein To compare morphology of the stomatal complex after digital corrections, epidermis was stained with Primulin, by vacuum infiltration, to contrast cell walls (Edwards et al, 1988, see also Verbelen et al, 1994) *In situ* fluorescence profiles were compared to mounting medium to assess specimen aberrations (Fig 7)

The magnitude of errors in multidimensional biological microscopy

In this paper, and other reports (Fricker and White, 1992, White *et al*, 1995), we describe spatial and photometric aberrations in 3D fluorescence microscopy Lateral spatial distortions are equivalent in conventional, confocal and two-photon microscopy, chromatic error can be expected up to 0.05% of the field (about 150nm for a Nikon 60x 1 4NA Plan apo objective), with a similar value for monochromatic blurring Axial chromatic shifts may be up to half the equivalent confocal PSF FWHM for the visible range, i e 300nm (60x 1 4 NA Nikon Plan Apo oil/water), 1μm (Zeiss 0 85 NA Plan-neofluar oil/water)) and 2μm (Nikon 10x Fluor, 0 6NA air/water) Uncorrected axial UV

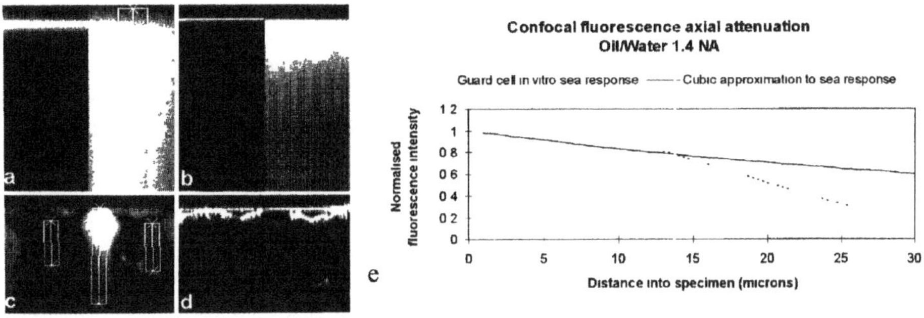

Figure 7. (x,z) sections of aqueous fluorescent 'seas'. 25x 0 8NA Zeiss Plan Neofluar glycerol immersion into glycerol (a) showed little attenuation, 60x Nikon '1 4 NA' Plan Apo Oil lens into aqueous medium (b) showed 50% attenuaticn over 100μm. Fluorescence (x,z) section through stomata of leaf epidermis (c) and in reflection (d) showed combined absorption, refraction and scattering Confocal attenuation responses expanded from parameters (e) 'sea' (solid line) and guard cell '*in situ* sea' (dotted line) The curves are not co-normalised but the slope (attenuation) is initially the same, increasing into the guard cell by sample refraction and optical sectioning

focusing may be from 5-10μm (350/360nm) or up to 50μm (325nm) from the focus of visible wavelengths Chromatic-based attenuation will be negligible for conventional and two-photon fluorescence, but up to 50% (lateral) or 75% (axial) for confocal. Monochromatic z-correction factors will be 0 84 - 0 88 (oil/water) depending on optical sectioning, and 1 4 - 1 33 for air/water Monochromatic axial attenuation for confocal and two-photon (100μm depth) will be up to 50% (oil/water), less than 4% (glycerol/glycerol) and above 50% for an oil lens with hydrated specimens

In all our models and empirical measurements, reducing confocality gives axial distortions

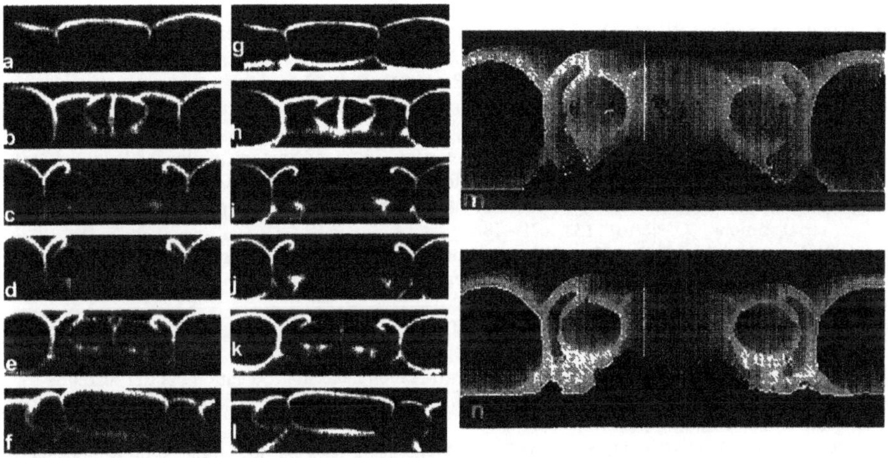

Figure 8. Axial distortion and attenuation degrades vertical (x,z) sections of hydrated, Primulin stained leaf epidermis (a-d) when using an oil lens Images are 'stretched out' by monochromatic focus error from mismatched immersion and mountant The same data (h-l) after our confocal correction of axial geometry (factor = 0 846) and attenuation using the *in situ* calibration, cell wall structure around the inner region of the guard cell is more faithfully represented in 3-D reconstructions (m) before and (n) after correction

approaching those of conventional optics, with a larger contribution from near axial compared to higher angle rays and a corresponding fall in attenuation Hell *et al*, (1993) show a somewhat greater axial attenuation than we have seen This may reflect differences in confocal geometry, confocality and/or objective lens performance Guilak (1993), showed attenuation into articular cartilage; nearly linear over 30μm then levelling out, in a similar manner to our quadratic response, over 100μm

DIGITAL IMAGE CORRECTION AND RESTORATION

Spatially varying transfer function components can be treated separately, allowing deblurring techniques that assume a constant PSF. All modes require two 3-D functions, a spatial lattice and PSF, but confocal and two-photon also have 3-D attenuation Specimen calibration allows images to be 'pre-corrected' (fig 8) before further restoration, e g by iterative deconvolution (Shaw, 1995), or 'blind deconvolution' (Holmes *et al, 1995)* where a PSF is restored from the image.

The availability of high NA water immersion lenses goes some way towards addressing aberrations in non-refractive fully hydrated samples. In some cases, an oil lens may still give the best images into thin or weakly refractive specimens For example, a '1 4 NA' oil lens has an effective NA of 1 33 for aqueous samples giving 1 23 times more signal than a 1 2 NA water lens With post correction of residual artefacts (e g Rigaut and Vassey, 1991), this extra efficiency may be realised.

ACKNOWLEDGEMENTS

NSW thanks the Dept of Plant Sciences, Oxford for support and is now a Royal Society Industry Fellow. RJE is supported by the Wellcome Trust JW acknowledges BBSRC for a Graduate Studentship NSW thanks Carlos Rubbi and Graham Hogg for many helpful discussions

REFERENCES

Agard, D A and Sedat, J W (1983) Three dimensional architecture of a polytene nucleus *Nature* **302**, 676 - 681

Born, M and Wolf, E (1991), *Principles of Optics*, Pergamon press, Oxford.

Brackenhoff, G J , Blom, P and Barends, P (1979) Confocal scanning light microscopy with high aperture immersion lenses. *J Microsc* **117**, 219 - 232

Carlsson, K (1991) Influence of specimen refractive index, detector signal integration, and non-uniform scan speed on the imaging properties in confocal microscopy *J Microsc.* **163**, 167 - 178

Carlsson, K and Liljeborg, A. (1989). A confocal laser microscope scanner for digital recording of optical serial sections *J. Microsc* **153**, 171 - 180

Cogswell, C.J and Larkin, K G (1995). The specimen illumination path and its effect on image quality *Handbook of Biological Confocal Microscopy*, Second Edition (Ed J B Pawley). pp 127 - 138 Plenum Press, New York

Denk, W., Strickler, J H and Webb, W.W (1990) Two-photon laser scanning fluorescence microscopy. *Science* **248**, 73 - 76

Denk, W , Piston, D.W and Webb, W W (1995) Two-photon molecular excitation in laser scanning microscopy. *Handbook of Biological Confocal Microscopy*, Second Edition (Ed J.B. Pawley). pp 445 - 458. Plenum Press, New York

Edwards, M C., Smith, G M and Bowling, D J.F. (1988) Guard cells extrude protons prior to stomatal opening - a study using fluorescence microscopy and pH electrodes *J Exp. Bot.* **39**, 1541 - 1547

Entwistle, A. and Noble, M. (1994). Optimising the performance of confocal point scanning laser microscopes over the full field of view *J Microsc* **175**, 238 - 251

Fricker, M.D and White, N S. (1992) Wavelength considerations in confocal microscopy of botanical specimens, *J Microsc* **166 (1)**, 29-42

Gahm, T. and Witte, S (1986) Measurement of optical thickness of transparent tissue layers *J Microsc* **141**, 101 - 110

Gu, M. and Sheppard, C.J R (1995) Comparison of three dimensional imaging properties between two-photon and single photon fluorescence microscopy *J Microsc* **177**, 128 - 137

Guilak, F. (1993), Volume and surface area measurement of viable chondrocytes *in situ* using geometric modelling of serial confocal sections. *J. Microsc.* **173(3)**, 245-256

Hell, S., Reiner, G., Cremer, C and Stelzer, H.K., 1993, Aberrations in confocal fluorescence microscopy induced by mismatches in refractive index *J Microsc* **169(3)**, 391-405.

Holmes, T J., Bhattacharyya, S , Cooper, J.A., Hanzel, D , Krishnamurthi, V., Lin, W.C., Roysam, B , Szarowski, D H. and Turner, J.N. (1995) Light microscopic images reconstructed by maximum likelihood deconvolution.). *Handbook of Biological Confocal Microscopy*, Second Edition (Ed. J.B Pawley). pp 389 - 402. Plenum Press, New York.

Inoué, S. (1986). *Video Microscopy* Plenum press, New York.

Inoué, S (1995). Foundations of confocal scanned imaging in light microscopy *Handbook of Biological Confocal Microscopy*, Second Edition (Ed. J.B. Pawley). pp 1 - 18 Plenum Press, New York.

Inoué, S and Oldenbourg, R. (1993). Optical instruments; Microscopes. In· *Handbook of Optics*. Second Edition (Ed. Optical society of America). McGraw-Hill, New York.

Jacobsen, H , Hanninen, P., Soini, E. and Hell, S W. (1994) Refractive index induced aberrations in two-photon confocal fluorescence microscopy *J. Microsc.* **176**, 226 - 230

Keller, H E. (1995) Objective lenses for confocal microscopy. *Handbook of Biological Confocal Microscopy*, Second Edition (Ed. J B Pawley) pp 111 - 126. Plenum Press, New York

Lacey, A J. (1989). *Light Microscopy in Biology, A Practical Approach*. IRL Press, Oxford.

Lucosz, W (1966) *J. Opt Soc Am.* **56**, 1463

Moss, R.A. and Loomis, W E. (1952). Absorption spectra of leaves. I The visible spectrum. *Plant Physiol* **27**, 370 - 391.

Pawley, J.B. (1995) Light paths of current commercial confocal light microscopes for biology. *Handbook of Biological Confocal Microscopy*, Second Edition (Ed J.B. Pawley) pp 581 - 598 Plenum Press, New York.

Rigaut, J P. and Vassey, J. (1991). High resolution three-dimensional images from confocal scanning laser microscopy. Quantitative study and mathematical correction of the effects due to bleaching and fluorescence attenuation in depth *Anal Quant Cytol Histol.* **13**, 223 - 232

Sandison, D R., Williams, R M , Wells, K.S., Strickler, J and Webb, W W (1995) Quantitative fluorescence confocal laser scanning microscopy (CLSM). *Handbook of Biological Confocal Microscopy*, Second Edition (Ed J B. Pawley) pp 39 - 54. Plenum Press, New York.

Shaw, P., J. (1995). Comparison of wide-field/deconvolution and confocal microscopy for 3D imaging *Handbook of Biological Confocal Microscopy*, Second Edition (Ed J B Pawley). pp 373 - 388 Plenum Press, New York

Shotton, D.M. (1989) Review Confocal scanning optical microscopy and its applications for biological specimens. *J. Cell Sci.* **94**, 175 - 206

Tao, L and Nicholson, C. (1995) The three dimensional point spread functions of a microscope objective in image and object space. *J Microsc* **178**, 267 - 271.

Taylor, D., L- and Wang, Y -L (1989a) *Fluorescent Analogues, Labelling Cells and Basic Microscopy* in: Methods in Cell Biology Vol 29. Acad Press, London.

Taylor, D , L- and Wang, Y -L (1989b) *Quantitative Fluorescence Microscopy - Imaging and Spectroscopy* in Methods in Cell Biology Vol 30 Acad. Press, London.

Verbeek, P , W. and van Vliet, L.J (1994) On the location of curved edges in low-pass filtered 2-D and 3-D images. *IEEE Trans. Pattern Anal. Machine Vision.* **16 (7)**, 726 - 733

Verbelen, J P and Stickens, D (1994). *In vivo* determination of fibre orientation in plant cell walls with polarisation CSLM. *J. Microsc.* **177**, 1 - 6

Visser, T D , Groen, F.C A , Brakenhoff, G J., 1991, Absorption and scaterring correction in fluorescence confocal microscopy *J. Microsc* **163**, 189-200.

Visser, T.D , Oud, J L. and Brakenhoff, G J , 1992, Refractive index and axial distance measurements in 3-D microscopy. *Optik.* **90**, 17-19

van Vliet, L J (1993). PhD thesis Gray - Scale Measurements in Multi-Dimensional digitized Images Delft University Press, Delft.

Weyers, J D.B and Travis, A J (1981) Selection and preparation of leaf epidermis for experiments on stomatal physiology. *J. Exp Bot* **32**, 837 - 850

White, N.S. (1995). Visualisation systems for multidimensional CLSM images. *Handbook of Biological Confocal Microscopy*, Second Edition (Ed. J.B Pawley). pp 211 - 254 Plenum Press, New York.

White, N.S., Errington, R.J., Fricker, M.D. and Wood, J.L. (1995). Aberration control in quantitative imaging of botanical specimens by multi-dimensional fluorescence microscopy *J Microsc* in press.

Wilson, T. (1990) *Confocal microscopy.* Academic Press, London

Wilson, T. and Sheppard, C.J.R (1984) *Theory and Practice of Scanning Optical Microscopy* Acad. Press, New York.

Wooley, J T (1971). Reflectance and transmittance of light by leaves *Plant Physiol* **47**, 656 - 662.

FLUORESCENT PROBES

Jan Slavík

Institute of Physiology, Czech Academy of Sciences
Vídenská 1083, CZ-142 20 Prague 4, Czech Republic

INTRODUCTION

Fluorescent probes represent one of the most widely developing fields in biology and medicine[1-4]. Their application potential is enormous. Probes are dyes employed act as "molecular reporters". The information about what is going on in their molecular neighbourhood can be derived from their fluorescence signal and their location inside cells can be determined from fluorescence microscopy images. All fluorescence parameters such as fluorescence intensity, quantum yield, emission spectrum, excitation spectrum, polarization (anisotropy), lifetime, time-resolved fluorescence intensity and time-resolved fluorescence polarization can be used for encoding what is happening in the close (i.e. molecular) neighborhood of the "spying" molecule. Interdependence of fluorescence parameters, many of them being related by physical equations, may serve as a check of the correctness of the experimental procedure. This is very important when working with a biological material.

The basic problem is to have the right dye at the right place in the cell. There are hundreds of fluorescent dyes listed in glossy catalogues. However, these catalogues are as alluring as Playboy or Penthouse magazines. Dyes are listed according to their application area (ion-sensitive, membrane potential, membrane fluidity, etc.) and list of literature references is included[4].

The immense progress and capabilities inherent to fluorescent techniques may be illustrated on the example of intracellular ion measurement. In the past, using ion-selective microelectrodes or destruction (permeabilization) of cells or pH - dependent distribution of weak acids or bases, each measurement yielded only one single value of ion concentration from the sample. Now, with ion-sensitive fluorescent dyes, maps of intracellular ion concentrations can be obtained in vivo, with spatial resolution of 200 nm, with millisecond temporal resolution, possibility of 3D-reconstruction, and ion concentration changes detected as small as 0.01 pH or 20 nM [Ca^{2+}]. The enormous amount of new data brought along some surprising results, e g., calcium waves moving across the cell cytoplasm, calcium spikes in individual cells, and various pH and calcium heterogeneities.

INTRACELLULAR IONIC COMPOSITION

Measurement and mapping of ionic composition can be done both in the extracellular and the intracellular space. The technique originates from the old idea of the litmus paper, gradually changing its color from red to blue with increasing pH. The ion-sensitive dyes exhibit mostly two fluorescent forms, one free and one with bound ion that can be spectroscopically distinguished from each other Distinction can be based on changes in "color" - i.e. fluorescence excitation and/or fluorescence emission spectra and/or on fluorescence lifetime (quantum yield). The main requirement on the dye is the stability of its pK (and pK*) and a sufficient selectivity for the chosen ion [5]. Some dyes require UV excitation (fura-2) which may bring difficulties with a sensitive biological material.

The most popular pH-sensitive dyes are based on fluorescein (carboxyfluorescein, BCECF, SNARF, SNAFL, NERF) and pyranine (DHPDS). They all exhibit pH-dependent spectral shifts in excitation, some of them also in emission, and can be used for ratio imaging.

For calcium and magnesium, there is a large number of dyes based on fluorescent derivatives of the calcium chelator BAPTA (quin2, indo-1, fura-2...). They all show spectral shifts suitable for ratio imaging, but they need UV excitation A new series of indicators has been developed matching the 488 nm excitation line of argon laser and thus convenient for confocal microscopy (or lifetime imaging) - fluo-3, fura red, calcium green, magnesium green.... They exhibit no spectral shifts, only a ion-concentration dependent lifetime/quantum yield changes.

The sodium-sensitive SBFI and potassium-sensitive PBFI dye are composed from a fluorescent benzofuran group similar to that in fura-2 attached to a crown with cavity size fitting either Na^+ or K^+ ions, respectively The size of the crown group determines the selectivity for the chosen ion. Both dyes show spectral shifts with Na^+ or K^+ and (with some difficulties) can be used for ratio imaging.

All commercially available probes for chlorides show only a chloride concentration dependent change in lifetime and quantum yield but no spectral changes This is because the chloride ion causes only collision quenching of dye fluorescence after the contact with the molecule. Therefore, these dyes can be employed only in lifetime imaging or in a semiquantitative way in the quantum yield measurements.

The same applies to metal probes There is a zinc probe TSQ and iron probes MA-DFO and NBD-DFO which show a change in lifetime selectively dependent on the concentration of Zn^{2+}, resp. Fe^{3+} ions. Heavy-metals can be followed by recently developed new probes APTA and APTRA which can be used for Cd^{2+}, Hg^{2+}, Pb^{2+}, Ba^{2+}, La^{3+}. These probes, however, show only a lifetime/quantum yield change with the concentration of the particular ion.

MEMBRANE POTENTIAL - SENSITIVE FLUORESCENT DYES

The dyes used for the measurement of the membrane potential are divided into two groups.

The first group of dyes is called distributive (also "slow"). After addition to the cell suspension, dye molecules are distributed across the plasma membrane according to the Nernst equation. This equilibration process takes generally minutes. The fluorescence of the internalized probe is supposed to be quenched and the magnitude of the membrane potential is calculated from the quenching of fluorescence intensity at the very beginning and after the equilibration. Unfortunately, the adequate theory of the quenching of slow dyes is missing and their behavior dubious. Some of the dyes are toxic when accumulated inside the cell. The most frequently used dyes are carbocyanines, oxonols, merocyanines, and rhodamines.

The latter group of dyes consists of dyes directly sensing the intensity of the electrical field across the membrane. These dyes are called electrochromic (also "fast") dyes and they are assumed to be incorporated in the membrane and respond directly to the electric field. Their response to the change in the electric field is almost immediate. The best of them are styryl dyes. One of the most popular electrochromic dye is ANEPPS, which shows a fluorescence intensity change of 8% per 100 mV accompanied by a spectral change both in excitation and in emission. These spectral changes are suitable for ratio imaging both in excitation and emission However, there are still difficulties with the proper incorporation of the dye in the membrane as the dye should stay only in the outer leaflet of the plasma membrane.

Usual calibration procedure of membrane-potential sensitive fluorescence dyes is based on the artificially imposed potassium diffuse potential and valinomycin addition. Sometimes an application of external electrical field across the cell or membrane is employed.

Both fast and slow dyes give only a semiquantitative data on the membrane potential and their application requires caution and patience

MEMBRANE FLUIDITY

Fluorescence offers three following ways of measurement of the membrane "fluidity" depending on the type of fluorescence parameter measured.

1. **Polarization** measurement provides information on the rotational mobility of the in the membrane-incorporated fluorescent dyes. The idea is based on the concept of a wobbling motion of the dye (most often DPH, TMA-DPH or anthroyl-labelled fatty acids) in a cone-restricted space. The technique can also be adapted for the measurement of cytosol viscosity.

2. **Excimer** fluorescence exhibited by pyrene and perylene and their derivatives can be employed for the measurement of translational mobility of in the membrane-incorporated fluorescent dyes. One can use either pyrene or perylene directly or a phospholipid derivative with an attached pyrene or perylene fluorophore.

3. **Bleaching** of dyes after a short strong laser pulse is used in the FRAP technique (Fluorescence Recovery After Photobleaching) The gradual recovery of fluorescence intensity within the measured bleached spot gives sufficient information for calculating the lateral diffusion coefficient. The procedure requires labelling of membrane proteins with a fluorescent dye (usually eosin) and then the bleaching pulse is applied using a special experimental laser set-up.

FLUORESCENT AND FLUOROGENIC SUBSTRATES

Mapping of enzymic activity can be easily done with fluorescent or fluorogenic substrates. For instance, peptidases make a pronounced red spectral shift in AA - aminoquinoline fluorescence, glucosidases turn nonfluorescent naphthofluorescein digalactoside into fluorescent naphthofluorescein, esterases turn nonfluorescent fluorescein diacetate into fluorescent fluorescein. The catalogues of fluorescent dyes show long lists of commercially available substrates[4].

IN VIVO LABELLING

Cells and their compartments can be *in vivo* selectively labelled. The labelling is based on empirical testing of thousands of different dyes and cells For example, mitochondria can be stained with rhodamine 123 or dihydrorhodamine, endoplasmic reticulum with carbocyanines, Golgi with NBD-ceramide, cytoskeleton with NBDphalloidin or phallacidin, DNA, RNA with DAPI, Hoechst 33258 and 33842, oligonucleotides with FISH technique (cf elsewhere in this book). There is vital staining with fluorescein diacetate or AM-fluorescein derivatives. Last but not least, forget labelling with fluorescent antibodies, namely with fluorescein isothiocyanate (FITC) and tetramethylrhodamine isothiocyanate (TRITC) plays a very important role in selective labelling. Endocytotic pathways can be visualized e g. by addition of fluorescent labelled dextrans (cf. Gregor et al., *this volume,* p 102)

HOW TO GET THE DYE AT THE RIGHT PLACE

The right position of the dye molecule is very important for its correct "reporting". According to the problem studied, fluorescent probes may be targeted toward the cell surface, cell cytoplasm, surface of organelles, organelles, outside the cell For measurement of intracellular ions it is important to have the dye freely moving inside the cytoplasm, for measurement of membrane fluidity or of membrane potential by an electrochromic dye it is necessary to have the probe incorporated into the membrane, etc. There is an elegant way of introduction of dyes through the plasma membrane barrier using chemically modified precursor probes (acetylated, AM forms) which penetrate the plasma membrane easily. In the cytoplasm they are then enzymically cleaved into the fluorescent form which is retained inside the cell. The mechanical permeabilization (microinjection, bead or scrape or scratch loading, osmotic shock, mild sonication) electroporation or chemical permeabilisation are not so elegant but often useful The behavior of real dyes inside the cells is complex and it usually includes some kind of sequestration, binding and leakage[3].

CONCLUSION

The future of fluorescence probes will certainly include syntetisation of better and more selective and more intensively fluorescent probes. The exciting area is, however, the application of several fluorescent probes simultaneously. Using selective excitation, selective detection, delayed fluorescence dyes and lifetime imaging, it will be easy to monitor simultaneously several physiological parameters such as membrane potential and pH or pH and calcium concentration or all three of them. The application of red and infrared dyes will broaden the usable spectral range and in some cases will enable to lower the background fluorescence signal. Confocal microscopy will become standard equipment in all laboratories and the supplementation of confocal microscope with lifetime imaging quite usual.

REFERENCES

1. B. Herman and J.J. Lemasters. *Optical Microscopy: Emerging Methods and Applications,* Academic Press, San Diego (1993)
2. W.T. Mason. *Fluorescent and Luminescent Probes for Biological Activity,* Academic Press, London (1993)
3. J. Slavík. *Fluorescent Probes in Cellular and Molecular Biology,* CRC Press, Boca Raton, Florida (1994)
4. R. P. Haugland. *Molecular Probes - Handbook of Fluorescent Probes and Research Chemicals,* Molecular Probes Inc., Eugene, Oregon (1992)
5. A. Kotyk and J. Slavik. *Intracellular pH Distribution and Its Measurement,* CRC Press, Boca Raton, Florida (1989)

FLOW CYTOMETRY *VERSUS* FLUORESCENCE MICROSCOPY

José-Enrique O'Connor
Departamento de Bioquímica y Biología Molecular, Facultad de Medicina,
Universidad de Valencia, Blasco Ibáñez 17, 46010-Valencia, SPAIN

INTRODUCTION

Since the pioneer work of the botanist Matthias Jakob Schleiden and the zoologist Theodor Schwann in 1839, and of Rudolph Virchow in 1859, cell research progressed in two opposite directions. Cell biologists focused their increasingly more powerful microscopies into the cell structure to reveal the great morphological complexity of the cytoplasm. A growing number of subcellular organelles thus challenged the early biochemists to discover their specific molecular features and their coordination to maintain an ordered cell life. The biochemists' answer to such a challenge consisted usually in tearing apart cells into their discrete components and obtaining information on molecules and pathways of each single part. The assembly of this jigsaw puzzle into an integrated view of a functional cell, and of such a cell within higher levels of organization, could not be achieved exclusively by pure biochemical methods. As early as in 1961, Jean Brachet wrote that "The cell biologist tries to explain in molecular terms what he sees under his microscope; he has become a molecular biologist. The biochemist in turn has become a biochemical cytologist, equally interested in the structure of the cell and the biochemical activity in which it is involved" (1).

More than thirty years later, current cell biology has enriched Brachet's reflection with two additional concepts. First, the accurate study of cell functions and metabolic responses demands, in many cases, the biochemical analysis to be undertaken on the cell as a whole. Second, the integration of cell functions and the asignment of specific functions to specific structures will benefit of the analysis of multiple parameters in the same cell or subcellular organelle.

Fluorescence microscopy and flow cytometry meet the criteria mentioned above. Both methodologies may preserve cell integrity or may be applied to isolated subcellular components. A great number of fluorescent markers specific for cell components or functions have been developed in the last years, either by chemical design or by taking advantage of the properties of natural molecules. A careful choice of fluorescent markers allows to visualize and quantify structures and functions by image cytometry within a single cell or in a few of them. Flow cytometry is usually applied to detect and quantitate multiple cell components and/or functions in more or less large, more or less heterogeneous cell populations.

Both methodologies have advantages and pitfalls and there is a current consensus on that both are complementary to each other. However, the microscopical tradition of cell biology (seeing is believing!) tends to put the weight on the image side of cytometry when one considers subcellular analysis. On the other hand, classical biochemists may disregard image techniques because they are more inclined to test tubes and numerical data. In that context, flow cytometry provides a compromise between both approaches. Specific fluorescent markers for most cell compartments may be used in whole cells or in isolated organelles. The multiparametric

Fluorescence Microscopy and Fluorescent Probes
Edited by J. Slavík, Plenum Press, New York, 1996

Table 1. TYPES OF FLUORESCENT MARKERS USED IN FLOW CYTOMETRY

Molecules with chemical reactivity
Molecules with structural specificity
Fluorescent indicators and chelators
Fluorogenic enzyme substrates
Fluorescent macromolecules and polymers

capabilities of flow cytometers allow simultaneous sensitive measurement of different components/functions. Finally, the nature of flow cytometric data makes possible quantitation (or, at least, semi-quantitation) of cellular properties, thus yielding objective results.

FLUORESCENCE MICROSCOPY AND THE DEVELOPMENT OF FLOW CYTOMETRY

Microscopy and image analysis preceded historically and indeed led eventually to the development of flow cytometry. Thus, sixty years ago, Andrew Moldavan suggested counting red blood cells as they were forced through a capillary tube on the stage of a microscope interfaced to a photodetector. Although it is unclear if Moldavan's instrument was ever built, it is considered the first step in the development of flow cytometry. Following this microscope-based approach, Kamentsky and Melamed developed in 1967 the first flow sorter, able to deviate one cell at a time out of the main stream, and Dittrich and Göhde, in 1969, built the first system in which fluorescence intensity histograms could be generated.

The Coulter principle gave origin to a wide family of hematological counters and analyzers: instruments that automatically record electrical signals from single cells or particles afer exiting a small orifice. Based on this approach, Crosland-Taylor took advantage of the properties of laminar fluids to develop hydrodynamic focusing of cells, achieved by injecting a cell suspension into the core of a capillary stream flowing faster. This concept was capital for the development of microscope-independent flow cytometers, such as those designed in the early 70's by Van Dilla and Herzenberg. Such systems, based on the orthogonal spatial disposition of illumination, flow and detectors, made posible the implementation of electrostatic sorting devices, based upon the Fulwyler and Sweet technique, originally conceived for ink jet computer printers. On the other hand, those instruments were already capable of multiparametric analysis, one of the capital features of the modern flow cytometers.

The technical development of flow cytometers is parallel to the development of fluorescent dyes and techniques to be applied specifically to reveal macromolecules and other cell components. The availability of fluorescent monoclonal antibodies allowed the immunophenotyping of cells by their surface antigens, as well as the detection of specific intracellular proteins. Finally, the design of fluorescent probes responding to variations in the cell milieu and to specific enzyme activities has incorporated cell biochemistry to the range of flow cytometric applications.

The advances in physics, chemistry, engineering, computer sciences, biology and medicine have contributed to the present expansion of flow cytometry. This has been accomplished, on one side, by the development of highly sophisticated instruments, usually restricted to research laboratories and, on the other, by the design of low-price, user-friendly systems, which are widely distributed in clinical laboratories. For better reviews of the history of flow cytometry, please read references 2 and 3.

In the other hand, the recent expansion of confocal microscopes and other complex image systems surpasses the growth of flow cytometry in the recent years. According to Robinson (4), this may be attributed both to the development of low-cost, high-capacity computers and the reduction in the cost of sophisticated video systems. In this way, altough conventional image analysis has been available for many years, it has only been economically feasible to the majority of scientists for about 5 years (4). A direct consequence of this is the growing interest of flow cytometrists in the implementation of image technologies and the evidence that both approaches, flow and image, are indeed complementary to each other. As it should become clear along this chapter, complementarity arises from the similar use of fluorescent markers to reveal cell functions and structures and from the essentially different ways in which such information is obtained and processed.

TECHNICAL BASES AND ANALYTICAL APPLICATIONS OF FLOW CYTOMETRY

Flow cytometry is basically an analytical method which allows the measurement of fluorescence emission and light scatter, induced by appropriate illumination of cells or microscopic biological particles, as they flow one by one at a high speed through a sensing area. In some systems, the individual cells or particles may be physically sorted according to the properties exhibited (3).

Flow cytometry takes advantage of the development of a substantial number of fluorescent markers, which bind especifically to cellular molecules, which are selectively accumulated in cell compartments or which modify their properties by specific biochemical reactions (3,5). In this way flow cytometry may reveal and quantitate structures and functions of single cells or particles at high speed and under a multiparametric approach, thus allowing to characterize cell populations on the basis of a wide range of biological properties.

Because of the increasing number of measurable parameters and the development of affordable cytometers interfaced to user-friendly, high-capacity computer systems, flow cytometry has currently a wide range of applications in many basic and clinical fields, wherever is needed to identify, to characterize the biological response and to separate single cells or particles.

COMPONENTS OF A FLOW CYTOMETER

The basic components of a typical flow cytometer are the fluidics system, the illumination system, the optics system, the electronics system and the data collection and analysis system.

a) Fluidics system: surrounds the sample stream with a particle-free laminar sheath fluid which moves the sample at a constant rate through a detection zone where the biological particles are exposed, one by one, to the illuminating beam.

b) Illumination system: delivers a light beam to illuminate the sample. Most cytometers use laser light because it is coherent, monochromatic, polarized, narrow, stable and of known intensity.

c) Optics system: focuses the illumination on the sample particles,detects the light scattered by the particles and selects the fluorescence emitted as the particles cross the laser beam.

d) Electronics system: provides a constant illumination intensity, detects and scales the response of particles in pulse analog form, converts analog signals into digital and generates sorting pulses.

e) Data collection and analysis system: in most modern cytometers it is based upon personal computers and common operating systems. It allows multiparametric data acquisition and analysis in real time or in list mode, as well as gated analysis. In addition, these data can be further processed off line with the aid of commercial and public domain software.

SPECIFIC FEATURES OF FLOW CYTOMETRIC ANALYSIS

As a matter of fact, any flow cytometer measures exclusively optical properties, i.e the light emitted by fluorochromes or scattered by the flowing particles, while flow cytometrists are usually interested in evaluating biological properties. In most cases, it is posible to link both kinds

TABLE 2. BIOLOGICAL PARAMETERS ANALYZABLE BY FLOW CYTOMETRY

Structures that do not need exogenous markers	**Functions that do not need exogenous markers**
Cell volume	*Redox state*
Cell texture	*Cell viability*
Fluorescent pigments	
Structures that need exogenous markers	**Functions that need exogenous markers**
Macromolecules	*Cell viability*
Low-weight metabolites	*Plasma membrane integrity*
Specific receptors	*Transport and internalization*
Specific regions in macromolecules	*Membrane potentials*
Subcellular organelles	*Specific enzyme activities*
	DNA synthesis
	Ionic movements

of properties by using appropriate fluorescent markers, as those indicated schematically in Table 1, and which may be applied in most cases in fluorescence microscopy. The adequate application of these molecules allows to detect and quantify structural sites or to follow dinamic functions in the cells. However, some structural or functional features can be studied without the need of incorporating exogenous markers, taking advantage of endogenous fluorescent molecules or of physical interactions between cells and light. According to the specific applications of fluorochromes, and following Howard Shapiro's well known classification (3), the types of biological parameters analyzable with most of flow cytometers are outlined in Table 2.

Obviously, the listed parameters may be studied through other analytical approaches - even using the same fluorescent markers. However, the technical features of flow cytometry allow to obtain a degree of information about these parameters in each single analysis, which is probably unique among biological methodologies. The biological information that may be obtained from a flow cytometric analysis, is summarized in Table 3.

TABLE 3. POTENTIAL INFORMATION GENERATED BY FLOW CYTOMETRIC ANALYSIS

Intensity of expression of one or more parameters in a cell population

Heterogeneity of expression of one or more parameters within a cell population

Relationship between multiple parameters within a cell population

Relationship between multiple parameters in single cells

Evolution of dynamic parameters

Correlation with parameters measured by other techniques after cell sorting

TABLE 4. CURRENT AREAS OF APPLICATION OF FLOW CYTOMETRY

Clinical routine

Clinical research

Basic research

Biotechnology

Environmental research

FIELDS OF APPLICATION OF FLOW CYTOMETRY

In the short -but eventful- history of flow cytometry (2,3), at the beginning there were the large cytometers and the small number of applications - easily separated as "clinical" or "basic". As technological developments occurred within and around the cytometers, smaller instruments and larger number of applications became the state-of-art and this makes more difficult to define categories. However, most current applications fall within either of the fields listed in Table 4. Specific applications could be located in the listed fields, keeping in mind that very few applications would belong exclusively to a single area. Thus, disciplines like immunology, hematology and oncology, in which flow cytometry is applied mostly for clinical routine, have increased notably their basic research aspects (6). On the other hand, applications developed in the basic laboratory, especially in molecular biology and cell physiology, are currently adapted to clinical or biotechnological purposes (3,7).

FLOW CYTOMETRY VERSUS (?) FLUORESCENCE MICROSCOPY

From what has been stated above, it should be clear that fluorescence microscopy -and subsequent image analysis - and flow cytometry are not opposite but complementary by necesity when exhaustive information about cell populations is required.

In first place, it is very important (but not always done!) to inspect visually the original samples or cell preparations in order to check cell integrity, staining, and other quality parameters, before running on the flow cytometer. This is specially mandatory when flow cytometric procedures are to be set up. In the other hand, cell sorting makes possible to visualize single cells separated physically, according to their flow cytometric properties. In this way, morphological characteristics of individual cells may be correlated to populational properties. Contrarily, such approach may be imprescindible to interpretate unclear or unexpected flow cytometric distributions.

Because of its nature, flow cytometry requires disgregation of supracellular structures such as biological solid tissues or experimental monolayers, thus losing substantial information on intercellular communication, but providing multiparametric information on large cell populations. Contrarily, fluorescence microscopy allow the visualization of small groups of cells in a more physiological context and the study of interactions between neighbour cells.

Both technologies generate information on a cell-by-cell basis. Flow cytometry typically processes up to thousands of cells per second and kinetic studies of dynamic or transient phenomena may be easily performed in most cases. However, fluorescence-microscopy based image techniques may allow the sustained visualization of single (or few) cells and, thus, kinetic measurements within the same cell are feasible.

The function of subcellular compartments may be also specifically studied both by flow cytometry and fluorescence microscopy. This is possible because of the existence of fluorescent probes that bind specifically to or are accumulated within subcellular compartments such as lysosomes, mitochondria, etc. While the use of fluorescence confocal techniques increases sensitivity to subcellular analysis within intact cells, flow cytometry allows the functional and structural study at the level of single subcellular elements in suspension (8-10).

It seems clear that the ideal situation for a laboratory involved in cell research is to implement both flow and image systems. In fact, this is refflected by the scientific organizations in the field of cytology, which include both methodologies under the common term of Analytical Cytology, such as the *International Society for Analytical Cytology* or just Cytometry, such as the *Deutsche Gesselschaft für Zytometrie, Gruppo Italiano di Citometria, Sociedad Ibérica de Citometría,* and other national societies.

REFERENCES

1. Brachet J. The living cell. In: The Living Cell, W.H. Freeman & Co., San Francisco. p. 13, 1961.
2. Melamed MR, Mullaney PF and Shapiro HM. An historical review of the development of flow cytometry instruments. In: Flow Cytometry and Sorting, 2nd Ed. (Melamed, Lindmo, Mendelsohn, Eds.), Wiley-Liss, New York, 1990, pp. 1-10.
3. Shapiro HM. Practical Flow Cytometry, 3rd. Ed., Alan R. Liss, Inc., New York, 1995
4. Robinson JP. Links between flow and image cytometry. Proceedings of the IV Meeting of the Iberian Society of Cytometry, Oviedo (Spain), 20-24 May 1995, pp. 32-33.
5. Slavik J. Fluorescence Probes in Cellular and Molecular Biology, CRC Press, Inc., Boca Raton, 1994.
6. Bauer KD, Duque RE and Shankey TV, eds. Clinical Flow Cytometry: Principles and Applications Williams and Wilkins, Baltimore, 1993.
7. Lloyd D, ed. Flow Cytometry in Microbiology. Springer, London, 1993.
8. O'Connor JE, Vargas JL, Kimler BF, Hernandez-Yago J, Grisolía S. Use of rhodamine 123 to investigate alterations in mitochondrial activity in isolated mouse liver mitochondria. Biochem Biophys Res Commun 151: 568-573, 1988.
9. Guasch RM, Guerri C, O'Connor JE. (199). Study of surface carbohydrates on isolated Golgi subfractions by fluorescent-lectin binding and flow cytometry. Cytometry 19: 112-118, 1995.

MULTICHANNEL FLUORESCENCE MICROSCOPY AND DIGITAL IMAGING - ON THE EXCITING DEVELOPMENTS IN FLUORESCENCE MICROSCOPY

Heinz Gundlach

Carl Zeiss Jena GmbH
Division of Microscopy
Jena, Germany

The last 15 years have seen an unexpectedly great renaissance and a partial revolution in light microscopy. The recent progress in light microscopy is due to new design in optics and instrumentation as well as improvement of optical contrast enhancement techniques and new developments in fluorescence methods in combination with the increase of number of fluorescence dyes.

Multicolor fluorescence- and fluorescence digital imaging microscopy today is of great importance by nonisotopic detection of multiple nucleid acid sequences for the analysis of cell components, chromosomes and genes.

OPTICS FOR FLUORESCENCE MICROSCOPY

The ICS optics (= Infinity Color Corrected System, Carl Zeiss 1986) is an example of recent state of the art in microscope imaging (Fig 1). The most outstanding feature of the ICS optics is that the objective with ist infinite image distance and the tube lens together form an optical system which produces a fully corrected intermediate image.

One of the most important features of the ICS optics is that the space of the parallel beam path between objective and tube lens can be optimally utilized to accommodate 3 Fl or 4 Fl reflector sliders as well as magnification changer, DIC slider, bertrand lens slider etc. without any need for further optics (Fig 1).

Dual and triple band pass filtersets are available now for the simultaneous detection of more than one fluorchrome (Fig. 2).

Multi-channel fluorescence microscopy is of great importance by nonisotopic in situ hybridization for simultaneous imaging of multiple nucleic acid sequences for the analysis of cell components, genes and chromosomes

For routine applications in the cytogenetic and cytopathology laboratory multi channel fluorescence filtersets have the following advantages:

- high speed because there is no filterchange
- high precision imaging without any image shift
- more space in the fluorescence filter slider for additional filter sets.

Fluorescence Microscopy and Fluorescent Probes
Edited by J. Slavík, Plenum Press, New York, 1996

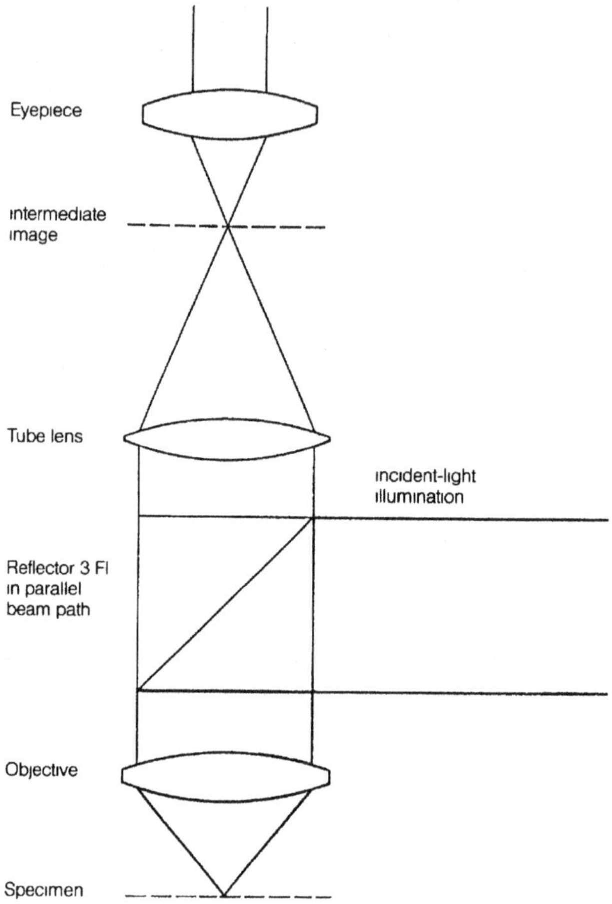

Figure 1 Imaging light path ICS optics together with epi fluorescence light path

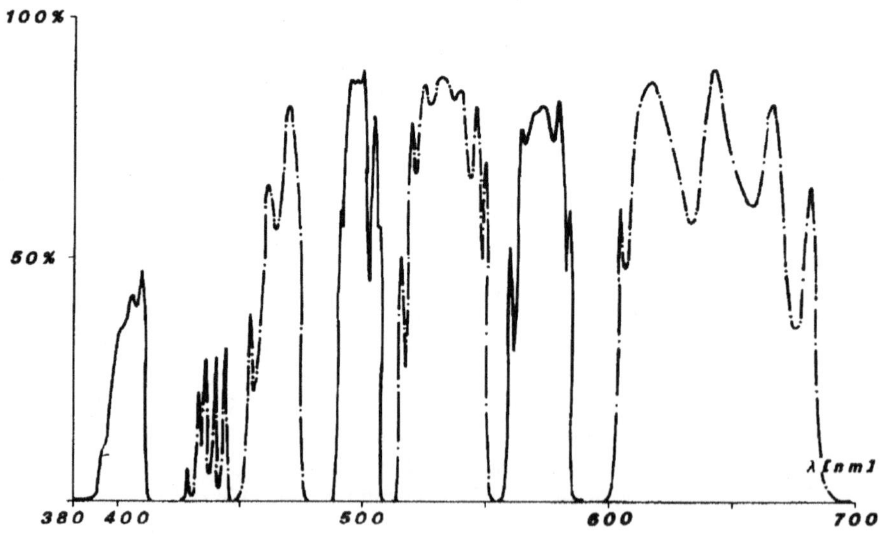

Figure 2 Triple band pass filter curve

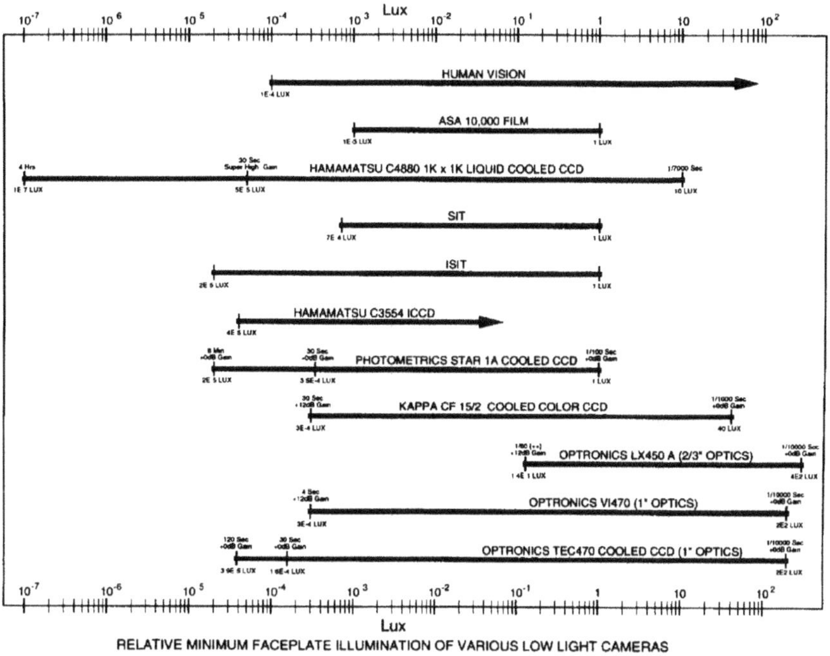

Figure 3 Overview of range of sensitivities of various CCD cameras

The application of those filtersets are influenced and limited by the distribution of the fluorescence and the intensities. Optimal results are produced, if the fluorescence signals have more or less the same intensity and the structures are not overlapping very much.

Because of the signal to noise ratio in conventional fluorescence microscopy and photomicrography the application of multilabeled fluorescence methods are limited. Digital imaging techniques are more and more being used to supplement conventional fluorescence microscopy. Very weak fluorescent signals can be documented more readily, image reproduction problems often occur when conventional multiexposure photomicrography is applied. The potential of image processing by means of a labeled specimen and quantitative data on signal intensities or measurements can be easily assessed. Digital imaging is performed using sensitive camera systems or laser scanning microscopy. For low light level applications, e. g. weak fluorescence or small FISH signals, cooled or integrating CCD cameras, black and white as well as color, are available now (Fig. 3). By means of the digitized images, simultaneous detection of multiple labeled objects and ratio imaging will supplement conventional fluorescence in a new dimension Confocal laser scanning microscopy is designed for obtaining optical sections through a labeled specimen e. g. for the analysis of fluorescent signals in interphase nucleus Three-dimensional reconstruction and -analysis can be complished by applying appropriate computer software to a stack of digitized images.

REFERENCES.

Arndt-Jovin, D.J. (1985) Science, 230-247
Gundlach, H. (1994) Computerized Cytology and Histology Laboratory, Chicago, 335-345.

Lengauer, Chr. (1993) Human Mol. Gen., Vol. 2, No.5 505-512.
Lichter, P. (1991) Gata 8 (1), 24-35. Lichter, P., T. Cremer (1992) Human Cytogenetics, Vol.I, 157-192.
Nederlof, P.M. (1989) Cytometry 10, 20-27.
Ried, Th. et al. (1992) Proc. Natl. Acad. Sci. USA, Vol. 89, 1388-1392.
Taylor, D.L. (1992) American Scientist, Vol. 80, 322-335

FLUORESCENCE LIFETIME IMAGING AND SPECTROSCOPY IN PHOTOBIOLOGY AND PHOTOMEDICINE

Herbert Schneckenburger,[1,2] Michael H Gschwend,[1] Karsten König,[3] Reinhard Sailer,[1] and Wolfgang S L. Strauß[1]

[1]Institut für Lasertechnologien in der Medizin und Meßtechnik and der Universität Ulm, Helmholtzstr. 12, D-89081 Ulm;
[2]Fachhochschule Aalen, Optoelektronik, D-73428 Aalen;
[3]Institut für Molekulare Biotechnologie, Beutenbergstr. 11, D-07706 Jena

INTRODUCTION

Time-resolving fluorescence imaging has become a valuable tool in photobiology and medical diagnostics[1]. Its techniques include two-dimensional laser scanning microscopy with signal detection in pre-selected time windows[2] as well as fluorescence microscopy and spectroscopy using pulsed laser excitation and time-gated image intensifying units[3-6]. In addition, experimental systems working in the frequency domain become more and more promising. In this case, the excitation light and the voltage of the image intensifier are modulated with a certain phase shift. This shift may be selected such that the detection of fluorescent components with specific lifetimes is enhanced, whereas other components are suppressed. The method has been applied to measure the distribution of specific markers[7,8] and of natural fluorophores[9] in various cellular environments.

The applications of time-resolving imaging techniques are manyfold and include the detection of intracellular calcium or oxygen concentrations[10] as well as pH values[10,11], the distinction between various intracellular fluorophores (e.g. free and protein-bound NADH[9]), and the discrimination between tumour-specific fluorescent markers and autofluorescence of the tissue[4,5]. In the present article an experimental setup for simultaneous detection of fluorescence spectra and images is described, and some present applications - ranging from measurements of respiratory function, caries detection, photosensitization and measurements of the impact of high light doses on the photosynthetic apparatus of spruces - are summarized.

MATERIALS AND METHODS

The experiments described below were carried out with cultivated BKEz-7 endothelial cells from calf aorta[12], human teeth, as well as needles from spruces of various damage classes which had been growing under different light conditions (Freudenstadt, Black Forest)[6]. For studies of photosensitization, BKEz-7 cells were incubated with various porphyrins.

The equipment recently developed in our laboratory (Fig.1) consists of a fluorescence microscope with an optical multichannel analyzer (OMA; Hamamatsu Photonics, IMD C4562)

Fluorescence Microscopy and Fluorescent Probes
Edited by J. Slavík, Plenum Press, New York, 1996

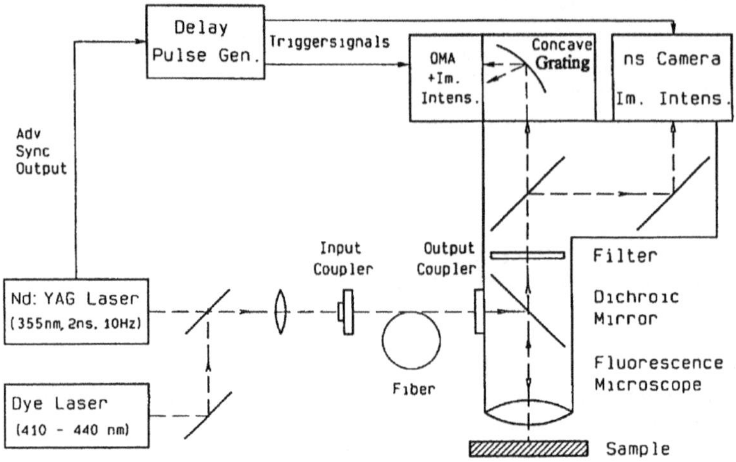

Figure 1. Experimental setup for time-gated microscopic imaging and spectroscopy.

- combined with a purpose-made polychromator - and a highly intensifying video camera (Proxitronic, NCA, sensitivity 10^{-6} lx in cw mode) for detection of fluorescence spectra and images, respectively. The image intensifiers of both detectors are time-gated using the electric output pulses of the exciting laser and adjustable synchronization electronics (for details see Ref. 6). Light pulses of the second or third harmonic of the Q-switched Nd:YAG laser (Spectra Physics; DCR 11; wavelength 532 or 355 nm, pulse duration 2 ns, repetition rate 10 Hz) or alternatively a dye laser (428 nm) - pumped by the Nd:YAG laser - are coupled to the microscope via glass fiber.

For most applications reported below a time gate of variable delay, but a constant width of 5 ns, and pulse energies ranging between 0.10 mJ (428 nm) and 1.2 mJ (532 nm) were used. These pulse energies were further attenuated in the microscope such that radiant exposures of 10 mJ/cm² or less were attained on the samples. After integration over 50-100 excitation pulses the light dose during one experiment was kept below 1 J/cm² which appeared to be tolerable for avoiding photochemical reactions. The integration occurred directly on the IMD detector (which was triggered repeatedly) during fluorescence spectroscopy, but was carried out off-line in the image acquisition unit when the NCA camera was used. For some *in vivo* measurements of teeth the video camera was taken off from the microscope and fixed on an optical table.

The appropriate time gates were selected on the basis of fluorescence decay measurements using a frequency doubled laser diode (Hamamatsu Photonics, PLP-01, wavelength 390 nm, pulse energy 10 fJ, repetition rate 1 MHz) and a single photon counting equipment (Tennelec), as described previously[13].

APPLICATIONS

Coenzymes of the Respiratory Chain

Autofluorescence of cells and tissues has been attributed to the proteins tryptophan and tyrosine (300-390 nm), the coenzymes NADH (nicotinamide adenine dinucleotide; 430-470 nm), NADPH (nicotinamide adenine dinucleotide phosphate; 440-470 nm) and flavins (510-530 nm), as well as the extracellular components collagen and elastin[14,15]. In addition, porphyrins are formed during synthesis of heme or cytochromes, which may be accumulated in

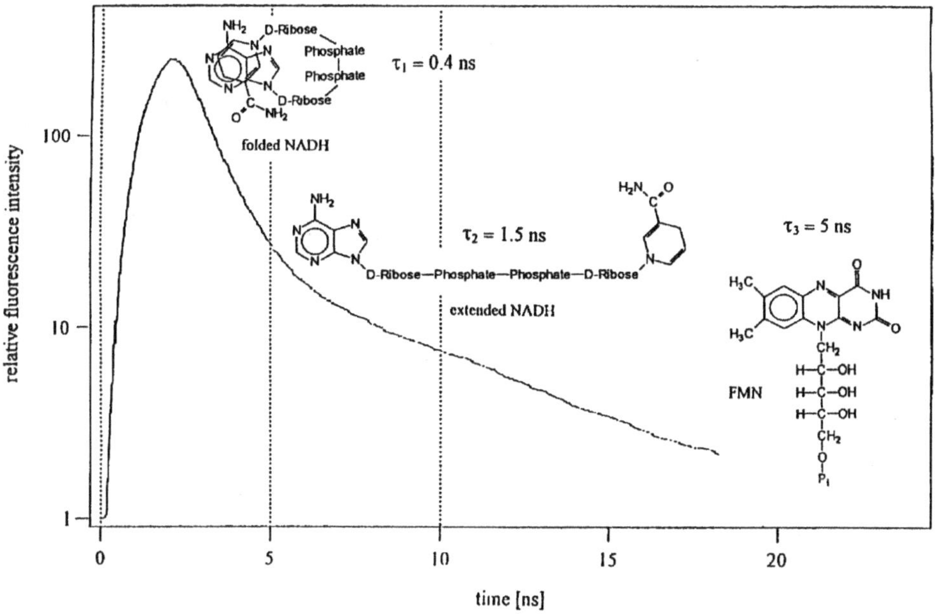

Figure 2. Decay kinetics of autofluorescence from BKEz-7 endothelial cells (λ_{ex} = 390 nm; $\lambda_{em} \geq$ 435 nm).

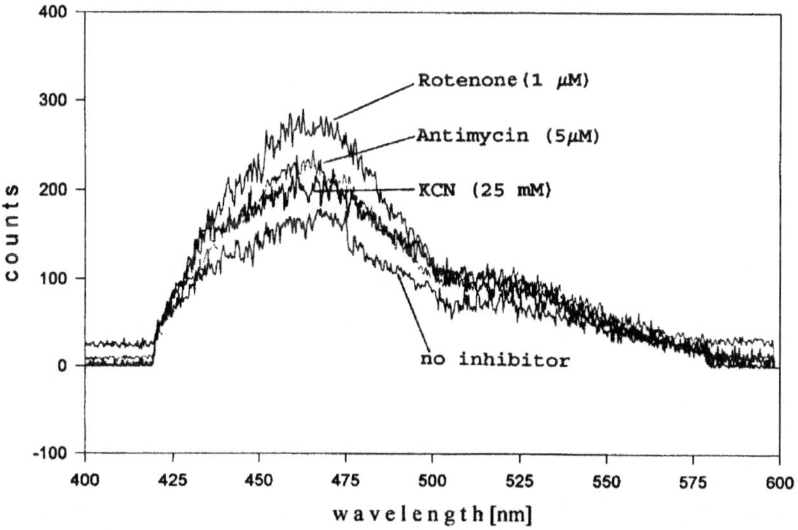

Figure 3. Emisison spectra of BKEz-7 endothelial cells incubated with inhibitors of the respiratory chain (time gate 0-5 ns; λ_{ex} = 355 nm).

certain bacteria[16]. Present studies are concentrated on mitochondrial NADH and flavins whose fluorescence intensity depends on their oxydation-reduction states[17,18] These coenzymes may therefore be used as indicators of respiratory function.

The fluorescence decay curve of cultivated endothelial cells from calf aorta is shown in Fig. 2. This curve can be described by a superposition of 3 exponentially decaying components with decay times of about 0.4 ns, 1.5 ns and 5 ns which - in agreement with literature[9,19,20] - may be attributed to free (folded) NADH, protein-bound (unfolded) NADH and flavin molecules, respectively. As recently demonstrated for endothelial cells as well as for yeast cells (*Saccharomyces*)[21], these components can be detected rather selectively within time gates of 0-5 ns, 5-10 ns and 10-15 ns after the exciting laser pulse, respectively. Fluorescence intensity of free NADH was correlated with oxygen concentration - and therefore with respiratory function - whereas the fluorescence of protein-bound NADH was almost independent from oxygen[21].

For more detailed studies of respiratory function, the electron transport chain of BKEz-7 endothelial cells was inhibited at various sites of the inner mitochondrial membrane using rotenone, antimycin, KCN or oligomycin for selective inhibition of the complexes I (NADH-coenzyme Q reductase), III (coenzyme QH_2-cytochrome c reductase), IV (cytochrome oxidase) or V (ATP synthase), respectively. Fig 3 shows the corresponding fluorescence spectra - dominated by free NADH - within the time gate 0-5 ns. In comparison with non-incubated reference cells, NADH emission increases after inhibition, in particular, if inhibition occurs in complex I (rather than in the terminal complexes IV or V). This increase is related to a higher amount of reduced NADH after disruption of electron transport and can be further used for detailed investigations of respiratory deficiencies.

Some relatively high background fluorescence may be due to cytoplasmatic NADH, to the coenzyme NADPH, as well as to further - so far unknown - fluorescent components. In an attempt to enhance the specifity of the measured signal from mitochondria, BKEz-7 cells were incubated with a specific mitochondrial marker, and energy transfer from NADH to the marker molecules was determined. As shown in a recent publication[22], the energy transfer rate increased considerably after inhibition of the complexes I or III of the respiratory chain.

Caries Detection

As previously demonstrated, intrinsic fluorescence from carious regions of teeth exhibits longer decay times than healthy parts[6,15] The dominating component of T = 17 ns (Fig. 4) is characteristic for porphyrins like protoporphyrin (see below) and may be due to porphyrin-producing bacteria. Although - at continuous (cw) detection - the emission of healthy enamel or dentin is often stronger than the fluorescence from carious regions, the caries signal is more pronounced at decay times above 25 ns with respect to the exciting laser pulse. In Fig. 4, the fluorescence of human incisors is shown at cw (4a) or time-gated (30-55 ns; 4b) detection, using the 546 nm band of a mercury lamp or the 532 nm line of the frequency doubled Nd:YAG laser, respectively. Unspecific emission from whole teeth is registered in the cw mode, whereas carious regions at the bottom of the teeth are detected selectively within the delayed time window. For diagnostic applications it may be of some interest that real time measurements of fluorescence intensity are possible without any calibration, image subtraction or further mathematical procedures.

Photosensitization

In view of photodynamic therapy of tumours and other diseases, measurements of uptake, distribution and light-induced reactions of various photosensitizers in cells and tissues are presently carried out[6,11]. These photosensitizers include the naturally occurring porphyrins uroporphyrin III (UP), coproporphyrin III (CP), protoporphyrin IX (PP) as well as the synthetic tetra-, tri-, di- and monosulphonated tetraphenylporphyrins (TPPSn with n = 4,3,2 and 1) Both groups of porphyrins exhibit decreasing hydrophilicity in this order.

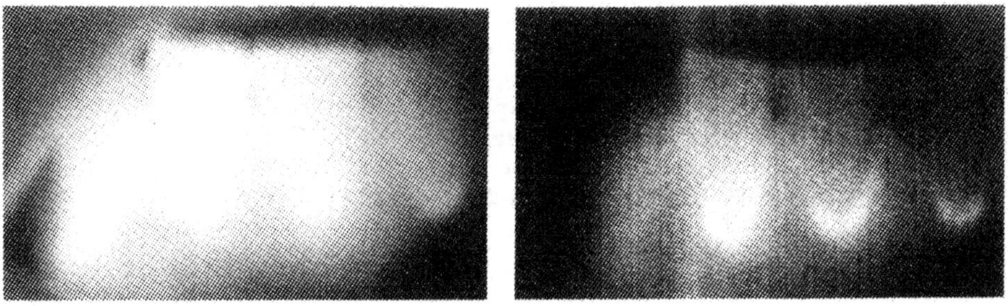

Figure 4. In vivo fluorescence of human incisors using cw (left) and time-gated (30-55 ns after laser pulse; right) detection (λ_{ex} = 532 nm, $\lambda_{em} \geq$ 590 nm).

From pH dependent fluorescence spectra and decay kinetics in solution[11,23] we deduced that the hydrophilic porphyrins UP, CP and $TPPS_4$ were composed by molecules of different degrees of protonation. The composition of these species in cultivated epithelial and endothelial cells reflected a pH value of the microenvironment around 5. The porphyrins were therefore localized within lysosomes, as also indicated by a few fluorescent spots in the perinuclear region.

The more hydrophobic PP showed a pronounced red shift of fluorescence spectra with decreasing polarity of the solvent (from 622 nm in aqueous buffer solution to 639 nm in benzene or octanol). The fluorescence spectrum of PP in endothelial cells (Fig. 5) was almost identical to the spectra of the least polar solutions. This result indicates that PP was preferentially localized within a microenvironment of low polarity ,i.e. in membranes. Pronounced photobleaching and the appearance of a photoproduct at 665-680 nm were detected after irradiation (Fig. 5) This indicates that a photochemical reaction occurs, which - by additional measurments of phototoxicity - was found to be lethal.

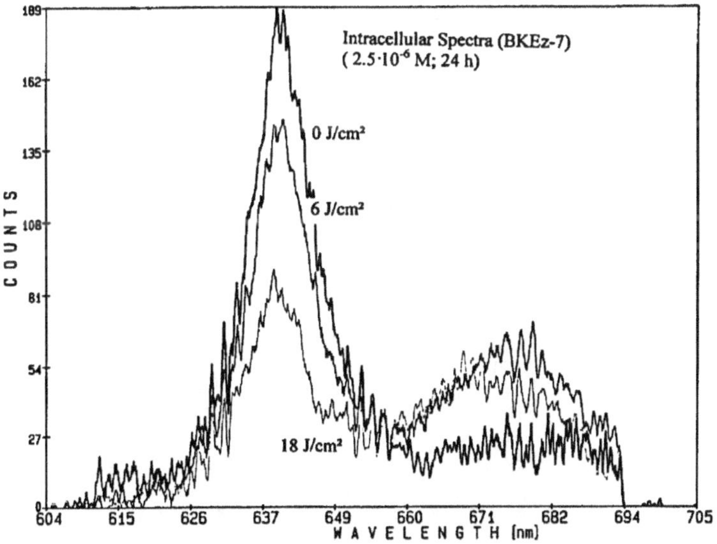

Figure 5. Time-gated emission spectra of intracellular protoporphyrin IX before and after light application (λ_{ex} = 532 nm; time gate 0-5 ns)

Time-gated fluorescence measurements as shown in Fig. 5, turned out to be advantageous for detection of short-lived photoproducts (T = 4-5 ns) and - in addition - allowed for a discrimination of short-lived aggregates ($T \leq 2$ ns) and longer-lived monomers (15-18 ns). Monomeric porphyrins were reported to have the best photosensitizing properties[24]. Time-gated experiments showed that monomers of PP and related substances are located preferentially in the plasma membrane, nuclear membrane and adjacent intracellular sites[6] which are supposed to correspond to further membranes. Therefore, porphyrin location in cellular membranes seems to play an essential role for the efficacy of photodynamic therapy.

Photosynthesis

Fluorescence decay curves and time-gated emission spectra were recorded for needles of the second and third age class of a declining (damage class 2) and an apparently healthy (damage class 0) spruce. These needles had been growing either in full sun light or in artificial shadow (using a wire mesh with about 20 % transmisson) or in natural shadow. From a 3-exponential fit of the decay curves, lifetimes of about 100-200 ps, 400-600 ps and 2.5-3.5 ns were obtained for chlorophyll fluorescence in the wavelength range of 690-800 nm. It turned out that the relative intensity of the long-lived component (I_3) was a measure for the number of those chlorophyll molecules whose energy transfer to the reaction centres was obstructed. In general, I_3 increased in summer up to about 5 % for the „healthy" and up to about 20 % for the declining spuce prior to decreasing in automn and winter[25]. Usually, for sun needles the increase was more pronounced and occurred earlier during the season than for shadow needles. This behaviour which was first detected in 1993[25], was confirmed by measurements in 1994 and 1995 (Fig. 6). This Figure shows an increase of I_3 of light exposed needles during the summer of the second year and afterwards some „stabilisation" on a rather high level. In contrast, the increase of I_3 of shadow needles was less pronounced and „reversible" during the first year. Only during the third year the high values of I_3 were maintained. A clear decrease of I_3 was detected after artificial shading (May 1994).

Preliminary conclusions of these results are: (i) I_3 can be regarded as an appropriate parameter for measuring defects to the photosynthetic apparatus; (ii) impairments of the photosystems depend on the dose of incident light and are most pronounced for pre-damaged spruces (e.g. by ozone, nutrient deficiencies, heat etc); (III) defects are probably located within the Photosystem II, as can be deduced from the time-gated spectra at 10-15 ns (where almost exclusively I_3 is detected).

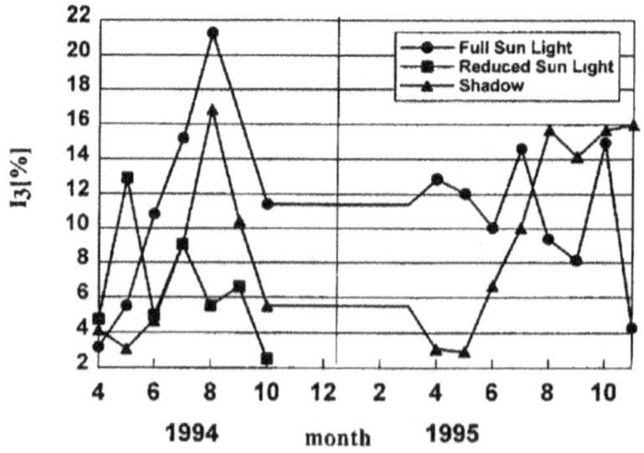

Figure 6. Biannual course of the long-lived component of chlorophyll fluorescence (needles of *Picea abies*, Freudenstadt, age class 1993, damage class 2).

CONCLUSION

We conclude that time-resolved and time-gated fluorescence spectroscopy and microscopy are powerful methods in various fields of medical diagnostics and environmental research.

ACKNOWLEDGMENT

The authors thank M. Bauer, R. Hahn and C. Hintze for their cooperation. Financial support by the German Ministry of Research (BMBF) and the German-Israeli Foundation for Scientific Research and Development (GIF) is gratefully acknowledged.

REFERENCES

1. H. Schneckenburger, Time-resolved fluorescence imaging, *J. Photochem. Photobiol. B:Biol.*24:63 (1994).
2. E.P. Buurman, R. Sanders, A. Draijer, H.C. Gerritsen, J J.F. van Veen, P.M. Houpt and Y.K. Levine, Fluorescence lifetime imaging using a confocal laser scanning microscope, *Scanning* 14:155 (1992).
3. X.F. Wang, T. Uchida, D.M. Coleman and S. Minami, A two-dimensional imaging system using a gated image intensifier, *Appl. Spectrosc.* 45:360 (1991).
4. R. Cubeddu, G. Canti, P. Taroni and G. Valentini, Time-gated fluorescence imaging for diagnosis of tumours in a murine model, *Photochem. Photobiol.* 57:480 (1993).
5. M. Kohl, J. Neukammer, U. Sukowski, H. Rinneberg, D Wöhrle, H.-J. Sinn and E.A. Friedrich, Delayed observation of laser-induced fluorescence for imaging of tumours, *Appl. Phys.* B56:131 (1993).
6. H. Schneckenburger, K König, T. Dienersberger and R Hahn, Time-gated microscopic imaging and spectroscopy in medical diagnosis and photobiology, *Opt. Eng.* 33:2600 (1994).
7 C.G Morgan, A.C. Mitchell and J.G. Murray, Prospects for confocal imaging based on nanosecond fluorescence decay time, *J Microsc.* 165 49 (1992).
8. T.W.J. Gadella, Jr, T.M. Jovin and R.M. Clegg, Fluorescence lifetime imaging microscopy (FLIM): Spatial resolution of microstructures on the nanosecond time scale, *Biophys. Chem.* 48:221 (1993).
9. J.R. Lakowicz, H. Szmacinski, K. Nowaczyk and M.L. Johnson, Fluorescence lifetime imaging of free and protein-bound NADH, *Natl. Acad. Sci. USA* 89:1271 (1992).
10. R. Sanders, Fluorescence lifetime as a contrast mechanism in confocal imaging, Doctoral Thesis, University of Utrecht (NL), ISBN 90-393-0648-6 (1995).
11. H. Schneckenburger, M.H. Gschwend, R. Sailer, A. Rück and W.S.L. Strauß, Time-resolved pH dependent fluorescence of hydrophilic porphyrins in solution and in cultivated cells, *J. Photochem. Photobiol. B:Biol.* 27:251 (1995).
12. W. Halle, W.-E. Siems, K.D Jentzsch, E. Teuscher and E. Göres, Die *in vitro* kultivierte Aorten-Endothelzelle in der Wirkstofforschung - zellphysiologische Charakterisierung und Einsatzmöglichkeiten der Zellinie BKEz-7, *Pharmazie* 39:77 (1984).
13. H. Schneckenburger, W. Strauß, A Rück, H.K. Seidlitz and J.M. Wessels, Microscopic fluorescence spectroscopy and diagnosis, *Opt. Eng.* 31:995 (1992).
14. S. Andersson-Engels et al., Fluorescence imaging and joint measurements of tissue, *Photochem. Photobiol.* 53:807 (1991).
15 K. König and H. Schneckenburger, Laser-induced autofluorescence for medical diagnosis, *J. Fluoresc.* 4:17 (1994).
16 B. Kjeldstad, A. Johnsson, B. Sandberg, Influence of pH on porphyrin production in *Propionibacterium acnes*, *Arch. Dermatol. Res.* 276:396 (1984).

17 J.-M. Salmon, E. Kohen, P. Viallet, J.G. Hirschberg, A.W. Wouters, C. Kohen and B Thorell, Microspectrofluorometric approach to the study of free/bound NAD(P)H ratio as metabolic indicator in various cell types, *Photochem. Photobiol.* 36:585 (1982).

18 T. Galeotti, G.D.V. van Rossum, D.H. Mayer, B. Chance, On the fluorescence of NAD(P)H in whole cell preparations of tumours and normal tissues, *Eur. J. Biochem.* 17:485 (1979).

19 P. Galland and H. Senger, The role of flavins as photoreceptors, *J. Photochem. Photobiol. B:Biol.* 1:277 (1988).

20. H. Schneckenburger and K. König, Fluorescence decay kinetics and imaging of NAD(P)H and flavins as metabolic indicators, *Opt. Eng.* 31:1447 (1992).

21. R.-J. Paul and H. Schneckenburger, Oxygen concentration and the oxydation-reduction state of yeast: Determination of free/bound NADH and flavins by time-resolved spectroscopy, *Naturwissenschaften*, in press.

22. M.H. Gschwend, W.S.L Strauß, H. Schneckenburger, R Steiner, Time-resolved fluorescence microscopy for probing mitochondrial metabolites, *Proc. SPIE*, Vol. 2628 (in press).

23. W.S.L. Strauß, R. Sailer, H. Schneckenburger, N Akgün, V. Gottfried, L. Chetwer, S. Kimel, Photodynamic efficacy of naturally occurring porphyrins in endohelial cells *in vitro* and microsvasculature *in vivo*, submitted to *J. Photochem. Photobiol.B:Biol.*.

24. J. Moan, S. Sommer, Action spectra for hematporphyrin derivative and Photofrin II with respect to sensitization of human cells in vitro to photosensitization, *Photochem. Photobiol.* 40:631 (1981).

25. H. Schneckenburger and W. Schmidt, Time-resolved chlorophyll fluorescence of spruce needles after different light exposure, J. Plant Physiol., in press.

A VERSATILE TIME-RESOLVED LASER SCANNING CONFOCAL MICROSCOPE

I.V. Eigenbrot, B. Crystall and D. Phillips

Department of Chemistry
Imperial College, London SW7 2AY U.K.

INTRODUCTION

Confocal microscopes are quickly becoming a *de facto* standard piece of equipment for the investigation of fluorescence distribution in microdomains, particularly in the medical sciences [1,2]. However, their complexity and the requirement for fast, highly linear scanners, high quality optical elements and at least one laser source means that the cost of such apparatus tends to be prohibitive

Our research group has a long-standing interest in the spatial distribution of fluorophores in biological systems, particularly in the field of photodynamic therapy (PDT) [3]. Many researchers are currently attempting to discover where PDT sensitisers localise in cells and it was decided to construct and utilise a laser scanning confocal microscope with the added advantage of temporal resolution provided by the utilisation of time correlated single photon counting [4].

Although first proposed in 1957 by Minsky [5,6] for neurological studies, it was several decades before the technology became available to enable the confocal microscope to be used routinely. The recently developed instrumentation is described here, together with the results of some preliminary studies. The scanning head is part of a commercial system, the HBH C900 [7-9]. However, instead of using a pinhole [5] or a fibre optic [10] as the spatial filter to eliminate out-of-focus glare, we have used a simple optical system enabling the avoidance of the complex alignment procedures normally associated with confocal scanning systems, particularly when more than one laser is used regularly with the system or the fluorescence is too weak to be passed through a monochromator.

EXPERIMENTAL

A schematic of the apparatus can be seen in Figure 1. The instrument is built around a conventional upright microscope, the Olympus BH-2. A dichroic beamsplitter (DB1) of a suitable wavelength transmits the excitation beam, which is transmitted to the scanning mirrors (MX ,MY) via a beamwaist. The excitation beam is then scanned over the beamsteering lens (BS), which ensures that the beam always fills the entrance pupil of

Fluorescence Microscopy and Fluorescent Probes
Edited by J. Slavík, Plenum Press, New York, 1996

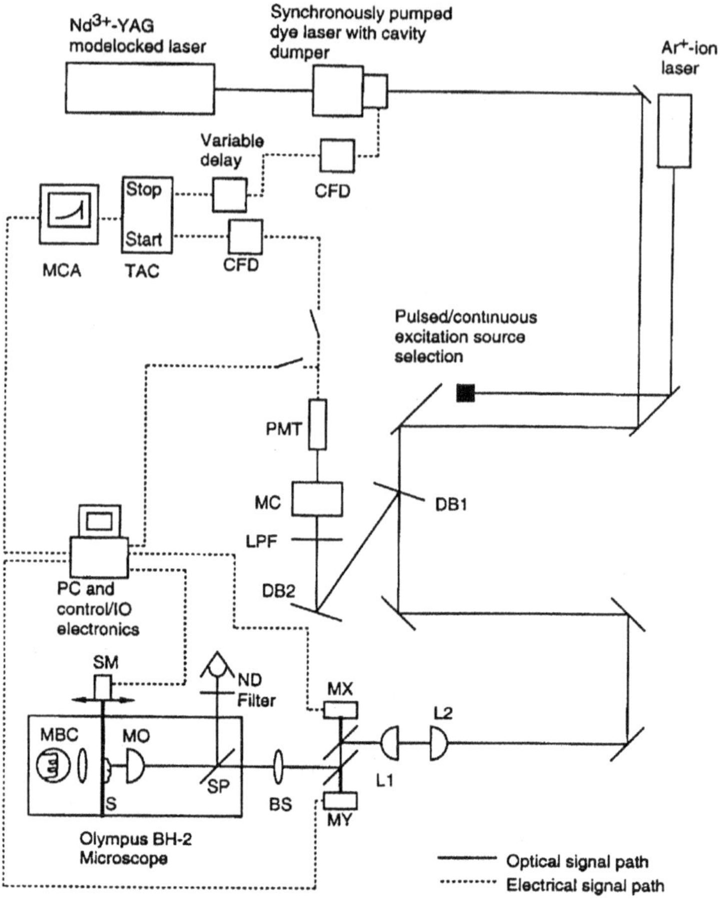

the microscope objective (MO) The galvanometers attached to the x- and y- axis scanning mirrors are operated by a computer and the excitation beam therefore performs a raster scan of the sample (S) The fluorescence from the individual voxels scanned travels back through the microscope along the same path, falls on the focusing element (L1), and the light coming from the focal plane is recollimated by the focusing element (L2) When this collimated fluorescence beam falls on the dichroic beamsplitter, it is reflected The beam is reflected from a second dichroic beamsplitter (DB2) and then passes through a longpass filter (LPF), which eliminates almost all remaining scatter from the excitation source. The emission then falls onto the slits of a monochromator (MC). The emerging spectrally filtered fluorescence is detected by a photomultiplier tube (PMT) For steady-state measurements, the photocurrent from the PMT is amplified, sampled by an analogue-to-digital converter at a rate precisely synchronised with the raster scan and passed to the computer In time-gated imaging mode, or for the collection of fluorescence decays, individual photon events are either counted by a gated counter, or passed to a time-to-amplitude converter (TAC) The collimating lens (L2) and the long fluorescence return path before the detection system act as a spatial filter, ensuring that very little light from out-of-focus planes reaches the detector. The axial resolution of this system is estimated to be ≈ 800 nm, with sub-micron lateral resolution, as the scanned excitation is effectively a diffraction-limited spot. The microscope stage can be moved under computer control by a stepper motor (SM), the minimum axial movement possible in the current configuration being ≈ 670 nm.

Laser excitation is provided by the output of a cavity-dumped dye-laser, synchronously pumped by a mode-locked CW Nd:YAG laser (Coherent Antares 76-s). Average excitation power at the sample is between 0.1 and 2 mW. For steady-state imaging, a tuneable Ar^+-ion (Spectraphysics 162), power at 488nm ~20mW, was used. Detection was carried out using a Hamamatsu R928 red-sensitive side-on PMT operated in photon counting mode Wavelength selection was achieved using cut-off filters and a Jarrel-Ash 1/4 metre monochromator.

Fluorescence lifetimes or time-gated images are collected as follows. Pulses form the PMT are routed via a Tennelec TC454 discriminator to an Ortec 467 TAC/SCA Stop pulses were provided by a synchronous signal from the cavity dumper driver. The TAC/SCA output is sent to either a Tennelc PCA-II MCA card or to an in-house designed 8-bit TTL-based counter which was interfaced to the PC via a modified slot-in card. The TAC/SCA is used to select those pulses falling within a predetermined voltage (time) range and only those pulses within the gate period were output to the MCA. This allows selection of gate timing and widths and the resulting output pulses are input to the TTL counter. The scanning of the sample then results in the collection of time gated images. Alternatively, complete fluorescence decays can be collected on the MCA.

RESULTS

The system was calibrated using a series of dye-filled microspheres obtained from Molecular Probes In order to test the lifetime capability of the apparatus, a number of dye/cell systems were measured. These are summarised in Table 1.

Table 1

Sample	Stain	Measured Decay Times
Murine mastocytoma cells	ethidium bromide	20.5ns
bovine kidney cells	Acridine orange	2.5ns / 10 5ns
e.coli	S_2Aluminium phthalocyanine	5ns

Ethidium bromide, incubated inside murine mastocytoma cells, exhibited an emission lifetime of 20.5ns, matching that of the dye in its bound, intercalated form whilst in the unbound state it is known to have a lifetime of approximately 1.8ns. Cells of e coli, incubated with disulphonated aluminium phthalocyanine, exhibit a lifetime of 5ns, indicating that the dominant decay was originating from the monomeric form of the dye [11] Previous work in leukaemic K562 cells [3] indicted that intracellular fluorescence consisted of two components, with a lifetime of 5.5ns and a minor one of 1ns. The latter component was assigned to the dye localised in a different environment, possibly in an aggregated form.

TIME GATED IMAGING

Figures 2a and 2b show two images of a pair of natural cotton fibres, one stained with 9,10 Anthracene dicarbonitrile and one with Rhodamine 6G. The images were

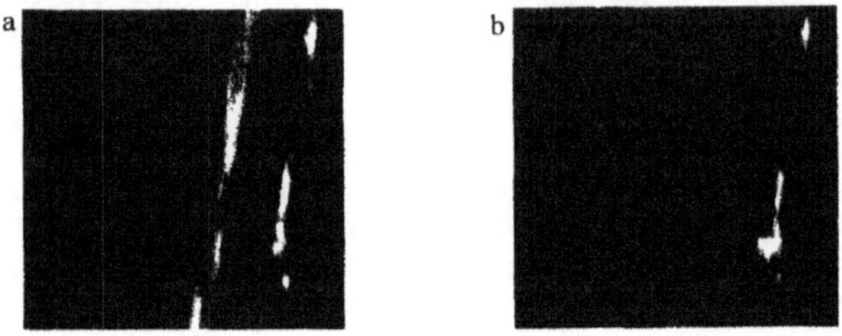

Figure 2a and b.
A pair of gated images of stained fibres taken at a) T=0-400ps and b) T=7000±200ps

obtained by laser excitation at 444nm and the emission was collected at 506nm, between the emission peaks of the two fluorophores The two images are of fluorescence collected at early (T=0-400ps) and later (T=7000ps ± 200ps) gates. The relative change in the fluorescence intensities between the two fibres is obvious from these images and is a result of the excited singlet state of 9, 10 Anthracene dicarbonitrile being longer lived than that of Rhodamine 6G The fluorescence intensity profiles of the two fluorophores on the fibres were extracted from an array of 16 images recorded in sequential gated images, each with window widths of approximately 400ps. The lifetime of the dyes on the fibres were then calculated using least squares fitting. The normalised resultant decays are shown in figure 3. The decays of the stained fibres were measured independently using TCSPC and both were found to be bi-exponential, with lifetimes of 2.37ns and 8.94ns (9, 10 Anthracene dicarbonitrile) and 2.62ns and 322ps (Rhodamine 6G). The decays shown in figure 3 match well the average values of the measured decays.

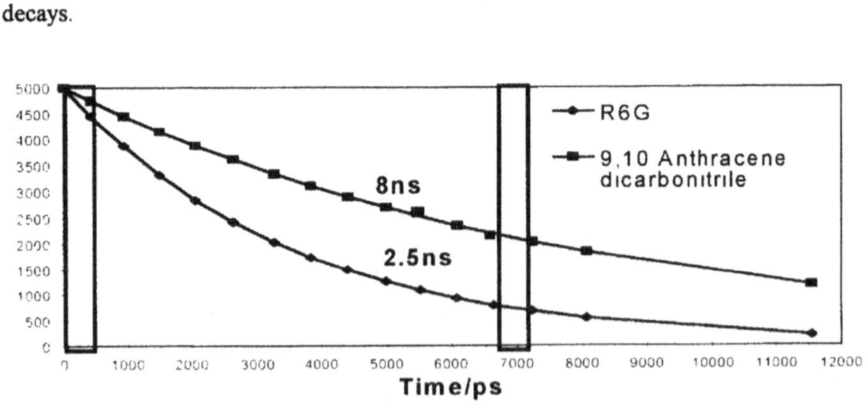

Figure 3 Normalised decays resulting from fitting of gated images Each point represents an image with a gate width of approximately 400ps The boxes illustrate the gate positions and widths for figures 2a and 2b

It will be possible to fit the decays more accurately if a greater number of gated images are recorded but photobleaching may pose a problem for this approach in some biological samples.

In conclusion, we have constructed a confocal microscope system which is capable of measurements with sub-micron spatial resolution along all three axes The optical system is usually complex and confocal fluorescence microscopes are therefore costly. The system described has been developed at low cost, yet has all the benefits of a conventional confocal system whilst adding temporal and spectral resolution to a conventional confocal microscope's capabilities.

REFERENCES

[1] S Paddock, Proc Soc.Exp. Biol. Med 198, 3 (1991)

[2] J.G White, W B. Amos, M. Fordham, Journal of Cell Biology 105, 41(1987)

[3] Ambroz, M., MacRobert, A J , Morgan, J , Rumbles, G , Foley, M.S.C., and Phillips, D. J. Photochem. Photobiol. B. Biol 22, 105-117, (1994)

[4] D.O'Connor and D. Phillips Time Correlated Single Photon Counting, Academic Press, London (1984)

[5] M. Minsky, *U.S. Patent #3013467* Microscopy Apparatus (1957)

[6] M Minsky, Scanning 10 128 (1988)

[7] M Gu, C.J.R. Sheppard, X. Gan, Journal of the Optical Society of America 8 1755 (1991)

[8] S. Kimura, T. Wilson, Applied Optics 30 2143(1991)

[9] J. Benschop, Appl Optics 30 1179(1991)

[10] P.M Delaney, M. Harris, R G. King, Clinical and Experimental Pharmacology and Physiology 20 197-198 (1993)

[11] Ambroz, M., Beeby, A., MacRobert, A.J , Simpson, M.S.C., Svenson, R. and Phillips, D. J. Photochem. Photobiol. B. Biol. 9, 87-95 (1991)

DISAPPEARANCE OF CYTOPLASMIC Ca^{2+} OSCILLATIONS IS A SENSITIVE INDICATOR OF PHOTODAMAGE IN PANCREATIC β-CELLS

Anders Tengholm and Eva Grapengiesser

Department of Medical Cell Biology
Uppsala University
Biomedicum, Box 571
S-75123 Uppsala
Sweden

INTRODUCTION

Epifluorescence microscopy is widely used for monitoring the cytoplasmic Ca^{2+} concentration ($[Ca^{2+}]_i$) in living cells. In our laboratory we have applied fluorometric techniques for measuring $[Ca^{2+}]_i$ in insulin-releasing pancreatic β-cells using conventional microscopy with fura-2 and confocal laser scanning microscopy with visible wavelength indicators (fluo-3, calcium green-1 and fura red). The β-cells were found to have an intrinsic ability of oscillating in response to glucose, the physiological stimulus of insulin release. When exposed to stimulatory concentrations of glucose the pancreatic β-cells reacted with influx of Ca^{2+}, resulting in large oscillations or a sustained elevation of $[Ca^{2+}]_i$ (Grapengiesser et al., 1989).

The ability of glucose to induce oscillations of $[Ca^{2+}]_i$ in individual β-cells have been confirmed by several investigators (Wang and McDaniel, 1990; Herchuelz et al., 1991; Theler et al., 1992; Pralong et al., 1994). However, differences exist regarding the proportion of cells showing oscillations. Analyzing the prerequisites for glucose generation of the oscillatory activity, we have found that this is critically dependent on gentle handling of the β-cells (Hellman et al., 1994). With improved techniques for cell isolation and efforts to minimize the exposure to the excitation light the percentage of mouse β-cells responding to 11 mM glucose with oscillations was found to be as high as 85 %. Indeed, disappearance of $[Ca^{2+}]_i$ oscillations represents a sensitive indicator for β-cell damage evoked by excitation light both in the UV and visible regions.

PHOTODAMAGE BY EXCITATION LIGHT OF 300-400 nm

Measurements of $[Ca^{2+}]_i$ in single β-cells were performed by dual wavelength microfluorometry with fura-2 adhering to a previously described protocol (Grapengiesser et al., 1989; Grapengiesser, 1993). The glucose-induced oscillations of $[Ca^{2+}]_i$ extended from a basal level of 60-90 nM to peaks exceeding 300 nM and had a frequency of 0.2-0.5 /min.

Fluorescence Microscopy and Fluorescent Probes
Edited by J. Slavík, Plenum Press, New York, 1996

The oscillations continued for periods exceeding 60 min (Figure 1) provided that the excitation energies at 340 and 380 nm were kept low.

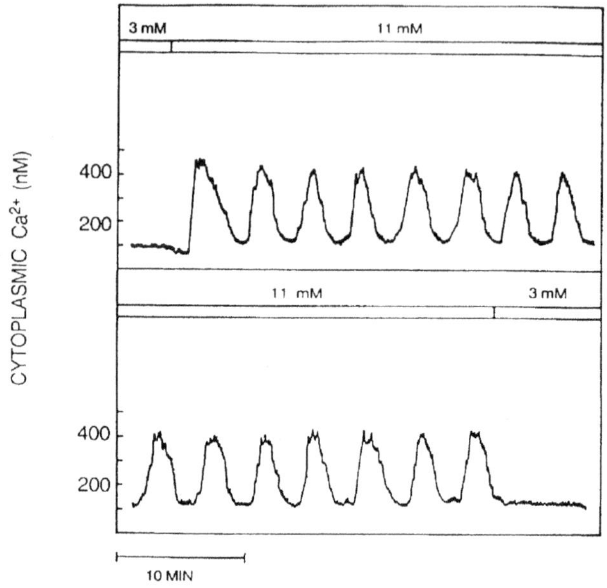

Figure 1. Oscillations of cytoplasmic Ca^{2+} induced by 11 mM glucose in an individual β-cell exposed to low intensity excitation (340 and 380 nm). The trace in the upper panel is continued in the lower panel.

Figure 2 shows the effects of tenfold increases of the excitation intensities. Within a few minutes the oscillations were changed into irregular fluctuations of $[Ca^{2+}]_i$, which returned to the basal value when glucose was lowered to 3 mM. Whereas similar effects were seen in 70 % of the β-cells, there was a complete loss of the glucose response in terms of raised $[Ca^{2+}]_i$ in 20 % of the cells. To investigate the effects of short-term exposure to UV light,

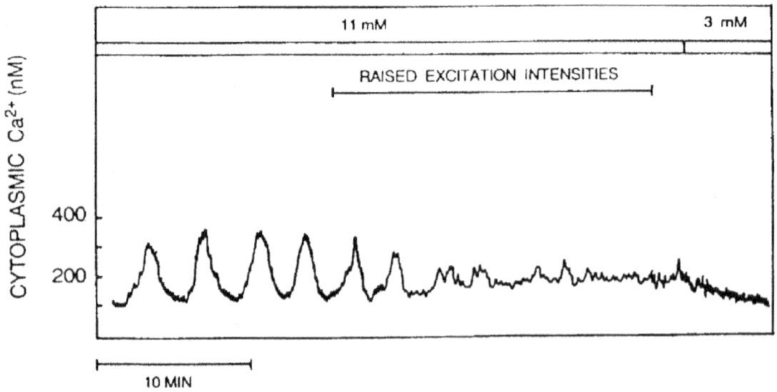

Figure 2. Effects of a tenfold increase of the excitation light intensity (340 and 380 nm) on glucose-induced Ca^{2+} oscillations in an individual β-cell. The duration of the intensity increase (removal of a neutral density filter) is indicated. The low noise during this period reflects the increase of the emitted fluorescence.

the β-cells were irradiated by <1 ms pulses of intense light produced by a xenon flash lamp assembly (Rapp and Güth, 1988) combined with a filter transmitting in the 300-400 nm region. Exposure to a UV flash with a radiant energy of 4-7 mJ/mm² resulted in a rapid disappearance of the oscillations (Figure 3). Repeated flashes induced more severe lesions manifested as an excessive and uncontrolled rise of $[Ca^{2+}]_i$, which is known to promote irreversible processes leading to cell death.

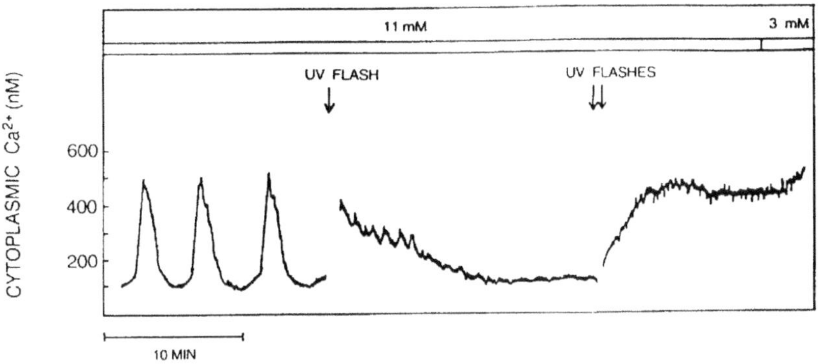

Figure 3. Effects of short pulses (< 1 ms) of UV light (300-400 nm) on the glucose-induced oscillations of cytoplasmic Ca^{2+} in an individual β-cell. The arrows indicate pulses with an energy of 4-7 mJ/mm² at focus.

In fura-2 measurements of $[Ca^{2+}]_i$, the intensity of the excitation light generated by a conventional xenon lamp (75 W) may easily cause photocyto-noxious effects, particularly during prolonged recordings. A considerable attenuation of the excitation light is consequently a prerequisite for maintaining an intact $[Ca^{2+}]_i$ response in the β-cells. To protect the β-cells from photodamage, the excitation intensities have to be decreased below a level generating 7 % photobleaching during the initial 10 minutes as determined in model experiments with cell-sized droplets containing fura-2 pentapotassium salt dissolved in an "intracellular" buffer (Grapengiesser, 1993).

The UV light used for excitation of fura-2 is in the region designated as UVA (320-380 nm). The adverse effects of UVA have been attributed to absorption by chromophores, such as flavins, porphyrins and nicotinamide coenzymes, resulting in the generation of free radicals (Black, 1987; Tyrell et al., 1990). Accordingly, the rapid interference of UV light with the oscillatory $[Ca^{2+}]_i$ response to glucose may reflect membrane lesions due to peroxidation of unsaturated lipids and crosslinking of proteins (Black, 1987).

PHOTODAMAGE BY EXCITATION LIGHT OF 488 nm

Although the UV absorbing, ratiometric Ca^{2+} indicators have many advantages, they have been of limited importance for confocal laser scanning microscopy. In this technique the most widely used light sources are visible wavelength laser systems, that are less expensive than the UV systems.

We have performed $[Ca^{2+}]_i$ measurements in a Noran Odyssey XL confocal laser scanning microscope equipped with a 300 mW multiline argon-ion laser providing 488 nm light suitable for the excitation of fluo-3, calcium green-1 and fura red. The excitation intensity could easily be varied by altering the diffraction efficiency of the acousto-optic deflector scanning the laser beam. The applied energies were estimated at the entrance aperture of the objective using an optical power meter (Newport Corporation, Fountain Valley, CA, USA). As expected, the visible wavelength Ca^{2+} indicators reported oscillations similarly to fura-2.

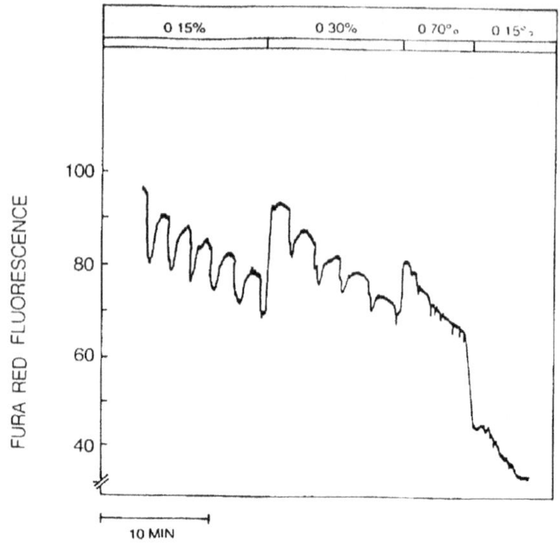

Figure 4. Glucose-induced oscillations of cytoplasmic Ca^{2+} recorded in a confocal microscope using fura red excited at 488 nm. (Note that Ca^{2+} binding to this indicator results in decrease of the fluorescence.) The gradual loss of fluorescence during the experiment is due to photobleaching and leakage of the dye from the cell. The bar indicates approximate laser illumination as per cent of maximum. Fluorescence intensity is expressed on an 8-bit gray-level scale.

More surprisingly, the glucose-induced $[Ca^{2+}]_i$ oscillations were sensitive also to 488 nm excitation light (Figure 4). Thus, already at 8 μW, corresponding to 0.7 % of maximum laser power, the oscillatory pattern disappeared.

The mechanisms by which visible light might interfere with cell function are not well understood. Fluorescent molecules in their excited state can react with molecular oxygen to produce highly reactive singlet oxygen (Tsien and Waggoner, 1995), mimicking the mechanisms in UV-induced damage. Possibly, two-photon effects may also occur at the high photon fluxes encountered in the focus spot (Svoboda and Block, 1994). Since the glucose-induced oscillations are known to be temperature-dependent, it is pertinent to consider whether production of heat contributed to the observed effects. However, according to estimates made by Denk et al. (1995), the heating would be negligible under the present experimental conditions.

CONCLUDING REMARKS

The oscillatory $[Ca^{2+}]_i$ response to glucose is believed to reflect a delicate balance between increased entry and stimulated removal of the ion from the cytoplasm (Hellman et al., 1992; 1994). The disappearance of regular $[Ca^{2+}]_i$ oscillations is a sensitive indicator of β-cell damage. Accordingly, cytotoxic agents have been found to mimic the effects of intense light on the glucose-induced oscillations of the β-cells (Grapengiesser et al., 1990).

The present results emphasize the importance of protection against photodamage in measurements with intracellular fluorescent probes. Caution is also required when using "caged" compounds, which release physiologically active agents after photolysis generated by a flash of UV light (Kaplan and Somlyo, 1989).

ACKNOWLEDGEMENTS

The skillful technical assistance of Mrs. Heléne Dansk is gratefully acknowledged. This study was supported by grants from the Swedish Medical Research Council (12I-562 and 12X-6240), the Swedish Diabetes Association, the Swedish National Board of Health and Welfare, the Novo Nordic Foundation, Novo-Nordisk Pharma AB, the Swedish Society for Medical Research and the Family Ernfors Foundation.

REFERENCES

Black, H., 1987, Potential involvement of free radical reactions in ultraviolet light-mediated cutaneous damage, *Photochem. Photobiol.* 46:213.

Denk, W., Piston, D.W., and Webb W.W., 1995, Two-photon molecular excitation in laser-scanning microscopy, *in*: "Handbook of Biological Confocal Microscopy", J.B. Pawley, ed., Plenum Press, New York.

Grapengiesser, E., 1993, Cell photodamage, a potential hazard when measuring cytoplasmic Ca^{2+} with fura-2, *Cell Struct. Funct.* 18:13.

Grapengiesser, E., Gylfe, E., and Hellman, B., 1989, Three types of cytoplasmic Ca^{2+} oscillations in pancreatic β-cells, *Arch. Biochem. Biophys.* 268:404.

Grapengiesser, E., Gylfe, E., and Hellman, B., 1990, Disappearance of glucose-induced oscillations of cytoplasmic Ca^{2+} in pancreatic β-cells exposed to streptozotocin and alloxan, *Toxicology* 62:263.

Hellman, B., Gylfe, E., Grapengiesser , E., Lund, P.-E., and Berts, A., 1992, Cytoplasmic Ca^{2+} oscillations in pancreatic β-cells, *Biochim. Biophys. Acta* 1113:295.

Hellman, B., Gylfe, E., Bergsten, P., Grapengiesser, E., Lund, P.-E., Saha, S., Berts A., Dryselius, S., Tengholm A., Liu, Y.-J., and Eberhardson, M., 1994, Pulsatile Ca^{2+} signalling and insulin release, *in*: "Frontiers of Insulin Secretion and Pancreatic β-cell Research", P. Flatt and S. Lenzen, eds., Smith-Gordon, London.

Herchuelz, A., Pochet, R., Pastiels, C., and van Praet, A., 1991, Heterogenous changes in $[Ca^{2+}]_i$ induced by glucose, tolbutamide and K^+ in single rat pancreatic B cells, *Cell Calcium* 12:577.

Kaplan, J.H., and Somlyo, A.P., 1989, Flash photolysis of caged compounds: new tools for cellular physiology. *Trends Neurosci.* 12:54

Pralong, W.F., Spät, A., and Wollheim, C.B., 1994, Dynamic pacing of cell metabolism by intracellular Ca^{2+} transients, *J. Biol. Chem.* 269:27310.

Rapp, G., and Güth, K., 1988, A low cost high intensity flash device for photolysis experiments, *Pflügers Arch.* 411:200.

Svoboda, K., and Block, S.M., 1994, Biological applications of optical forces, *Annu. Rev. Biophys. Biomol. Struct.* 23:247.

Theler, J.-M., Mollard, P., Guérineau, N., Vacher, P., Pralong, W.F., Schlegel, W., and Wollheim, C.B., 1992, Video imaging of cytosolic Ca^{2+} in pancreatic β-cells stimulated by glucose, carbachol, and ATP, *J. Biol. Chem.* 267:18110.

Tsien, R.Y., and Waggoner, A., 1995, Fluorophores for confocal microscopy. Photophysics and photochemistry. *in*: "Handbook of Biological Confocal Microscopy", J.B. Pawley, ed., Plenum Press, New York.

Tyrell, R.M., and Keyse, S.M., 1990, New trends in photobiology. The interaction of UVA radiation with cultured cells, *J. Photochem. Photobiol. B* 4:349.

Wang, J.-L. and McDaniel, M.L., 1990, Secretagogue-induced oscillations of cytoplasmic Ca^{2+} in single β and α-cells obtained from pancreatic islets by fluorescence-activated cell sorting, *Biochem. Biophys. Res. Commun.* 166:813.

ACKNOWLEDGMENTS

[illegible]

DISTRIBUTION OF INDIVIDUAL CYTOPLASMIC pH VALUES IN A CELL SUSPENSION

Petr Cimprich and Jan Slavík

Institute of Physiology, Czech Academy of Sciences
Vídeňská 1083, CZ-142 20 Prague 4, Czech Republic

INTRODUCTION

The application of pH-sensitive fluorescent probes represents both a progressive and an easy way of *in vivo* determination of cytoplasmic pH values. The ratio microscopy can deliver pH values of a large number of cells simultaneously. This illustrates the tremendous progress of the pH-measuring techniques, as ^{31}P NMR, ion-selective microelectrodes, distribution of weak acids or destruction (permeabilization) of cells yielded only one single value from each measurement procedure.

MATERIALS AND METHODS

Fluorescence Measurement

Several pH-sensitive dyes have been used, namely fluorescein, 5(6)-carboxyfluorescein, 2',7'-bis-(carboxyethyl)-5(6)-carboxyfluorescein (BCECF) and carboxy-SNARF-1. Cells were loaded after 30 min incubation in a 10μM solution of the acetate or acetoxymethyl ester (AM) form of the above mentioned dyes.

Pairs of fluorescence images were taken by a usual way, i.e. either by a conventional epifluorescence microscope (Zeiss, Germany) equipped with CCD camera and microchanel image intensifier, using double excitation 435 and 490 nm, emission at 520 nm or by a confocal microscope (Biorad MRC 600) with argon excitation 488 nm and emission separated for wavelength above 600 nm (one picture) and below 600 nm (second picture) by a split mirror. The resulting digitized images (Matrox) were processed with Nikon LUCIA software.

The pH-calibration was done in vitro (in a series of buffers) and in vivo (using nigericine). The yeast cells *Saccharomyces cerevisiae* were placed on a Petri dish filled with an approximately 1 mm layer. Methylene blue staining was used as a check for cell viability. The results shown in this paper are those obtained by BCECF. Other two dyes have been employed as a control and showed similar results[1]

RESULTS

The distribution of pH values of a cell population appears in all cases to fit a curve close to the Gaussian distribution as shown in Fig. 1. Cells suspended in water exhibit typically the mean pH value 5.6 and the standard deviation 0 17 pH unit. Cells incubated in a strong buffer (15 min in 0.2 M triethanolamine-phthalate) show a shift of the mean pH value with the pH of the buffer but the half-width of the distribution curve remains similar to that in water. It decreases only slightly from 0.40 to 0 36 pH unit (Fig. 2)

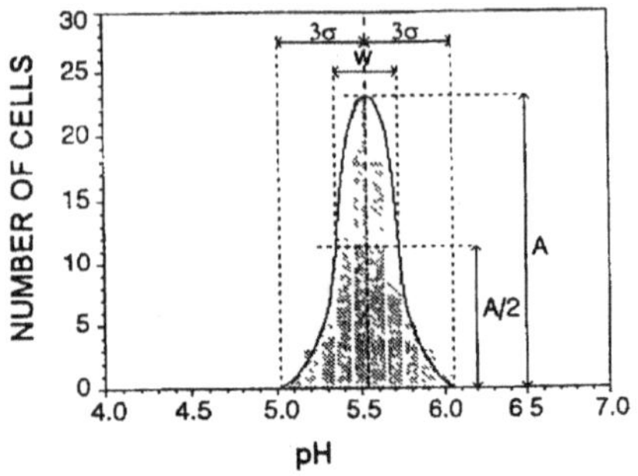

Fig. 1. Distribution of intracellular pH of yeast cells population suspended in a water

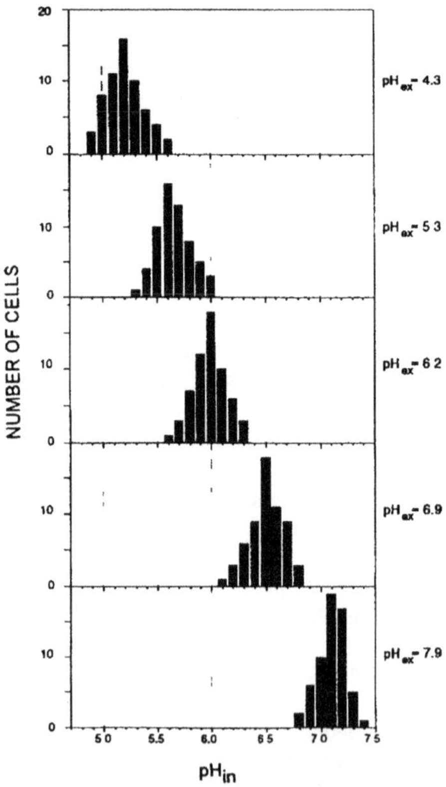

Fig. 2. Distribution of intracellular pH of yeast cells population suspended in a series of buffers (0 2 M triethanolamine-phthalate)

DISCUSSION

The average pH values both in water and in buffers tally with the literature data [2-6]. The existence of individual differences in cytoplasmic pH among cells is new. Preliminary results obtained with other cell types show that the phenomenon is quite general, rather than a specialty of the yeast *Saccharomyces cerevıesıae*. Individual differences have been by flow cytometry in lymphocytes[7-8], in macrophages and foreign body giant multinucleate cells[9] and in plant tissues[10]. If one realizes that cytoplasmic pH in a cell depends on a large number of independent factors[11], this may explain the Gaussian shape of the distribution curve.

ACKNOWLEDGEMENT

The work was supported by the grant #202/94/1712 of the Grant Agency of the Czech Republic

REFERENCES

1. P. Cimprich, J. Slavík, J. and A. Kotyk, Dıstrıbution of individual cytoplasmic pH values in a populatıon of the yeast Saccharomyces cerevisiae, *FEMS Mıcrobıology Letters* 130: 245 (1995)
2. A. Kotyk, Intracellular pH of baker's yeast, *Folıa Mıcrobıol.* 8:, 27-31 (1963)
3. J. Slavík, Intracellular pH of yeast cells measured with fluorescent probes, *FEBS Lett.* 140:22 (1982)
4. J. Slavík and A. Kotyk, Intracellular pH distribution and transmembrane pH profile of yeast cells, *Bıochım. Bıophys. Acta* 766:679 (1984)
5. G.W.F.H. Borst-Pauwels and P.H.J Peters, Effect of medium pH and the cell pH upon the kinetical parameters of phosphate uptake by yeast, *Bıochım Bıophys. Acta* 643:572 (1977)
6. A. Kotyk and G. Georghiou, Protonmotive force ın yeast pH, buffer and species dependence, *Bıochem. Int.* 24:641 (1991)
7. D F. Gerson, Determination of intracellular pH changes in lymphocytes with 4-methylumbelliferone by flow microfluorometry, in: *Intracellular pH: Its Measurement, Regulatıon, and Utılızatıon ın Cellular Functions,* R. Nuccıtelli and D.W.Deamer, eds., Alan R. Liss, New York, (1982) p. 375
8. O.Seksek, N. Henry-Toulmé, F. Sureau and J. Bolard, SNARF-1 as an ıntracellular pH ındicator in laser microspectrofluorometry: a critical assessment, *Analyt Bıochem.* 193:49 (1991)
9. J. Slavík and K. Smetana Jr. Intracellular pH of macrophages and foreign body giant cells colonizing an implant surface, *Journal of Fluorescence* (in press), (1996)
10. J.T.P. Albrechtová, J. Slavík and E. Wagner, Use of fluorescent probes and CLSM for pH-monitoring in the whole plant tissue, *thıs volume*, p 132
11. R. Nuccitelli and J.M. Heiple,. Summary of the evıdence and dıscussion concerning the involvement of pHi in the control of cellular functions, in *Intracellular pH Its Measurement, Regulation, and Utılızatıon ın Cellular Functıons*, R. Nuccitellı and D.W Deamer, eds., Alan R. Liss, New York (1982), p.567

THE EFFECT OF LYSOSOMAL pH ON LACTOFERRIN-DEPENDENT IRON UPTAKE IN *Tritrichomonas foetus*

Martin Gregor[1], Jan Tachezy[2] and Jan Slavík[3]

[1]Department of Physiology and Developmental Biology
[2]Department of Parasitology, Faculty of Science, Charles University, Viničná 7, CZ-128 44 Prague 2, Czech Republic
[3]Institute of Physiology, Czech Academy of Sciences, Vídeňská 1083, CZ-142 20 Prague 4, Czech Republic

INTRODUCTION

Tritrichomonas foetus is a parasitic protozoan which causes a sexually transmitted disease of cattle. The establishment of infection depends on the ability of *T. foetus* to acquire iron from the host as was demonstrated in experimentally infected mice[1]. Most of the iron available in mucosal secretions, the environment colonised by this parasite, is rather firmly bound to host iron-binding proteins such as lactoferrin and transferrin. Trichomonads as well as other pathogens therefore evolved specific mechanisms which allow them to withdraw iron from these proteins[2,3,4]. It was shown recently that lactoferrin is specifically bound to the surface of *T. foetus,* endocytosed and transported into hydrolase containing lysosome-like organelles[5]. Although there is no direct evidence for release of iron within this cell compartment, the low pH in the lysosomes might provide a suitable environment for such process. Since information about the intracellular pH of protozoa is rather limited[6,7] and since no data are available on the pH of the endo/lysosomal compartment in trichomonads, we attempted (1) to determine the pH of the cytoplasm and the lysosome-like organelles of *T. foetus,* (2) to monitor pH changes in these organelles in trichomonads treated with agents inhibiting endo/lysosomal acidification, and (3) to investigate whether a pH increase in the lysosomal-like organelles influences the iron uptake from lactoferrin by *T. foetus.*

MATERIAL AND METHODS

An axenic culture of *T. foetus* clone Lub-1 MIP was used in all experiments. The original stock of Lub was isolated in 1965 by Stepkowski (Lublin, Poland) from a bull (*Bos taurus*) It was axenized, cloned and cryopreserved by Kulda in 1970 (personal communication). The trichomonads were maintained in TYM medium[8] supplemented with 10% inactivated horse serum. The cells used in all experiments were harvested in logarithmic phase and washed 3 times in ice cold phosphate buffer saline (PBS; pH 7.4)

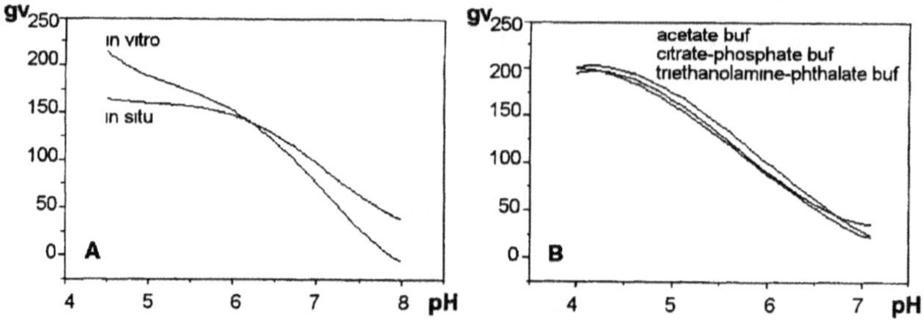

Figure 1. (a) Standard curves for BCECF-AM. Both *in vitro* and *in situ* standard curves were constructed The differences between the curves were negligible in the measured pH values range. The distortion of *in situ* curve in the more acidic and basic areas was probably caused by incomplete permeability of the plasma membrane to TEPA standard buffer[21,22]. gv is the gray value of the ratio image.

(b) Standard curve for FITC-lactoferrin Calibration was prepared in acetate (upper curve), citrate-phosphate (middle curve) and triethanolamine-phthalate (lower curve) buffers

To determine the pH of *T. foetus* cytoplasm, $4x10^6$ cells in 0.1 ml PBS were incubated for 30 minutes at 37°C with 10 μM 2′,7′-bis-(carboxyethyl)-5 (or 6)-carboxyfluorescein pentaacetoxymethyl ester (BCECF-AM). After incubation, the cells were washed 3 times with PBS, resuspended in 10 μl PBS and immediately examined by fluorescence excitation imaging microscopy (microscope JENALUMAR, Zeiss Germany, objective lens 100x, excitation wavelengths 435 nm and 490 nm, emission 520-530 nm, CCD camera, microchanel image intensifier, Matrox card digitizer, grabbing of 50 images, 255 grey scale results after rationing, image analysis software LUCIA-NIKON[9]) The pH calibration for BCECF-AM was done both *in vitro* and *in situ* (Fig 1a) using potassium triethanolamine-phtalate (TEPA) buffer solutions with pH ranging from 4 5 to 8.0 pH. *In situ* (intracellular) standard curve were prepared by incubation of the cells in the presence of 15 μM nigericin for 10 minutes at 37°C to allow equilibration of the trichomonads cytoplasm with the buffer.

The pH of lysosome-like organelles was determined by the same procedure using fluorescein isothiocyanate (FITC) labelled lactoferrin (60 μg/100 μl cell suspension)[10] In experiments monitoring the effect of agents inhibiting endo/lysosomal acidification, the cells were pretreated for 30 minutes with different concentrations of monensin, chloroquine or ammonium chloride in PBS as indicated in Fig 4 FITC-lactoferrin was subsequently added to the cell mixture, and the cells were incubated for another 30 minutes at 37°C. Finally, the cells were washed 3 times with PBS containing tested concentrations of inhibitors and examined. Standard curves for FITC-lactoferrin were made *in vitro* using series of buffers (acetate, citrate-phosphate, triethanolamine-phthalate) in the pH range from 4 to 7. No significant effect of the buffer composition on the calibration curve was found (Fig. 1b)

Iron uptake from lactoferrin by *T. foetus* was determined using ^{59}Fe-saturated ligand[11]. The cells ($5x10^6$), pretreated with inhibitors of acidification or control cells without the inhibitor treatment, were incubated in 250 µl of PBS with 25 µg ^{59}Fe-lactoferrin for 1 hour at 37°C. After incubation the cell-associated radioactivity was determined.

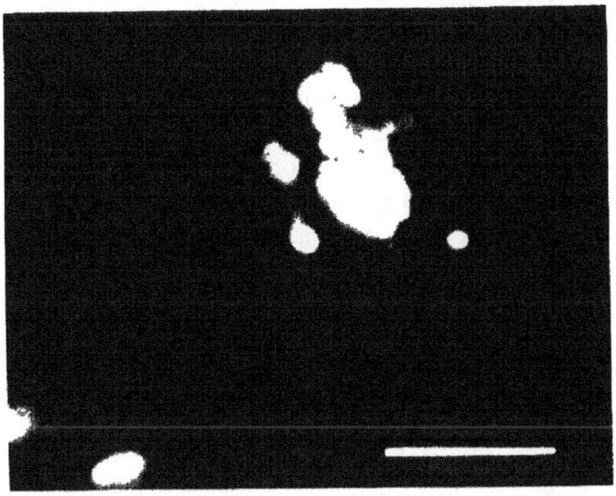

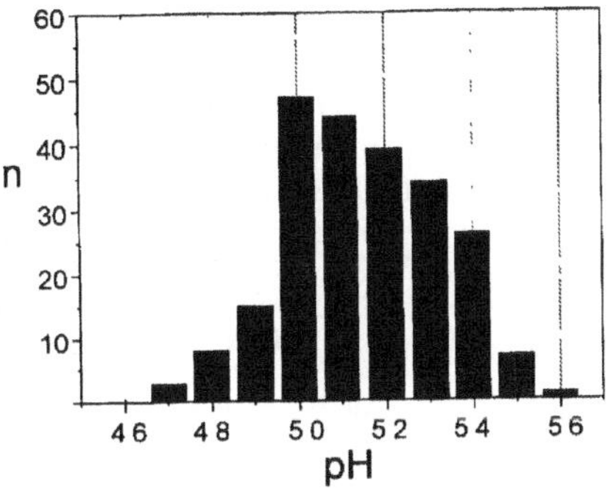

Figure 2. Accumulation of FITC-lactoferrin in lysosome-like organelles of *T. foetus.* The clusters of organelles were apparent in the central portion of the cell Bar, 10 µm

Figure 3. Distribution of lysosomal-like organelles pH in 224 cells loaded with FITC-lactoferrin, n is a number of individual cell.

RESULTS AND DISCUSSION

A weakly acidic pH was found in the cytoplasm of *T. foetus* (pH 6.28±0.2 [SE], range 5.8-6.7 pH, n=44). Such a low intracellular pH has only been reported in a specific subpopulation of *Dictyostelium discoideum* amoebae[12] In the majority of cell lines, a neutral or weakly basic pH has been determined[13,14,15]. Although the pH of the *T. foetus* environment within the host is not known, it was found that the trichomonads rapidly acidify culture media up to pH 4.2[16] The acidic pH of the *T. foetus* cytoplasm might be therefore related to its potential to acidify its environment

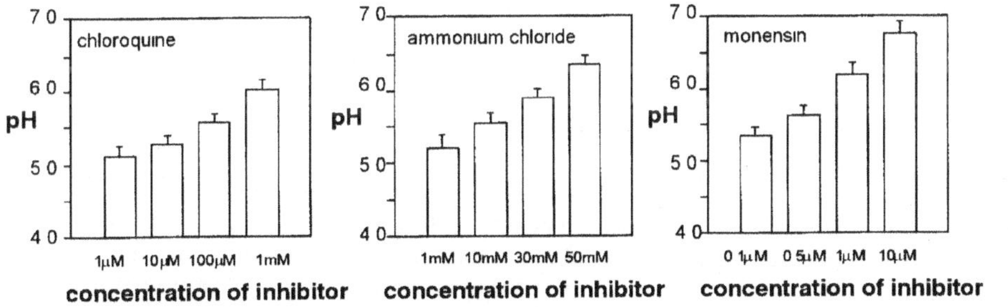

Figure 4. Effect of weak bases and ionophore on pH of lysosome-like organelles. The cells were pretreated with indicated concentrations of chloroquine, ammonium chloride and monensin before addition of FITC-lactoferrin. Bars represent standard errors, n=75.

To measure the pH of lysosome-like organelles of *T. foetus,* FITC-labelled lactoferrin was used as a pH-sensitive probe. The trichomonads accumulated this protein in vesicles of various sizes distributed throughout the cytoplasm (Fig. 2). The average number of fluorescent vesicles was about 30 per cell. The pH of the vesicles measured in 224 cells by means of fluorescence ratio imaging microscopy, revealed a mean pH value 5.2 ±0.18 The individual pH values fit the Gaussian curve ranging from 4.5 to 5.6 (Fig. 3). Comparable pH values of lysosomes were recently found using FITC-dextran and FITC-labelled silica particles in *Dictyostelium discoideum*[6] and rabbit alveolar macrophages[17], respectively. Our pH estimation of lysosome-like organelles in *T. foetus* is also consistent with biochemical data of Searle and Müller[18] They reported a pH optimum of 5.2 for acidic inorganic pyrophosphatase, an enzyme present in lysosome-like organelles of *T. vaginalis* and probably also in other protozoa utilising inorganic pyrophosphate as a phosphoryl donor[19].

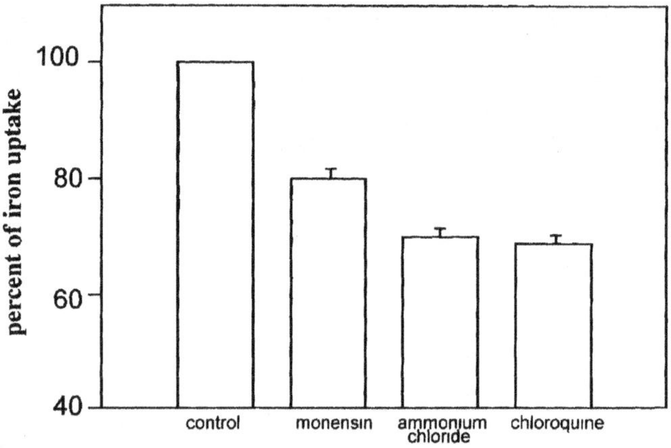

Figure 5. Inhibition of iron uptake from ^{59}Fe lactoferrin in the cells pretreated with ammonium chloride (50 mM), chloroquine (1 mM) and monensine (10 μM). Bars represent standard errors, n=3.

The effect of agents inhibiting endo/lysosomal acidification are summarised in Fig. 4. Both weak bases (ammonium chloride and chloroquine) and the ionophore monensin caused a pH increase in lysosome-like organelles of *T. foetus.* The highest concentrations of ammonium chloride (50 mM), chloroquine (1 mM) and monensine (10 μM) caused an increase of the pH value to 6.35 ±0.12, 6 02 ±0.16 and 6.76 ±0.17, respectively. The same inhibitor concentrations were used in the iron uptake experiments. As apparent from Fig. 5, treatment of trichomonads with ammonium chloride, chloroquine and monensin caused a decrease of iron uptake from ^{59}Fe-lactoferrin to 63%, 70% and 80% respectively, when compared with the untreated cells.

Specific binding and endocytosis of iron-containing lactoferrin by *T. foetus* was found by Affonso et al.[5] and Tachezy et al.[20]. However, it is difficult to demonstrate directly whether iron is released from lactoferrin within the endo/lysosomal compartment or at the parasite surface as was suggested for *T. vaginalis*[3] Our findings confirmed that lactoferrin is internalised and accumulated in the acidic cell compartment of *T. foetus.* The correlation between a pH increase of lysosome-like organelles and a decrease of iron accumulation from ^{59}Fe-saturated lactoferrin suggests an active involvement of the acidic cell compartment in the iron release from lactoferrin in *T. foetus.*

REFERENCES

1 J. Kulda, M. Budilová , The effect of ferric ammonium citrate on the multiplication of *Tritrichomonas foetus* in the peritoneal cavity of mice, *J. Protozool.* 24, 51A (1977)
2. C.S. Voyiatzaki, K.P. Soteriadou, Identification and isolation of the *Leishmania* transferrin receptor, *J. Biol. Chem.* 267:9112-9117 (1992)
3. K.M. Peterson, J F. Alderete, Iron uptake and increased intracellular enzyme activity follow host lactoferrin binding by *Trichomonas vaginalis* receptors, *J. Exp. Med.* 160:398-410 (1984)
4. I. Coppens, F.R. Opperdoes, P.J. Courtoy, P Baudhuin, Receptor-mediated endocytosis in the bloodstream form of *Trypanosoma brucei*, *J. Protozool.* 34·465-473 (1987)
5. A.L. Affonso, M. Benchimol, K.C. Ribeiro, U. Lins, W. de Souza, Further studies on the endocytic activity of *Tritrichomonas foetus, Parasitol. Res.* 80· 403-413 (1994)
6. L. Aubry, G. Klein, J.-L. Martiel, M. Satre, Kinetics of endosomal pH evolution in *Dictyostelium discoideum* amoebae, *J. Cell Sci.* 105· 861-866 (1993)
7. R. Allen, A. Fok, Nonlysosomal vesicles (acidosomes) are involved in phagosome acidification in *Paramecium, J. Cell Biol*.97: 566-570 (1983)
8. L. Diamond, The establishment of various trichomonads of animals and man in axenic cultures, *J. Parasit.* 43:488-490 (1957)
9. P. Cimprich, J. Slavík, A. Kotyk, Distribution of individual cytoplasmic pH values in a population of the yeast *Saccharomyces cerevisiae, FEMS Microbiol. Let.* 130:245-252 (1995)
10. D. Legrand, J Mazurier, P. Maes, E. Rochard, J. Montreuil, G. Spik, Inhibition of the specific binding of human lactotransferrin to human peripheral blood phytohaemagglutinin-stimulated lymphocytes by fluorescein labelling and location of the binding site, *Biochem. J* 276:733-738 (1991)
11. G.W Bates, J. Wernicke, The kinetics and mechanism of iron(III) exchange between chelates and transferrin, *J. Biol. Chem.* 246.3679-3685 (1971)
12. R. Furukawa, J.E. Wampler, M. Fechheimer, Cytoplasmic pH of *Dictyostelium discoideum* amebae during early development: identification of two cell subpopulations before the aggregation stage, *J. Cell Biol.*, 110:1947-1954 (1990)
13 S. Bassnett, L. Reinisch, D. C. Beebe, Intracellular pH measurement using single excitation-dual emission fluorescence ratios, *Am. J. Physiol.* 258 C171-C178 (1990)
14. G R. Bright, G. W. Fisher, J. Rogowska, D L. Taylor, Fluorescence ratio imaging microscopy: temporal and spatial measurements of cytoplasmic pH, *J. Cell Biol.*, 104:1019-1033 (1987)
15. O. Seksek, N. Henry-Toulmé, F Sureau, J. Bolard, SNARF-1 as a intracellular pH indicator in laser microspectrofluorometry: a critical assessment, *Anal. Biochem.* 193:49-54 (1991)
16. F. F. Pindak, W. A Gardner, jr , M. M. de Pindak, Growth and cytopathogenicity of *Trichomonas vaginalis* in tissue cultures, *J. Clin. Microbiol* 23:672-678 (1985)
17. K. Nyberg, U. Johansson, A. Johansson, and P. Camner, Phagolysosomal pH and location of particles in alveolar macrophages, *Fundam. Appl. Toxicol.* 16:393-400 (1991)
18 S.M. Searle, M. Müller, Inorganic pyrophosphatase of *Trichomonas vaginalis, Mol. Biochem. Parasitol.* 44:91-96 (1991)
19. E. Mertens, E. Van Schaftingen, M. Müller, Presence of a fructose-2,6-bisphosphate-insensitive pyrophosphate: fructose-6-phosphate phosphotransferase in the anaerobic protozoa *Tritrichomonas* foetus, *Trichomonas vaginalis* and Isotricha prostoma, *Mol. Biochem. Parasitol.* 37:183-190 (1989)
20. J. Tachezy, J. Kulda, I. Bahníková, P. Suchan, J. Rázga, J. Schrevel, Tritrichomonas foetus: iron acquisition from lactoferrin and transferrin, *Exp. Parasitol.*, in press
21. P Cimprich, J. Slavík, Fluorescent measurement of intracellular pH, *J. Fluorescence*, in press
22. CH. S. Owen, Comparison of spectrum-shifting intracellular pH probes 5′(and 6′)-carboxy-10-dimethylamino-3-hydroxyspiro [7H-benzo[c] xanthene-7,1′(3′H)-isobenzofuran]-3′-one and 2′,7′-Biscarboxyethyl-5(and 6)-carboxyfluorescein, *Anal. Biochem.* 204:65-71 (1992)

ON THE PROTEIN-ERROR OF THE CALCIUM-SENSITIVE FLUORESCENT INDICATOR FURA-RED

N. Opitz , T. Porwol , E. Merten , H. Acker

Max-Planck-Institute for Molecular Physiology
Rheinlanddamm 201 , 44139 Dortmund , Germany

INTRODUCTION

The relatively new Ca^{2+}-sensitive fluoroprobe FURA-RED (Molecular Probes, Inc.) is recommended as a fluorescent indicator (FURA-II analog) for quantitative determination of intracellular free Ca^{2+}-concentrations on the basis of dual excitation / single emission wavelength measurements (excitation ratio). In comparison to FURA-II this new fluoroprobe comprises the advantage of excitation within the visible spectral range with a fluorescence emission > 570 nm and may therefore be applied in instruments where UV-excitation is not provided (e.g. confocal laser scanning microscopes). In addition, FURA-RED may be suited in combination with the pH sensitive indicator BCECF to perform simultaneously intracellular pH and Ca^{2+} measurements, since both indicators largely overlap with their main excitation bands, but differ significantly with regard to their fluorescence emission.

In a detailed study (predominantly based on absorption measurements) Kurebayashi et al. [1] first reported the use of FURA-RED in frog skeletal muscle fibres for determination of intracellular calcium. From FURA-RED's slow diffusion coefficient in myoplasm the authors conclude that about 85 % of the indicator molecules are bound to muscle constituents of large molecular weight (e.g. aldolase, an abundant myoplasmic protein) elevating the intracellular K_d value to about 1.1 - 1.6 µmol/l. Moreover, first fluorescence measurements revealed substantial changes of the intracellular extreme ratios (R_0, R_1) as compared to in vitro calibration solutions lacking muscle proteins. However, the myoplasmic alterations of the indicator could be taken into account so that Kurebayashi and colleagues could reliably determine intracellular calcium concentrations under various experimental conditions. This result encouraged us to use the indicator in conjunction with a suitably supplemented confocal laser scanning microscope [2] (Biorad, MRC 600) for cytoplasmic Ca^{2+}-measurements in human melanoma cells (IGR) grown in monolayer culture.

Fluorescence Microscopy and Fluorescent Probes
Edited by J. Slavík, Plenum Press, New York, 1996

METHODS AND RESULTS

Prior to intracellular measurements FURA-RED was characterized in vitro using a Perkin-Elmer spectrofluorometer (LS 50). Fig.1 shows fluorescence excitation spectra of the indicator in dependence on three different Ca^{2+}-concentrations (using a calcium calibration kit provided by Molecular Probes). The excitation spectra exhibit an isosbestic point at 465 nm enabling to interpret the Ca^{2+}-dependence of the spectra as a sum of two fundamental spectral components (corresponding to the completely dissociated and associated indicator forms) weighted by the respective dissociation (α) and association degree ($1-\alpha$). The completely associated spectral component comprises a broad absorption band from 380 to 470 nm with a peak-absorption at about 410 nm, whereas the entirely dissociated component shows a peak at about 460 nm centered within the absorption band from about 420 to 500 nm. Evaluation of such spectra at the wavelengths 442 nm and 488nm (corresponding to the emission lines of the Ar^+- and He-Cd-laser of the confocal microscope) and subsequent ratio formation of the signals results in calibration curves as depicted in Fig.2 for two different temperatures. As demonstrated by the best fits (solid lines), the experimental data can be well approximated according to the equation :

$$R(pCa) = R_o - (R_o - R_1) / (1 + 10\char`^(pK_d^* - pCa)) \qquad (1)$$

with : R_o, R_1 = extreme ratios,
$pK_d^* = pK_d - \log(Q)$ = apparent pK_d value , Q = spectral parameter
$pCa = -\log((Ca^{2+}))$

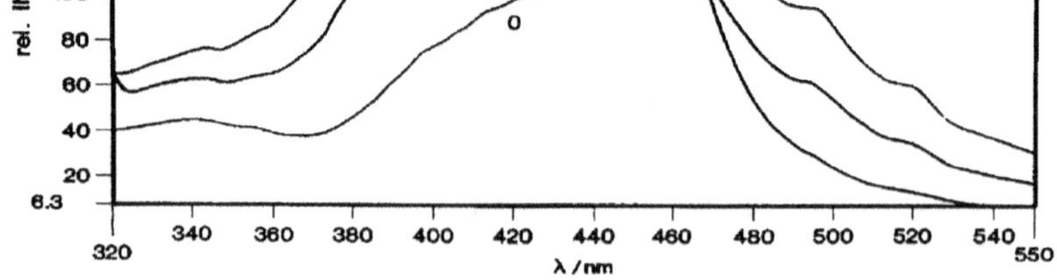

Fig.1 : Fluorescence excitation spectra (λ_{em}=620 nm) of FURA-RED in dependence on three different Ca^{2+}-concentrations (pH = 7.20 , ionic strength = 100 mmol/l KCl , ambient temperature)

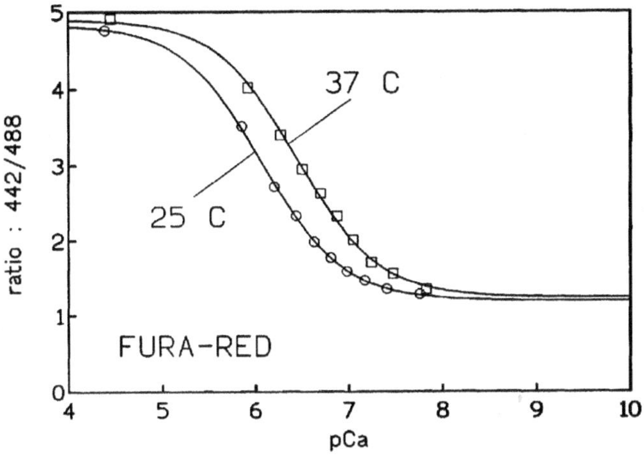

Fig.2 : In vitro calibration curves of FURA-RED (best fits of the data points according to eq.1) for two different temperatures plotted as ratio versus pCa as monitored by a Perkin-Elmer spectrofluorometer (LS 50)

Remarkably, the temperature sensitivity of the calibration curves can be mainly explained by the temperature dependence of the K_d value of the indicator and the spectral parameter Q since the extreme ratios seem to be virtually unaffected by a temperature rise from 25 ^oC to 37 ^oC. Thus, the distribution of the fundamental spectral components does not seem to change within this temperature range.

The K_d values are estimated from these measurements to about K_d (25 ^oC) = 240 nmol/l and K_d (37 ^oC) = 90 nmol/l (pH=7.20 ; ionic strength = 100 mmol/l KCl). Similar results as shown in Fig.2 are also obtained in case of measurements performed with the confocal laser scanning microscope (CLSM). Consequently, reproducible measurements can be performed with this indicator in vitro independent of the respective instrument used.

Human melanoma cells, incubated with the acetoxymethyl ester form of FURA-RED (10 µmol/l) (with 25 % w/w Pluronic in DMSO), were recorded with the modified CLSM [2] by alternating excitation with the 442 nm line of the He-Cd-laser and the 488 nm line of the Ar^+-laser. Fluorescence was measured in both cases at about 640 nm via a bandpass filter in front of the photomultiplier. Intracellular calibration of the cells (under administration of ionomycin 10 µmol/l) was performed according to a dynamic calibration procedure described in detail elsewhere [3,4] . Briefly, the intracellular Ca^{2+}-transient (induced by a well defined extracellular Ca^{2+}-jump, i.e. from 39.8 µmol/l to 1 mmol/l) is monitored during short time intervals (every 30 s) as shown in Fig.3 (left panel). Afterwards this transient is kinetically analyzed and transformed into a calibration curve as displayed in Fig.3 (right panel) in comparison to an in vitro calibration characteristic denoted cuvette (both curves are normalized to R_o). As can be recognized from this figure, cytoplasmic Ca^{2+}-measurements in human melanoma cells using FURA-RED are meaningless due to the drastically diminished intracellular Ca^{2+}-sensitivity of the indicator.

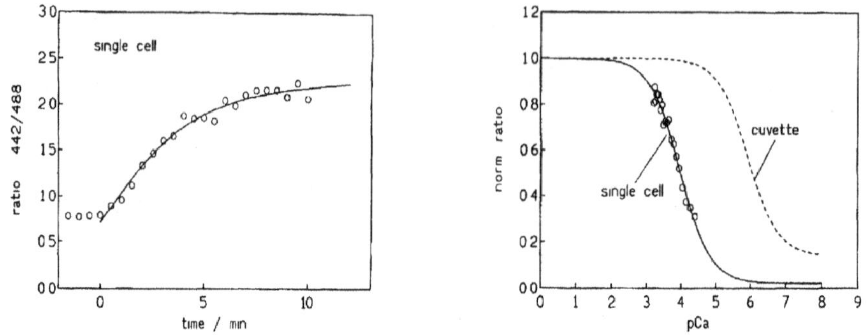

Fig.3 Intracellular Ca^{2+}-transient (left panel) of a single cell (under administration of ionomycin, 10 µmol/l) induced by an extracellular Ca^{2+}-jump from 39.8 µmol/l to 1 mmol/l (monitored by the CLSM with a time interval of 30 sec) and corresponding calibration curve (right panel) of the cell resulting via transformation of the transient after nonlinear regression analysis of the experimental data.

To investigate whether this huge sensitivity shift (from nmol/l to µmol/l range) is attributable to intracellular interactions of indicator molecules with intracellular substances such as proteins or other biomolecules, we studied the influence of protein addition (bovine serum albumin, 1 %) upon the spectral distribution of the excitation and emission spectra as well as upon the in vitro calibration characteristic of the fluoroprobe. The results of these investigations as depicted in Fig.4 show a drastical red-shift of the excitation spectra of approximately 20 - 60 nm (depending on the dissociated or associated indicator component, respectively) along with an about 60 nm shift of the isosbestic point towards shorter wavelengths, whereas the distribution of the fluorescence emission was only slightly affected (not shown).

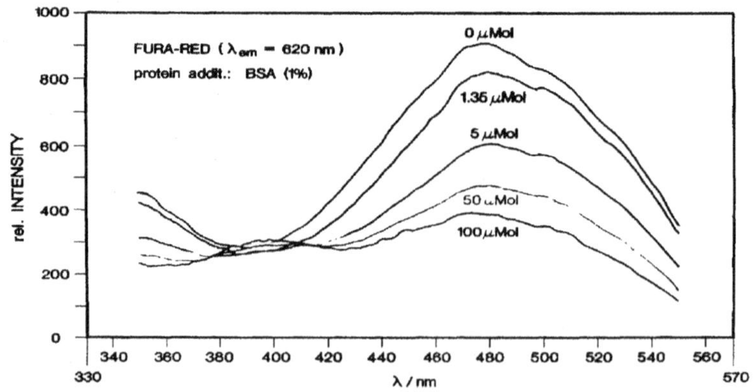

Fig.4 : Influence of protein addition (bovine serum albumin, 1 %) upon the spectral distribution of the excitation spectra of FURA-RED for different calcium concentrations recorded by the Perkin-Elmer spectrofluorometer (LS 50)

In addition, a considerable increase of the fluorescence intensity (about 10-fold) was observed presumably due to an enhanced fluorescence quantum efficiency since light scattering of the protein-indicator-solution and protein-autofluorescence were negligible. Moreover, a sensitivity change (expressed in terms of the K_d value) from about 240 nmol/l (in vitro) to approximately 6.6 µmol/l (after protein addition) was estimated via a best fit of the experimental data of a multipoint calibration characteristic.

DISCUSSION AND CONCLUSION

In accordance with Kurebayashi et al. [1] substantial changes of the calibration characteristic of FURA-RED were found intracellularly. However, in contrast to absorption measurements in frog skeletal muscle fibers, these huge changes in human melanoma cells monitored via the fluorescence of the indicator could not be taken into account for reliable determinations of intracellular calcium within the physiological range due to a drastical decrease of the intracellular Ca^{2+}-sensitivity (see Fig.3 : shift of the intracellular calibration curve by about 2 pCa-units towards higher calcium concentrations). As demonstrated by the in vitro investigations, such alterations can be simply induced by the addition of bovine serum albumin (1 %) suggesting that intracellular interactions of indicator molecules with proteins and other biomolecules are probably affecting the calibration characteristics.

Preliminary investigations using other frequently applied Ca^{2+}-sensitive fluoroprobes (e.g. FURA-II, INDO-I, FLUO-3, etc.) with regard to their protein dependence further suggest that the protein-error of fluorescent indicators is often the major interference during intracellular measurements. Unfortunately, although this error is known since decades [4], little attention is paid to it. As a consequence, quantitative cytoplasmic ion determinations have to be considered sceptically unless they are based on intracellular calibrations. However, even an in situ calibration may be regarded cautiously if it's based on a two-point calibration (as usually performed to determine the extreme ratios) assuming further an intracellularly unaffected K_d value. Thus, intracellular multipoint calibration procedures are necessary prerequisits for reliable quantitative determinations of intracellular ion concentrations and distributions [6].

Other publications concerning FURA-RED report a combination of the dye with the Ca^{2+}-sensitive indicator FLUO-3 enabling emission ratioing and, therefore, the application in confocal microscopy [7,8,9,10]. However, an emission ratio mode based on two different fluoroprobes demands several criteria to be fulfilled. Above all, both indicators have to distribute similarly within the cells and should possess largely identical photostabilities as well as low leakage rates. Although Lipp and Niggli [7] reported that the above requirements are largely fulfilled in heart muscle cells, deviations in different cell lines could possibly occur. In addition, variations of the degree of intermolecular interactions of the fluoroprobes with cytoplasmic constituents may result in considerable changes of the ratios indistinguishable from those caused by $(Ca^{2+})_i$.

ACKNOWLEDGEMENTS

We are indebted to B. Bölling for providing the human melanoma cells (IGR) grown in monolayer culture. Financial support by grants DFG (Ac37/9-1) and EC-project (BMH-CT94-1068) is appreciated.

REFERENCES

1. N. Kurebayashi, A.B. Harkins, S.M. Baylor, Use of FURA-RED as an intracellular calcium indicator in frog skeletal muscle fibers. Biophys.J.64: 1934-1960 (1993)
2. N. Opitz, M. Weinlich., F. Mooren, F.-K. Keinemann, G. Weidemann, R. Kinne, Dual wavelength excitation in confocal laser scanning microscopy : intracellular pH measurements with a fluorescent indicator. Biomed.Tech. 36: 132-133 (1991)
3. N. Opitz, E. Merten, H. Acker,Evidence for redistribution-associated intracellular pK-shifts of the pH sensitive fluoroprobe c-SNARF-1. Pflügers Arch.427 : 332-342 (1994)
4. N. Opitz,E.Merten,H.Acker,A dynamic calibration method for fast multipoint calibration of intracellularly entrapped ion sensitive fluoroprobes.Biomed.Tech.38:309-310 (1993)
5. Int.Series of Monographs in Anal.Chemistry, Vol.51 (Indicators), ed. E. Bishop, Pergamon Press Oxford, NY, Toronto, Sydney, Braunschweig (1972)
6. N. Opitz, T. Porwol, E. Merten, H. Acker, Cytoplasmic ion imaging : evidence for intracellular calibration heterogeneities of ion-sensitive fluoroprobes, this volume
7. P. Lipp, E. Niggli, Ratiometric confocal Ca^{2+}-measurements with visible wavelength indicators in isolated cardiac myocytes. Cell Calcium 14: 359-372 (1993)
8. M. Burnier, G. Centeno, E. Burki, H. Brunner, Confocal microscopy to analyze cytosolic and nuclear calcium in cultured vascular cells. Am.J.Physiol.266: C1118-C1127 (1994)
9. E.J.Novak,P.S.Rabinovitch,Improved sensitivity in flow cytometric intracellular calcium measurements using fluo-3 fura-red fluorescence ratios. Cytometry17 : 135-141 (1994)
10. D. Schild, A. Jung, H.A. Schultens, Localisation of calcium entry through calcium channels in olfactory receptor neurons using a laser scanning microscope and the calcium indicator dyes fluo-3 and fura-red. Cell Calcium 15: 341-348 (1994)

CYTOPLASMIC ION IMAGING : EVIDENCE FOR INTRACELLULAR CALIBRATION HETEROGENEITIES OF ION-SENSITIVE FLUOROPROBES

N. Opitz , T. Porwol , E. Merten , H. Acker

Max-Planck-Institute for Molecular Physiology
Rheinlanddamm 201, 44139 Dortmund, Germany

INTRODUCTION

Ion sensitive fluoroprobes such as BCECF and carboxy-SNARF-1 are frequently used indicators for pH determinations in single cells and subcellular compartments as well as for monitoring intracellular pH (pH_i) topographies via pseudocolour encoded fluorescence ratio imaging from pixel to pixel [1,2,3]. Regardless the intracellular degree of spatial pH resolution, reliable pH_i determinations essentially depend on accurate calibration of the intracellularly entrapped fluoroprobes. The calibration procedure is usually performed according to the nigericin method [4], which allows clamping of the intracellular pH to the extracellular (pH_o). Calibration is either established on the basis of a whole cell or within a more or less extended subcellular compartment and the characteristics are displayed as pH encoded pseudocolour bar within the image frame. This calibration is assumed to be valid for other cellular compartments and, in case of ion-imaging, it is even thought to be valid for every single pixel of the complete pixel field.

However, the assumption of a topographically invariant intracellular calibration requires a reliable behaviour of the intracellularly applied indicator. Concerning carboxy-SNARF-1, this intracellular integrity of the dye does not seem to exist, since calibration curves vary considerably from one cell to another as well as in comparison to an in vitro calibration characteristic (despite emission ratioing of the fluorescence signals). Moreover, calibration varies within different subcellular compartments of a single cell suggesting variations of the calibration characteristic even on a pixel basis. In case of carboxy-SNARF-1 the heterogeneity of the calibration is likely attributable to pH-dependent intracellular redistribution of the protonated form, as reported previously [5]. However, in case of BCECF, similar variations of the calibration were observed from one cell to another and also on a subcellular level. Investigations suggest that this heterogeneity is predominantly due to cytoplasmic interactions of BCECF molecules with intracellular substances, e.g. proteins, affecting the spectral distribution of the fundamental spectral components resulting in

Fluorescence Microscopy and Fluorescent Probes
Edited by J. Slavík, Plenum Press, New York, 1996

considerable changes of the extreme ratios and, therefore, in subcellular variations of the calibration characteristic. As a consequence, considerable restrictions are resulting in particular with regard to quantitative imaging of intracellular ion concentrations as well as in view of the spatial and temporal resolution of intracellular ion activities.

METHODS AND RESULTS

Intracellular pH measurements were performed in human malignant glioma cells (U 118 MG) grown as multicellular spheroids using confocal laser scanning microscopy (CLSM) and microspectrofluorometry (MSF). Extracellular pH (pH_o) was varied via suitable equilibration of the perfusion solution with different $CO_2/N_2/O_2$ gas mixtures provided by a gas mixing pump. pH_i was clamped to pH_o according to the nigericin method[4] and measured during intracellular steady state. Fig.1 shows mean cellular calibration curves of the indicators (depicted as normalised ratios R/R' versus pH) along with best fits of the experimental data in comparison to in vitro calibration characteristics denoted cuvette (experimentally verified). Nonlinear regression analysis of the experimental data was performed according to the equation :

$$\mathbf{R/R' = (1\text{-}a/(1+10\text{^}(pK^*\text{-}pH_o)))/(1\text{-}a/(1+10\text{^}(pK^*\text{-}6.77)))} \qquad (1)$$

with : $$\mathbf{pK^* = pK_i - \log(Q) \;;\; a = 1 - R_1/R_o} \qquad (2)$$

and: $pK_i = pK_a$ value of the indicator; Q = spectral parameter; R_o, R_1 = extreme ratios

As can be recognised from Fig.1 mean intracellular calibration curves considerably deviate from in vitro calibration characteristics for both pH indicators. In case of BCECF [6] a parameter determination of the mean cellular calibration curve (i.e. pK^* , $a=1-R_1/R_o$ and the correlation coefficient r) suggests that these distortions are predominantly attributable to intracellular interactions of the indicator molecules affecting the spectral distributions of the fundamental spectral components and, thus, resulting in considerable changes of the extreme ratios and a slightly altered pK^* value.

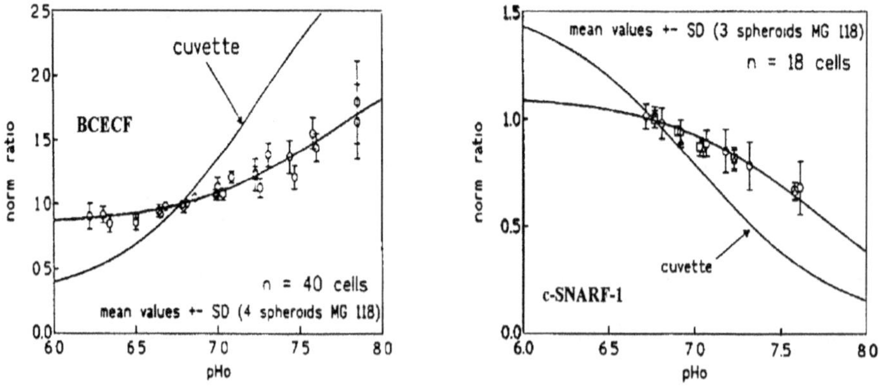

Fig.1 : Mean intracellular calibration characteristics of the pH indicators BCECF and carboxy-SNARF-1 in comparison to in vitro calibration curves (experimentally verified) denoted cuvette .

In contrast, the best fit of the experimental data in Fig.1 (carboxy-SNARF-1) can be interpreted in terms of a considerable change of the apparent pK value, pK*, whereas the extreme ratios remain almost unaffected. Hence , microspectrofluorometrical measurements were performed to investigate the spectral distributions of the fundamental spectral components of the intracellularly entrapped indicators. These measurements revealed a significant red shift of the uncorrected fluorescence emission spectra of BCECF in comparison to in vitro spectra supporting the assumption of intracellular interactions of the indicator molecules with proteins and/or other biomolecules. Conversely, almost identical spectral distributions of carboxy-SNARF-1 were observed independent of its microenvironment. However, a considerable change of the intracellular pH sensitivity of the indicator seems to exist in accordance with the observed pK shift of the apparent pK value, pK*, as determined via the best fit in Fig.1. As reported previously [5], this pK shift can be mainly explained by intracellular redistribution of the protonated component of carboxy-SNARF-1 between cytosol and lipophilic cell compartments (e.g. plasma membrane). A theoretical derivation of the redistribution associated intracellular pK shift results in [5] :

$$\Delta pK = \log(1 + q * P) \qquad (3)$$

with P = partition coefficient, q = factor, which contributes for different mean layer thicknesses of cytosol and plasma membrane .

To demonstrate the variability of the calibration curves from cell to cell, Fig.2 depicts the pH dependent courses of the normalised ratios for different single cells. In case of BCECF the variations of the calibration characteristics may be interpreted in terms of different extents of intracellular interactions resulting in different extreme ratios and pK* values from one cell to another, whereas the changes in case of carboxy-SNARF-1 may be explained by different degrees of intracellular redistribution, i.e. by varying q and P values from cell to cell.

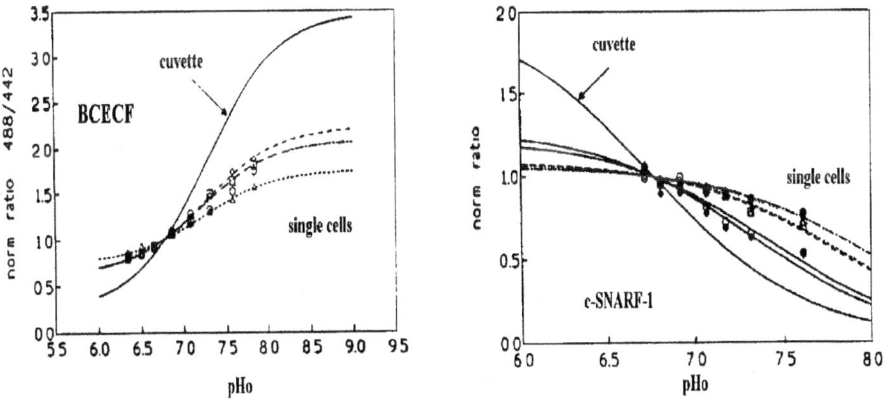

Fig.2 : Variation of the calibration curves of single cells in comparison to the in vitro calibration characteristic (experimentally verified) denoted as cuvette .

The heterogeneity of the calibration can be also observed on a subcellular level as exemplified in Fig.3 by means of the evaluation of 5 subcellular compartments of a single glioma cell loaded with BCECF (similar results as shown in Fig.3 were also found in case of the indicator carboxy-SNARF-1). This intracellular heterogeneity of the calibration in different subcellular compartments suggests variations of the calibration characteristic even on a single pixel basis.

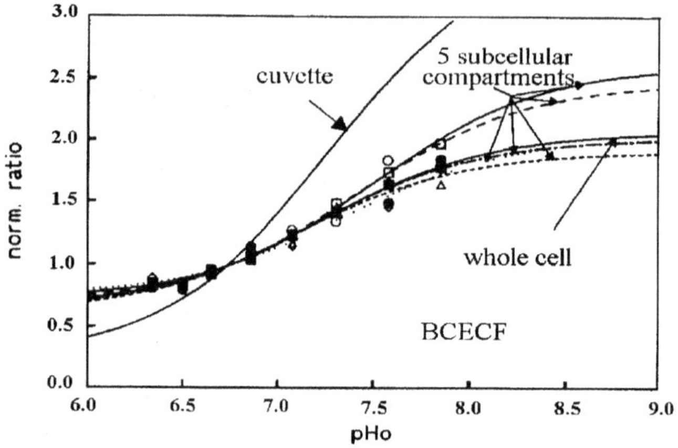

Fig.3 : Evaluation of 5 subcellular compartments (pixel field : 20*20) of a single glioma cell (loaded with BCECF) at various pH (picture sequence) resulting in calibration characteristics of the different compartments in comparison to an in vitro calibration curve (denoted cuvette) and a calibration curve of the whole cell (pixel field : 100*100) .

DISCUSSION

Evidence is presented in view of intracellular calibration inhomogeneities of the pH indicators BCECF and carboxy-SNARF-1 based on multipoint in-situ calibration procedures enabling the distinct determination of the complete set of calibration parameters, i.e. extreme ratios, apparent pK value and the correlation coefficient. In contrast to the usually performed two point calibration method (assuming intracellularly unaffected pK (or K_d) values despite frequently occurring changes of the extreme ratios), this multipoint calibration procedure can be considered adequate to describe the relative complex intracellular interactions of indicator molecules with cytoplasmic constituents and the associated variations of the calibration.

In case of carboxy-SNARF-1 the calibration heterogeneity may be attributed to pH dependent intracellular redistribution phenomena of the protonated indicator component between cytosol and lipophilic cell compartments resulting in varying pK^* shifts of the pH indicator in different subcellular regions (depending on the lipid content of the respective compartment) and, thus, in heterogeneity of the intracellular calibration.

Surprisingly, even in case of the frequently applied intracellular pH indicator BCECF considerable deviations of the intracellular calibration in comparison to cell free solution are observed. These distortions may be attributed to intracellular interactions of the indicator molecules with intracellular substances resulting in considerable changes of the extreme ratios and slightly altered pK^* values. In addition, it could be shown that calibration curves vary from one cell to another and even within different subcellular compartments suggesting variations of the calibration characteristic from pixel to pixel within the complete pixel field of the imaged cells .

To analyse intracellular calibration characteristics on a single pixel basis further investigations with the aid of a workstation are in progress in order to display the spatial distribution of the extreme ratios as well as the apparent pK values and corresponding correlation coefficients as function of the particular intracellular position.

CONCLUSION

Intracellular redistribution as well as intracellular interactions of indicator molecules affect the calibration characteristics of BCECF and carboxy-SNARF-1 to various extents in different cells and subcellular compartments. As a consequence, pH_i determinations are restricted to the same pixel field (and to the same vertical resolution in case of confocal microscopy) of the corresponding calibration curve. In practice, the spatial resolution of intracellular ion activities is limited to such extensions of the intracellular compartments (or corresponding pixel fields), which can be reproducibly identified within sequences of picture pairs monitored during calibration and measurement.

Moreover, since intracellular calibration probably varies from pixel to pixel, pH_i imaging on the basis of these indicators have to be regarded with reservation. The same may hold for other ion sensitive indicators such as the Ca^{2+}-sensitive fluoroprobe FURA-RED, which also suffers considerable distortions of its spectral distribution and Ca^{2+}-sensitivity due to interactions with proteins and other cytosolic substances [7,8]. Hence, quantitative intracellular imaging of concentration distributions using other ion sensitive fluoroprobes has to be considered sceptically unless the intracellular integrity of the dye is not clearly proved. To do this, systematic investigations of the physicochemical and spectroscopic properties of the indicators (in vitro <u>and</u> intracellularly) are essential prior to physiological measurements and interpretations.

Finally, intracellular interactions of indicator molecules with proteins and other cytoplasmic macromolecules may influence the temporal resolution in case of indicators such as BCECF and FURA-RED due to time-dependent variations of the protein distribution in different subcellular compartments caused for instance by intracellular plasma streaming and exo- or endocytose processes under physiological conditions in contrast to a rather instantaneously recorded calibration topography under nigericin or ionomycin administration.

ACKNOWLEDGEMENTS

We are indebted to B. Bölling for providing the glioma cells grown in multicellular spheroid or monolayer culture as well as to G. Holtermann for technical assistance regarding the microspectrofluorometric investigations. Financial support by grants DFG (Ac 37/9-1) and EC-project (BMH-CT94-1068) is appreciated.

REFERENCES

1. R.W. Tsien, M. Poenie, Fluorescence ratio imaging : a new window into intracellular ionic signaling, Trends Biochem.Sci.11: 450-455 (1986)
2. D. Weiner, L.L. Hamm, Use of fluorescent dye BCECF to measure intracellular pH in cortical collecting tubule, Am. J. Physiol.256: F957-F964 (1989)
3. G.R. Bright, G.W. Fischer, J. Rogowska, D.L. Taylor, Fluorescence ratio imaging mycroscopy : temporal and spatial measurements of cytoplasmic pH, J.Cell Biol.104: 1019-1033 (1987)
4. J.A. Thomas, R.N. Buchsbaum, A. Zimniak, E. Racker, Intracellular pH measurements in Ehrlich ascites tumor cells utilising spectroscopic probes generated in situ, Biochem.18: 2210-2218, (1979)
5. N. Opitz,E. Merten,H. Acker, Evidence for redistribution-associated intracellular pK-shifts of the pH-sensitive fluoroprobe carboxy-SNARF-1. Pflüg.Archiv 427:332-342 (1994)
6. N. Opitz, M. Weinlich, F. Mooren, F.-K. Keinemann, G. Weidemann, R.K.H. Kinne, Dual wavelength excitation in confocal laser scanning microscopy : intracellular pH measurements with a fluorescent indicator, Biomed. Tech. 36: 132-133 (1991)
7. N. Opitz, T. Porwol, E. Merten, B. Bölling, H. Acker, Cytoplasmic Ca^{2+} measurements : problems encountered with the Ca^{2+}-sensitive fluorescent indicator FURA-RED. Pflüg.Archiv 429, Suppl.6, R83 (1995)
8. N. Opitz, T. Porwol, E. Merten, H. Acker, On the protein-error of the calcium sensitive fluorescent indicator furared, this volume

THE EFFECT OF PROTEIN BINDING ON THE CALIBRATION CURVE OF THE pH INDICATOR BCECF

Jaromír Plášek[1], Jan Jaap ter Horst[2], Marcel Ameloot[2] and Paul Steels[2]

[1] Institute of Physics of the Charles University, Ke Karlovu 5, 121 16 Prague, Czech Republic
[2] Limburgs Universitair Centrum, Universitaire Campus, B-3590 Diepenbeek, Belgium

INTRODUCTION

The fluorescent dye BCECF (2',7'-bis-(2-carboxyethyl)-5-(and 6-)-carboxyfluorescein) is very popular as a dual excitation probe to monitor intracellular pH. The spectral characteristics are known to be different *in vitro* and *in vivo*. Upon introduction into the cellular environment its absorption shows a red shift of about 5 nm.[1] This spectral shift may exhibit a pronounced effect on pH *vs* BCECF fluorescence calibration curves obtained in ratiometric mode. It is not clear whether the spectral shift of BCECF is due to binding to cellular proteins or to other environmental effects.[2] Measurements of the translational mobility of BCECF in the cell cytoplasm provide contradictory results. It has been reported that BCECF is transiently bound to intracellular components of low mobility[3] although this was not observed by others.[2]

It is not unlikely that the charged pH indicator is bound to the high protein content of the cytosol. To evaluate this hypothesis an *in vitro* study was undertaken using bovine serum albumin. We investigated the effect of protein binding on the fluorescence of the pH sensitive dye BCECF and on the shape of pH *vs* BCECF fluorescence calibration curves, in particular.

MATERIALS AND METHODS

BCECF Solutions

Different Ringer solutions, specific for Malpighian tubules[4] (*Formica*) have been used: (i) the solution, further referred to as 113Na 143Cl, contained (in mM) 113 Na^+, 2 Ca^{2+}, 13 Mg^{2+}, 143 Cl^-, 8 citrate^{3-}, 17 fumarate^{2-}, 3 alanine, 139 glucose, 12 maltose, 10 trehalose and 12 Hepes; its osmolality was 375 milliosmole/kg H_2O and pH was adjusted to 7.2 with NaOH, (ii) the solution, further referred to as 113K 143Cl, is obtained when Na^+ is replaced by K^+ in the previously indicated Ringer solution, and (iii) either 113Na 143Cl or 113K 143Cl solution is diluted 1:1 (v:v) with distilled water. To study the assumed effect of BCECF binding to cytosolic proteins, bovine serum albumin, Fraction

V (Sigma Biochemicals, USA) was added to the above solutions. The maximum albumin concentration in our samples was 9% (w:w), i.e. approximately 1.5 mM.

A stock solution (1 mM) of BCECF (Molecular Probes, USA) in DMSO was added to a particular medium to a final BCECF concentration of 1 μM. Then 5 ml aliquots were taken, and their pH adjusted with KOH or NaOH or HCl to selected values. Usually, the calibration sets of twelve different pH values were prepared, their pH ranging from 5 to 9. For pH values between 6.5 and 7.5 (i.e. within a range close to the BCECF pK_a value), the pH in individual solutions of the set increased in steps of 0.1 pH units.

Fluorescence Measurements and Analysis of the Experimental Calibration Curves

BCECF fluorescence in the calibration solutions was measured with a Perkin-Elmer spectrofluorimeter, model LS-5B, controlled with a PC. The collected steady-state spectra were processed so that the fluorescence intensities integrated over the bands at 450 nm (*F450*) and 490 nm (*F490*) to simulate the transmission characteristics of the two excitation filters used in the microscope setup (Omega Optical 440DF20 and 490FD20). Then the ratio $R = F490/F450$ was assessed as a quantity comparable with a result of the microfluorimetric ratiometric assay. The following equation

$$R = \frac{AB\,10^{-pH} + C}{10^{-pH} + A} \tag{1}$$

was used as a fitting function in the experimental plots of *R vs pH*. This is only a general fitting function, suitable for any curve exhibiting a sigmoidal shape. A rigorous treatment including the dye binding with albumin would require a more complicated expression. Therefore, no real physical meaning can be attributed to the calculated fit parameters in our calibration experiments, i.e. to *A*, *B* and *C*.

Corrections for the Inner Filter Effect and Blank Fluorescence Due to Albumin

Albumin solutions at concentrations of about 4 - 9% can absorb part of the excitation light in BCECF experiments. Moreover, they emit a weak fluorescence when excited at a wavelength within the range of BCECF excitation spectra. Corrections are therefore required for both the above mentioned effects, to avoid artificial differences between calibration curves measured in pure aqueous media and in albumin solutions, respectively.

The correction for the inner filter effect in the ratio assessment is just multiplicative. This stems from the fact that the experimental ratio *R* of observed fluorescence signals at two excitation wavelengths is given by

$$R = F490/F450 = f([H^+]) \times I490_{exc}/I450_{exc} \tag{2}$$

where $I490_{exc}$ and $I450_{exc}$ are the respective excitation light intensities at the wavelength of 490 and 450 nm and where $f([H^+])$ is a function of pH as described by Grynkiewicz et al.,[5] and Van den Bergh et al.[6] This multiplicative correction factor for the albumin inner filter effect was assessed by two different methods. First, the BCECF fluorescence excitation spectra were measured in Ringer medium without albumin. An experiment was then performed under standard experimental conditions, as described above. Subsequently, a cuvette (1 cm) containing a corresponding albumin solution was placed into the excitation light beam, directly in front of the sample cuvette, and the measurement of the BCECF fluorescence excitation spectrum was repeated. The ratio of the fluorescence intensities measured in these two experiments were used to assess the correction factor. Second, the excitation spectra of albumin fluorescence were measured at six different albumin concentrations, ranging from 0.1% to 9% (w:w). The ratio *F490/F450* was determined and plotted against the known concentration of albumin solution. This plot exhibited a plateau at albumin concentrations below 1% (w:w). Then the *F490/F450* values measured at high albumin concentrations were normalized relative to the *F490/F450* value of the plateau, and used as the correction factors for the albumin inner filter effect. The results obtained with the two different methods were practically identical. The values of the multiplicative correction factor are 0.97 and 0.90 for 4.5% and 9% albumin solutions, respectively. The fluorescence spectra of BCECF in the presence of albumin were corrected for the intrinsic fluorescence of albumin by subtracting the spectrum of albumin without BCECF.

RESULTS AND DISCUSSION

The Effect of Protein Binding on the Excitation and Emission Spectra of BCECF Fluorescence

Considerable changes in both excitation and emission spectra of BCECF fluorescence were observed upon adding albumin (9%, w:w) to aqueous media and were found to be pH dependent (Fig. 1). At pH 8.5 albumin caused a red shift in both excitation and emission spectra of about 7 nm relative to spectra in pure aqueous media. In addition, the fluorescence intensity in the main peak of the excitation spectrum decreased (by about 40%), while the fluorescence intensity in the emission spectrum remained practically unchanged. The dependence of the fluorescence intensity on the concentration of bovine serum albumin has also been reported by others.[7] At pH 5.0 it was observed, in the comparison with the spectra at pH 8.5, that (i) the albumin induced spectral shift in the excitation spectrum was more pronounced, while no spectral difference between corresponding emission spectra was found, (ii) the decrease in fluorescence intensity observed on going from pure aqueous media to solutions with albumin was much larger, both in the excitation and emission spectrum.

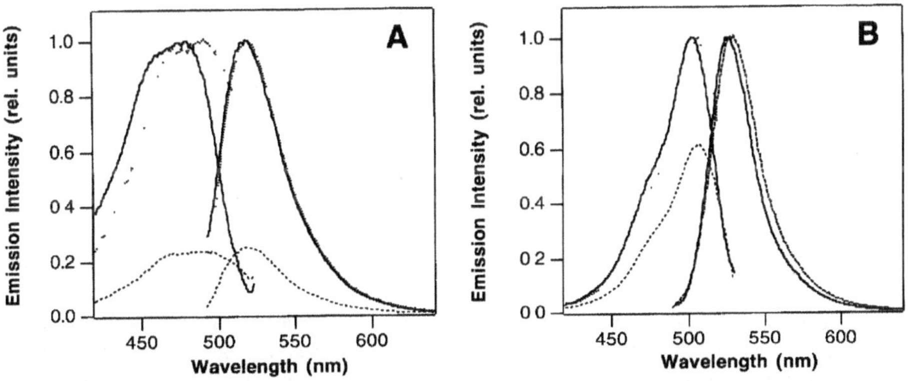

Figure 1. BCECF fluorescence excitation and emission spectra in pure 113Na 143Cl buffer and albumin solution: (A) pH = 5.0, (B) pH = 8.5; full lines - spectra in pure buffers (peak-height normalized), dotted lines - spectra in 9% (w:w) albumin solutions (peak-height normalized), dashed lines - spectra in 9% (w:w) albumin solutions (in scale to their counterparts measured in pure buffers; at pH 8.5 the "in scale" and the normalized emission spectrum for the albumin containing solution coincide). Excitation spectra were measured with λ_{em} = 535 nm, emission spectra with λ_{exc} = 480 nm; the spectral width of slits was 10 and 2.5 nm for the excitation and emission monochromator, respectively. BCECF concentration in all solution was 1 μM.

The spectral differences between BCECF fluorescence excitation spectra measured in pure buffers and albumin solutions resulted in concomitant differences in the probe fluorescence response to changes in pH of the solution. Examples are shown in Fig.2. On going from pure aqueous buffers a significant shift occurs in the position of the isoexcitation point (the excitation wavelength at which the emission intensity is independent of the solution pH; note that the isoexcitation point is different from the isosbestic point because the former depends on the relative quantum efficiencies of the species involved and on the wavelength of observation[8]).

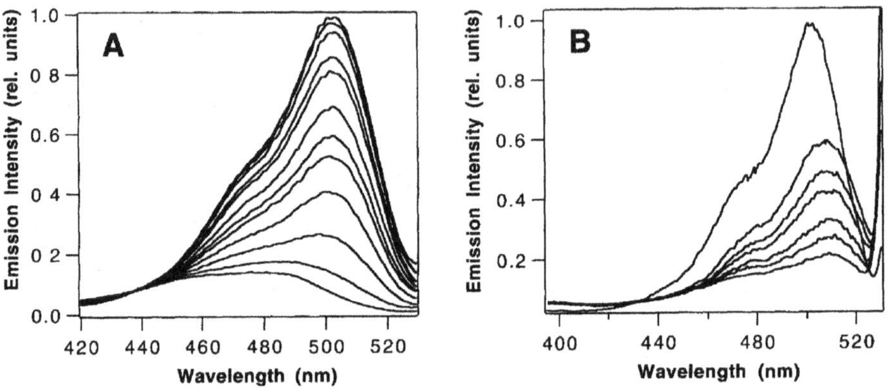

Figure 2. pH dependence of BCECF fluorescence excitation spectra: (A) 1 μM BCECF solution in pure aqueous buffer (113Na 143Cl), pH of BCECF solutions varied in decreasing order from pH = 9.02 (the highest curve) to 8.67, 8.07, 7.55, 7.31, 7.08, 6.88, 6.75, 6.50, 6.11, 5.66 and 5.02 (the lowest curve); (B) albumin solution, containing 5% albumin (w:w) in 51K 143Cl buffer and 0.22 μM BCECF, pH of BCECF solutions varied in decreasing order from pH = 7.53 (the highest curve) to 7.26, 7.09, 6.70, 6.50, and 6.30 (the lowest curve). The excitation spectra were measured with λ_{em} = 535 nm; the spectral width of excitation monochromator slits was 10 (A) and 5 nm (B), for the emission monochromator the width of the slits was 2.5 nm.

The Effect of Protein Binding on the Calibration Curves

In vitro calibration curves obtained in aqueous solutions with and without albumin were compared using measurements at twelve different pH values ranging from pH 5 to 9 (Fig.3). Also solutions of different ionic composition were used ranging from low K^+ concentration (113 Na 143Cl) to high K^+ concentration (113K 143Cl). The albumin concentration was 9% (w:w) which is close to a typical protein concentration in the cytosol of bacteria and mammalian cells (about 20%), e.g. Alberts et al.[9]

Albumin had very little effect on the slope of the linear portion of the calibration curve. Similar results were obtained by Musgrove et al.[10] who compared calibration curves obtained from BCECF measurements on detergent-lysed cells and by the nigericin method. However, the calibration curves for BCECF in media with albumin were shifted about 0.4 pH units towards the more alkaline values as compared to the results of calibration performed in pure aqueous media. Moreover, the respective calibration curves for BCECF in albumin solutions and in pure aqueous media exhibit slightly different profiles at the acid region (below pH = 6.0 - 6.2). From the titration curve of albumin[11] it is known that the net charge, and thus the binding capacity of albumin, does not change substantially within the pH range where BCECF calibration curves are nearly linear. In the acid region, however, the profile of calibration curves for BCECF in albumin solutions reflects not only the pH dependent properties of the fluorochrome, but changes in the net charge of albumin as well. Small, but not negligible, lateral shifts between calibration curves were also observed if the ionic strength of the calibration solutions varied (data not shown). In pH scale, the magnitude of such shifts was about 0.1 - 0.2 pH units which is about the detection limit of this fluorescent probe technique.[12]

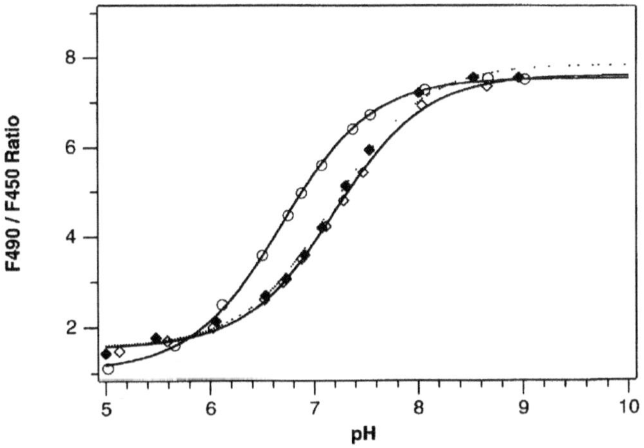

Figure 3. pH Calibration curves in pure buffer and albumin solutions: Full line and circles - pure aqueous buffer (113K143Cl), full line and diamonds - 9% (w:w) albumin solution in 113K143Cl buffer, dotted line - 9% (w:w) albumin solution in 113K143Cl buffer. BCECF concentration in all solution was 1 μM, excitation spectra were measured with λ_{em} = 535 nm, the spectral width of slits was 10 and 2.5 nm for the excitation and emission monochromator, respectively.

An interesting similarity exists between the results of our model experiments performed with BCECF and the pH sensitive properties of another fluorescein derivative, 4',5'-dimethyl-5-carboxy-fluorescein (Me_2CF). When the latter dye was used as a pH indicator exhibiting pH-dependent changes in its absorption spectrum, *in vitro* (measured with aqueous Me_2CF solutions in cuvettes) and *in vivo* (measured in isolated, perfused salamander proximal tubules) calibration curves were mutually shifted by about 0.4 pH units.[13] This similarity suggests that the binding of fluorescein and its derivatives to various proteins, including the intracellular ones, is a general phenomenon which precludes the use of *in vitro* calibration curves (obtained with aqueous BCECF solutions) in cell studies.

CONCLUSIONS

The interaction of BCECF with albumin has a pronounced effect on BCECF fluorescence and the corresponding calibration curves. The spectral shifts are in the same direction and of the same order of magnitude as has been observed for BCECF in cells. This seems to support the model in which spectral shifts are due to protein binding and provides further evidence that *in vitro* calibration curves with pure buffers cannot be used in cell studies as an easy alternative to the time consuming *in vivo* calibration procedures which are usually based on the use ionophores such as nigericin or monensin.[14]

Acknowledgement. The above work was carried at Limburgs Universitair Centrum and was supported by COST PECO grant No 07612 and the Nationaal Fonds voor Wetenschappelijk Onderzoek (Belgium).

REFERENCES

1. T.J. Rink, R.Y. Tsien and T. Pozzan, Cytoplasmic pH and free Mg^{2+} in lymphocytes, *J. Cell Biol.* 95: 189 (1982).
2. G.R. Bright, G.W. Fisher, J. Rogowska and D.L. Taylor, Fluorescence ratio imaging microscopy: temporal and spatial measurements of cytoplasmic pH, *J. Cell Biol.* 104:1019 (1987).
3. H.P. Kao, J.R. Abney and A.S. Verkman, Determinants of the translational mobility of a small solute in cell cytoplasm, *J. Cell Biol.* 120:175 (1993).
4. E. Van Kerkhove, R. Weltens, N. Roinel and N. De Decker, Haemolymph composition in *Formica (Hymenoptera)* and urine formation by the short isolated malpighian tubules: electrochemical gradients gradients and ion transport, *J. Insect Physiol.* 35:991 (1989).
5. G. Grynkiewicz, M. Poenie and R.Y. Tsien, A new generation of Ca^{2+} indicators with greatly improved fluorescence properties, *J. Biol. Chem.* 260:3440 (1985).
6. V. Van den Bergh, N. Boens, F.C. De Schryver, M. Ameloot, P. Steels, J. Gallay, M. Vincent and A. Kowalczyk, Photophysics of the fluorescent Ca^{2+} indicator Fura-2, *Biophys. J.* 68:1110 (1995).
7. P.A. Valant and D.H. Duncan, Calibration methods and avoidance of errors in measurement of intracellular pH (pHcyt) using the indicator bis(carboxyethyl)-5(6)-carboxyfluorescein (BCECF) in human platelets, *J. Fluorescence* 2:191 (1992).
8. R.Y. Tsien, Fluorescent indicators of ion concentration, in: *Methods in Cell Biology,* vol. 30, D.L. Taylor and Yu-Li Wang eds., Academic Press, San Diego (1989).
9. B. Alberts, D. Bray, J. Lewis, M. Raff, K. Roberts and J. D. Watson, *Molecular Biology of the Cell,* 2nd edition, Garland Publ., New York (1986).
10. E. Musgrove, C. Rugg and D. Hedley, Flow cytometric measurement of cytoplasmic pH: a critical evaluation of available fluorochromes, *Cytometry* 7:347 (1986).
11. T. Peters, Jr., Serum albumin, in: *The Plasma Proteins*, vol. I, 2nd edition, F. W. Putnam ed., Academic Press, New York (1975).
12. D.J. Pheasant and P.K. Hepler, Intracellular pH change is not detectable during the metaphase/anaphase transition in Tradecentia stamen hair cells, *Eur. J. Cell Biol.* 43:10 (1987).
13. J.R. Chaillet and W. Boron, Intracellular calibration of a pH-sensitive dye in isolated, perfused salamander proximal tubules, *J. Gen. Physiol.* 86:765 (1985).
14. J.A. Thomas, R.N. Buchsbaum, A. Zimniak and E. Racker, Intracellular pH measurements in Ehrlich ascites tumor cells utilizing spectroscopic probes generated in situ, *Biochemistry* 18:2210 (1979).

ARTIFACTS IN FLUORESCENCE RATIO IMAGING

Petr Cimprich and Jan Slavík

Institute of Physiology
Academy of Sciences of the Czech Republic
Vídeňská 1083
142 20 Prague 4
Czech Republic

INTRODUCTION

Successful development of reliable fluorescent ratio indicators for ion concentration, membrane fluidity and membrane potential and steady improvement of the imaging equipment (microscopes, detectors, computers) have made fluorescence ratio imaging microscopy a widely used method in biomedical research (Slavík, 1994) An increasing number of papers deal especially with Ca^{2+} and pH measurements. The proper interpretation of experimental data requires not only a profound knowledge of the fluorescent probe used and a reliable calibration standard curve but one must be also aware of artifacts inherent in the opto-electronical level of the technique These artifacts can arise from the following causes blurring of the image due to the involvement of out-of-focus light, background fluorescence, improper adjustment of the detection part of experimental setup, movement of the sample, and bleaching of the dye

ARTIFACTS CAUSED BY OUT-OF-FOCUS LIGHT

Fluorescence images taken using a classical (now called "wide-field") microscope contain not only the desired information but also a contribution from out-of-focus planes Glare around fluorescent objects arises from the projection of the light from fluorescence above and beneath the focus plane and the objects seem to be larger than they are in reality (Silver et al , 1992) As the fluorescence intensity of the glare is usually very low, ratio values from this area are computed by the division of small integers Consequently, confusing concentric rings can appear around the object edges in the computed ratio image These rings are, of course, completely meaningless In order to separate useful from useless optical information the ratio image must be thresholded or be compared to a bright field image and covered with a binary mask

The effect of the glare can be suppressed by use of a thin sample or by confocal scanning laser microscopy To some extent it can be also corrected (or at least compensated) by digital image processing The simplest technique is the nearest neighbours deblurring, a method employing additional images, one from above and one from beneath the focus plane of the sample More sophisticated techniques are based on deconvolution and require the determination of the point spread function of the imaging system (Buil, 1991)

ARTIFACTS CAUSED BY BACKGROUND LIGHT

The crucial requirement of the ratio methods is complete removal of the background light In the case of fluorescence ratio imaging the background light can include autofluorescence of the sample (light emitted by all fluorophores different from the dye employed for ratio measurement) and false light (outside light not arising from the measured sample - e g , from computer monitor)

The background light can be more or less compensated using the background subtraction As background image one should employ an image of the identical sample under the same conditions but without the fluorescent probe This procedure is also not flawless, as the fluorescence dye added may cause the quenching of the autofluorescence Consequently, the level of signal subtracted as background may be too high, because in the absence of the dye autofluorescence is higher

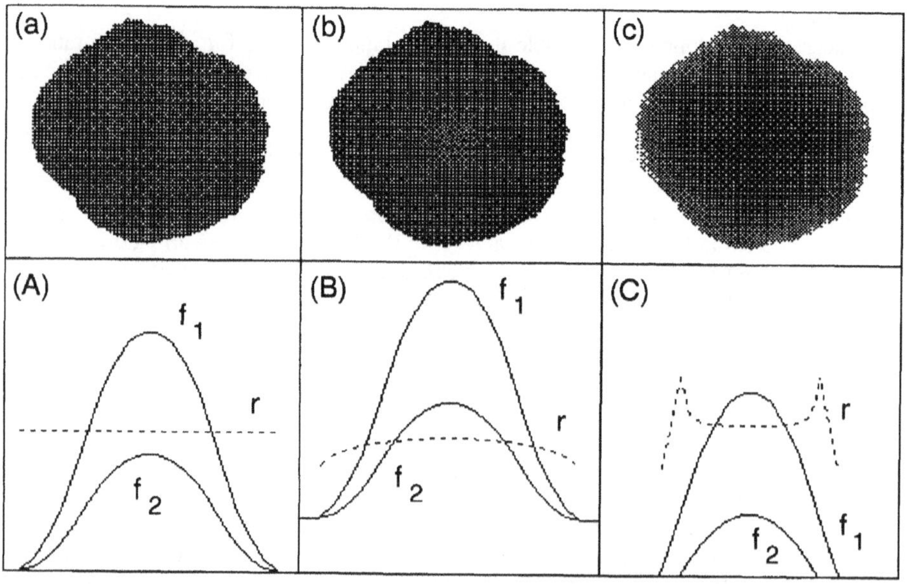

Fig. 1. Effect of background fluorescence (or improper offset adjustment) on computed ratio images Panels below (A, B, C) show profiles of model fluorescence images (f_1, f_2 = $f_1/2$) and their ratio (r = $f_1 \backslash f_2$) (A) No background signal present (or offset equals to zero), the ratio is independent on fluorescence intensity (B) Underestimated background subtraction (or positive offset), the ratio increases with increasing fluorescence intensity (C) Overestimated background subtraction (or negative offset), the ratio decreases with increasing fluorescence intensity Panels above (a, b, c) show the ratio images corresponding to profiles below

If the background light contributes to raw fluorescent images, the intensity of the resulting ratio image will depend not only on the studied parameter but also on fluorescence intensity (Fig 1) A small error in the removal of the background can cause a large error in the ratio A typical manifestation of imprecise background subtraction is a false gradient across the cytoplasm The centre of cells which, due to their spherical shape, fluoresces more than the edges (using wide-field microscopy), is brighter than the periphery in the ratio image and a false concentric distribution of the studied parameter appears

ARTIFACTS CAUSED BY IMPROPER ADJUSTMENT OF THE DETECTOR

Generally, detectors and digitizers transform the light intensity to the brightness of the digital image (expressed in shades of gray, often called "gray value") according to the following equation

$$\text{gray value} = \textit{gain} \times \text{light intensity}^{\gamma} + \textit{offset} \qquad (1)$$

The setting of the *gain* parameter allows to use all the dynamic range of the detector for a measurement This parameter causes no complications to ratio computing and simply must not be changed during the reading of the image couple and between the measurement and the calibration

The γ parameter should be keep equal to one (to ensure the linearity of the detector) but a slight deviation would not have a fatal consequences If it will not be changed during the experiment and the calibration, the resulting ratio image will be the exponential transformation of the image read by the linear detector

Unlike the *gain* and γ, the *offset* (*black level*) parameter is extremely important for the ratio imaging The *offset* must be equal to zero, what means, in other words, that no extra electric signal can be added during the detection An incorrect setting of the *offset* would blur the image with electronic background (Fig 1) The electronic background can be mostly compensated in the same way as the light background but for some detectors (for example for photo multipliers of confocal microscopes) also negative *offset* values may occur and background subtraction is then not possible On top of it, both positive and negative *offset* values are indicated using positive numbers on a control panel of the photo multiplier In this case it is necessary to calibrate the *offset* and to determine which "offset value" corresponds to the true zero *offset*

ARTIFACTS CAUSED BY MOVEMENT OF THE SAMPLE

To obtain a meaningful ratio image it is essential to calculate the ratio using two geometrically identical fluorescence images Corresponding pixels of both images must represent the fluorescence intensity signal from the same spot of a sample

As the fluorescence images are often read successively with a filter exchange between the readings, a movement of the fluorescent object can distort the ratio image In this case the movement of the entire sample (cell) results in appearance of bright and dark bands on opposite sides of the fluorescent object in the ratio image (Fig 2) Shape and area of the bands depend on the direction and magnitude of the movement The movement inside the sample (for example cytoplasmic steaming inside the cell) is more difficult to recognise and a slight internal movement can be easily misinterpreted as a false ionic gradient inside the fluorescent object

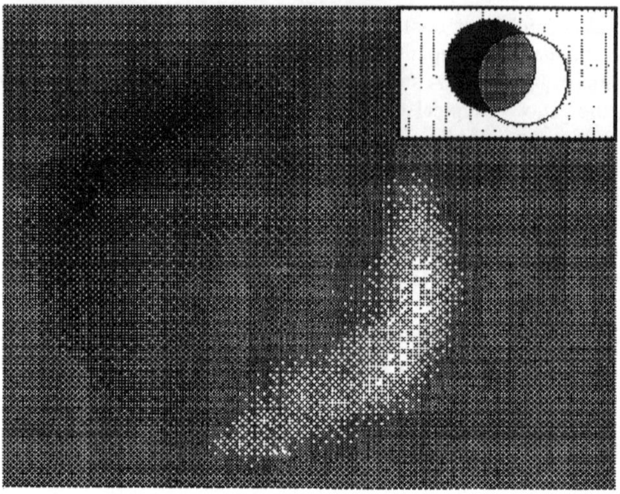

Fig. 2. Effect of movement of the entire sample Bands darker and brighter than the rest of the image appear on opposite sides of the fluorescent object

The movement artifacts can be overcome by use of a dual emission fluorescent probe and parallel double-channel detection In classical wide-field fluorescence microscopy it means a simultaneous detection with two properly aligned CCD cameras In confocal fluorescence scanning microscopy double-channel photomultiplier is used In the excitation ratio all what can be done is to minimize the time between the reading of fluorescent images for computing the ratios A digital image correction using geometrical transformation of one image from the pair is in principle possible but not simple, tiresome and not very reliable

ARTIFACTS CAUSED BY BLEACHING OF THE DYE

Fluorescent dyes bleach more or less during the measurement The amount of dye capable of fluorescence diminishes and the fluorescence fades during the excitation This is why in the ratio imaging with consecutive reading of images it must be taken into account which image of the pair was taken as the first and which as the second one

The fluorescence bleaching is directly proportional to the intensity of the excitation light and if the excitation intensity is constant, the fluorescence decay is linear in time Therefore if images of the pair are read in the same order and with a constant interval between the readings, fading can be included in the calibration curve and eliminated with a reasonable precision Another way how to compensate the bleaching is to take the images in sequence *first - second - first* and to compute the ratio using the average of both the "first" images Besides, the bleaching effect can be minimized by decreasing down the intensity of the excitation light (gray filters)

CONCLUSION

Commercially available imaging systems and fluorescent ratio probes have made fluorescence ratio imaging microscopy accessible for a large number of researchers The method became routine but one must still keep in mind the danger of artifacts leading to misinterpretation of results As described above, the most convenient approach to fluorescence ratio imaging is a combination of dual-emission ratiometric probe with confocal laser scanning microscopy and dual-channel detection This approach eliminates the artifacts caused by out-of-focus light and movement of the sample The bleaching of the dye does not distort the results due to the simultaneous detection but the high-intensity light source often causes a rapid irreversible fading of the excited area of the sample

REFERENCES

Buil, Ch., 1991, *CCD Astronomy*, Willmann-Bell, Richmond

Silver, R A., Whitaker, M., Bolsover, S R., 1992, Intracellular ion imaging using fluorescent dyes artifacts and limits to resolution, *Pflugers Arch* 420 559

Slavík, J., 1994, *Fluorescent Probes in Cellular and Molecular Biology*, CRC Press, Boca Raton

ACKNOWLEDGEMENT

The work was supported by the grant #202/94/1712 of the Grant Agency of the Czech Republic.

USE OF FLUORESCENT PROBES AND CLSM FOR pH-MONITORING IN THE WHOLE PLANT TISSUE. pH-CHANGES IN THE SHOOT APEX OF *Chenopodium rubrum* RELATED TO ORGANOGENESIS

Albrechtová, Jolana T.P.[1*], Slavík, Jan[2], and Wagner, Edgar[1]

[1]Institut für Biologie II., Albert-Ludwigs-Universität, Freiburg, Germany
[2]Institute of Physiology, Academy of Sciences of Czech Republic, Praha, Czech Republic
[*]On leave from: Institute of Experimental Botany, Academy of Sciences, Praha, Czech Republic

INTRODUCTION

Cytoplasmic pH changes were reported to be related to many life processes (Felle 1989). In many organisms pH changes specifically during the cell cycle and during the life time: activation and juvenility being associated with pH-increasing, and quiescent state and senescence with pH-decreasing (Felle 1989, Ross 1992). Data obtained in animal cells point to an important role of the cytoplasmic pH in regulation of metabolic processes (Busa and Nuccitelli 1984, Nucitelli and Deamer 1982) and in the activation of enzymes playing a role in signal transduction cascades (Felle 1989) There are not many studies carried out on plant material; however, all of them suggest the importance of intracellular pH in regulation of cell cycle and development (Felle and Bertl 1986, Gendraud and Lafleuriel 1983, Roos and Slavík 1987, Tort and Gendraud 1984). Local pH-changes are related to changes of membrane potential and proton fluxes (Blatt and Slayman 1987, Felle 1987, Felle and Bertl 1986, Frachisse et al. 1988), which lead to a hypothesis of pH playing a role as a second messenger in signal transduction between plasma membrane and cytoplasm (Felle 1989).

Organogenesis persists whole life in plants, and enhanced cell division, preceeded by cell activation, occurs on the sites of new organ initiation in a regular manner. Specialised site(s) of organ formation is (are) shoot meristem(s), i.e apical meristem and axillary

Corresponding author: Dr. J.T.P Albrechtová, Albert-Ludwigs-Universität, Biologie II./ Botanik, Schänzlestr. D-79104 Freiburg im Breisgau, Germany, Tel.: 0761-203-2637, Fax: 0761-203-2840

Abbreviations: NMR - nuclear magnetic resonance
SNARF - Seminaphthorhodafluor
CLSM - confocal laser scanning microscopy

meristems. Together with the change of the morphogentic program, e.g. from the vegetative development to the flowering, pattern of the cell division and the cell activation changes and leads to the changed pattern of the organogenesis (Albrechtová and Wagner 1995, Bernier et al. 1981, Lyndon and Battey 1985). An important role in organogenesis is ascribed to cells in superficial layer, i.e. tunica (Green 1992, Lyndon 1994).

As local cytoplasmic pH-changes may play a role as a second messenger in plant cell activation (Albrechtová et al. - in press, Albrechtová and Wagner 1995, Felle 1989, Kurkdjian and Guern 1988), specific cytoplasmic pH-changes in the site of organogenesis - apical meristem, could be involved in maintaining of the specific pattern of organ formation.

To investigate relationships between a pattern of pH and organogenesis on the apical meristem, we have used a method of pH-estimation by means of a fluorescent, pH-sensitive probe (Roos 1992, Slavík 1983), which enabled us to obtain pH-maps of whole sections throughout a meristem and of the superficial cell layer of the meristem. Pattern of cytoplasmic pH was compared with a pattern of apoplastic pH, in order to find out relationships of the cytoplasmic pH changes with a proton transport on the plasma membrane.

MATERIAL AND METHODS

MATERIAL

Chenopodium rubrum (strain 374, Ullmann et al. 1985) plants were used. They were grown in continuous light in growth chambers until the time of the photoperiodic treatment. One short dark period (20 h light / 4 h darkness) was applied before observation in order to synchronize plants. Samples were taken always one hour after the end of the dark period. Plants were observed in different developmental stages from seedlings (just after initiation of the second pair of leaves) to the initiation of sixth pair of leaves.

TISSUE PREPARATION

Fresh longitudinal sections were obtained from the central plane of each apex using a Vibroslice TSL (manual tissue slicer, World Precision Instruments). Sections were incubated at least 15 min in the dye solution in MES-buffer (25 mmol/l, pH 6) at a final concentration of 10^{-5} mol/l (stock solution. 1 mmol/l dye in DMSO) and then observed in the same dye solution.

No differences were detectable in the distribution of the dye in the tissue after 15 min of incubation and examination up to several hours The samples were monitored by confocal microscopy 2 and 3 h after the end of the dark period.

In some experiments, whole apices were used instead of the sections. The mode of incubation and observation was the same as for the sections.

METHOD OF pH-ESTIMATION

Cytoplasmic pH-probe carboxy SNARF-1 (Molecular Probes, Eugène, USA) was used for estimation of cytoplasmic pH, as described by Albrechtová et al. (in press). The probe was loaded into the cells in AM-form (acetoxymethyl esther), with subsequent cleaving of the AM group by cytoplasmic estherases. The dye without AM group may convert into two forms (basic or acidic, depending on the actual environmental pH), with different fluorescent properties (Slavík 1983). The Leica TCS 4D confocal laser scanning unit attached to a Leitz DM RBE fluorescent microscope with argon/krypton laser was used in ratio imaging arrangement. The two forms were scanned in a simultaneous dual-channel detection mode with excitation wavelength 488 nm and emission 520 nm for the acidic form and excitation

568 nm with emission 590 nm for the basic form, respectively. For every picture, a serie of the 8 optical sections was taken in regular intervals during the whole first cell layer and put together before ratios were calculated. The ratio of acidic versus basic form was computed to obtain the resulting maps of pH-values independent on the local concentration of the dye. No fading was detectable during the whole time of observation.

Every resulting ratio image was compared with an image of the one form of dye (mostly basic form, which was clearest in our filter arrangement). Due to the distribution of the dye in the cells, it was possible to evaluate, whether or not differences in the cell structure (size of vacuoles, intercellular spaces, position of the nucleus) influenced the differences of the fluorescence intensity.

The pH-values of the leaves (except very young primordia) and more differentiated cells on the stem could not be estimated, because of interference from the strong autofluorescence of chlorophyll. Due to the calculation of the ratio of both dye forms for every picture, the cells with chlorophyll autofluorescence were in some cases presented as darker ones and in some cases as lighter ones then the cells without chlorophyll. No regularity or relation to a treatment was detectable

Semiquantitative pH-marker primulin (Aldrich-Chemie, Steinheim, Germany) was used for the visualization of the apoplastic pH with excitation wavelength 488 nm and emission 520 nm (Loyall 1995). No ratiometric imaging can be done with this dye

CALIBRATION

The calibration of the relationship between the measured intensities of fluorescence ratios and cytoplasmic pH was done *in vitro* in a serie of buffers, as described by Roos (1992). The sensitivity of the dye was shown in the pH range between 5.0-8.5 with the highest sensitivity between 6.0-7.0, in accordance with the literature (Roos 1992) It was not possible to obtain precise calibration curves because of the insufficient accuracy of the ratio counting of the software. Therefore a control calibration was done using image analysis software "Lucia" (Laboratory Imaging, Praha), with the pictures of the basic and acidic forms of the dye as obtained from the confocal microscope, where we could reach the accuracy up to 0 1 pH-unit (cf. Albrechtová et al. - in press).

Relationship between the intensity of fluorescence of the primulin and the apoplastic pH was estimated using tissue sections incubated in a serie of buffers (Loyall 1995). The sensitivity of the dye to pH-changes was sufficient in the range between pH 3.5 - 7.5 with the highest sensitivity between 3.5 and 6 5

RESULTS AND DISCUSSION

CYTOPLASMIC pH

The dye carboxy SNARF-1 was used for studies on pH-changes in different unicellular organisms (Haworth and Fliegel 1993, Pönitz and Roos 1994, Roos 1992, Roos and Slavík 1987). pH-related changes of fluorescent signal were almost exclusively ascribed to cytoplasmic pH-changes (Haworth and Fleiegel 1993, Roos 1992), partially because of the detection limit of the dye (pH 5.5), and vacuoles being typically below pH 5.6 (see Roos 1992). The dye accumulates mostly in the cytoplasma. The proportion of fluorescence of the dye loaded in microsomal and mitochondrial compartments is less then 5% of the total fluorescence (Haworth and Fliegel 1993, Pönitz and Roos 1994). In accordance with this statement, we have found changes of intensity of fluorescence mostly in cytoplasma, using a serie of pictures with higher enhancement and estimated using the software "Lucia" again. It was confirmed, that most of the dye was loaded into the cytoplasma, and that the pH-differences stayed always in the range of cytoplasmic pH (6.0-7.5). Fluorescence of vacuoles and cell walls stayed always near to the detection limit of the dye. The amount of vacuolization is low in the apical meristem.

Differences in the intensity of fluorescence of cell nucleoli were also detectable, without any clear dependence on the treatment. Intensity of fluorescence of the nuclei showed only very small fluctuations in all treatments, being relatively low, i.e. pH about 6.0 (Albrechtová et al. - in press). This value could be related to the actual phase of the cell cycle (see further).

APOPLASTIC pH

Primulin was used for monitoring of apoplastic pH-changes in leaf epidermis of *Commelina communis* (Edwards et al. 1988) and on the shoot sections of maize (Loyall 1995), however, its use is rather limited, because it is very difficult to obtain realy quantitative results. A fluorescence intensity of the dye depends not only on the actual pH, but also on the mode of the binding (i.e. on the structure of cell wall, Duffus et al. 1984, Streiblová and Beran 1963) and on the actual concentration of the dye (Loyall 1995) However, with these limitations, it is possible to visualize differences in apoplastic pH in the plant tissue. The dye does not penetrate living cells (Edwards et al. 1988), and therefore there is no interference with underlying cytoplasm. Sensitivity of the dye is in the optimal range for monitoring pH in cell walls (Loyall 1995; cf. Methods)

PATTERNS OF pH-DISTRIBUTION

In the apical meristem the most basic cytoplasmic pH was detected in cells in the tunica layer (superficial layer). All other parts of the apex had no detectable or clearly pronounced differences (Albrechtová et al - in press). The degree of differences between cytoplasmic pH of cells in the tunica and corpus (inner tissue) changed with developmental state of the plants (Albrechtová et al. - in press).

Walls of tunica cells showed slightly more basic pH Cells of the future bud primordium in one axil of the two youngest leaf pair possessed usualy a more basic cytoplasmic and apoplastic pH. This is in accordance with a spiral phylotaxis of this plant One leaf primordium in every " leaf pair" therefore develops faster and its axillary bud primordium is initiated earlier. The apices were never symmetrical in left-right pH distribution, even when anatomical picture seemed to be symmetrical. In the area between the two youngest leaf primordia, a group of cells with more basic cytoplasmic pH was visible in the tunica layer on the site of the next leaf primordium, which was not formed yet at the time of observation.

DISCUSSION OF THE RELIABILITY OF THE METHOD

The methods used in the present study have allowed us to monitor cytoplasmic and apoplastic pH in living cells directly. However, despite the use of confocal microscopy, it was necessary to make sections through the meristem to observe cells inside It was not possible to see more than one cell layer (probably because of the high optical density of the cell walls). Sectioning of the tissue could bring about a wound reaction, which can change cytoplasmic pH

However, the features of the pH-distribution were the same in all apices observed, on the sections as well as on the whole apices. The degree of the differences between the most acidic and most basic cells was different and absolute pH-values were not the same in all apices. However, the differences between apices in different developmental stage were clear cut. As the absolute pH-values could not be determined precisely (cf. Methods), the differences could be due to limitation of confocal image analysis software. It cannot be excluded, that the absolute pH-values were shifted by a wound reaction, probably to the more acidic values, but the pH-values stayed always in the range stated in the literature for plant cells.

The absolute pH-values reported in the literature differ with different methods of estimation (e.g. microeletrodes, ^{31}P-NMR spectroscopy, fluorescent probes). In general pH-values shift towards the more acidic area when fluorescent probes are used, probably due to heterogenity of the cytoplasma (6.5-7.4 pH units, up to 60 in periferal cytoplasma and near the vacuoles; Roos 1992, Slavík 1983) Furthermore, direct effect of wounding by glass microelectrode pronounced as a shift to the more basic pH near the site of insertion of the microelectrode was observed using a fluorescent probe (Roos - personal communication). Mean pH-values stated in literature vary between pH 7.0-7.6 depending on the material (Felle 1991, Herrmann and Felle 1995, Walker et al. 1995), as measured with microelectrodes or ^{31}P-NMR spectroscopy. The lowest mean cytoplasmic pH-values were reported for plant cells with specific function requiring very precise turgor regulation (e.g. pH 6 26 in epidermal cells of the bending zone of the circumnutation shoots of *Phaseolus*; Millet et al. 1987). The highest pH-values were described in animal cells after activation, up to pH-values of 7.44-7.95 (fertilized frog egg; Webb and Nuccitelli 1982), but values in resting state were in average about 0,4 pH units lower. It seems very probable from the comparision of the data reported in literature, that plant cells have generally more acidic cytoplasmic pH-value than animal cells.

It was shown in different models, that there are cytoplasmic pH changes during the cell cycle in the range of 0,4 toward an acidification and 0,25 toward an alkalinization to compare with a resting state (Felle 1989) As cell division is highly synchronised in the shoot apical meristem of *Chenopodium rubrum* (King 1975), the differences in absolute pH-values between the apices could be also affected with an actual phase of cell cyclus at the time of taking a sample. Although the samples were taken always at the same physiological time (one hour after the end of the dark period), the individual time might have been slightly out of synchronisation before the beginning of the dark period, as the plants were grown several days in the continuous light prior to the dark treatment (cf Albrechtová et al., in press).

It was not possible to determine absolute pH-values in apoplast (cf. Loyall 1995). However, the differences between cell walls in different parts of the apical meristem were clear pronounced and reproducible. A good correlation between sites of pH-shifts in cell walls and cytoplasm strongly supports reliability of the results.

A shift to more basic pH occurred at specific sites, where enhanced cell division will follow (initiation of bud and leaf primordia). This is in accordance with the suggested role of the increased cytoplasmic pH in cell activation in different organisms (e.g. Barton et al. 1980, Gillies and Deamer 1979, Lee et al. 1983). Role of apoplastic pH is stated mostly for cell growth (Cleland et al. 1991, Rayle and Cleland 1970), in relation with extensibility of cell wall (Roberts and Haigler 1994, Taiz 1984, Vian et al. 1993) However, apoplastic pH-changes are necessary coupled with cytoplasmic changes on the base of ion and/or electron transport on the plasma membrane. In accordance with this statement, we have found a tight relationship between patterns of apoplastic and cytoplasmic pH. Apoplastic pH would therefore involve in cell activation as well. Furthemore, cell wall extensibility is supposed to change specifically in relation to the pattern of organogenesis (Lyndon 1994) Interactions between cell wall and cytoplasma and ion transport events on cell membranes (plasma membrane, tonoplast, nuclear envelope, etc.) thus can be a key to the understanding of the processes of cell activation and differentiation.

ACKNOWLEDGEMENTS

This work was sponsored with a stipendium of DAAD (Deutscher Akademischer Austauschdienst) and also partially by the grant #202/94/1712 of the Grant Agency of the

Czech Republic. Authors are grateful to Prof K.-F Fischbach for the kind permission to work on his confocal microscope.

REFERENCES

Albrechtová, J.T.P., and Wagner, E. - The response of the target cells in the shoot apex of *Chenopodium rubrum* to different photoperiods. Biol. Rhythm Res. 26 (1995), 360.

Albrechtová, J.T.P., Slavík, J., and Wagner, E. - Confocal pH-topography in the shoot apex of *Chenopodium rubrum* in relation to photoperiodic flower induction - Bot. Acta, in press.

Barton, J.K., Den Hollander, J A , Lee, T.M , MacLaughlin, A , and Shulman, R.G. - Measurement of the internal pH of yeast spores by ^{31}P nuclear magnetic resonance. Proc. Natl. Acad. Sci. USA 77 (1980), 2470-2473

Bernier, G., Kinet, J.-M., and Sachs, R M - The Physiology of Flowering. Vols.I,II, CRC-Boca Raton, 1981

Blatt, M.R. and Slayman, C.L. - Role of "active" potassium transport in the regulation of cytoplasmic pH by non animal cells Proc Natl Acad Sci USA 84 (1987), 2737-2741.

Busa, W.B. and Nuccitelli, R. - Metabolic regulation via intracellular pH. Am. J. Physiol 246 (1984), 409-438.

Cleland, R.E., Buckley, G., Nowbar, S., Lew, N.M., Stinemetz, C., Evans, M.L., and Rayle, D.L. - The pH profile for acid-induced elongation of coleoptile and epicotyl sections is consistent with the acid-growth theory. Planta 186 (1991), 70-74

Duffus, J.H., McDowell, W., and Manners, D.J. - The use of primuline to indentify the septum polysaccharide of the fission yeast *Schizosaccharomyces pombe*. Stain Technol. 59 (1984), 79-82.

Edwards, M.C., Smith, G.N., and Bowling, D.J F - Guard cells extrude protons prior to stomatal opening - a study using fluorescence microscopy and pH micro-electrodes. J. Exp. Bot. 39 (1988), 1541-1547

Felle, H. - Proton transport and pH control in *Sinapis alba* root hairs. A study carried out with double-barreled pH-microelectrodes J. Exp. Bot. 38 (1987), 340-354.

Felle, H - pH as a second messenger in plants In: W F. Boss and D.J Morré, eds., Second Messengers in Plant Growth and Development, pp. 145-166. Liss-New York, 1989.

Felle, H. - The control of cytoplasmic levels of Ca^{2+} and H^+ in plants In: C. Penel and H Greppin, eds., Plant Signalling, Plasma Membrane and Change of State, pp. 79-104 Imprimerie Nationale-Geneva, 1991

Felle, H. and Bertl, A - Light-induced cytoplasmic pH changes and their interrelation to the activity of the electrogenic proton pump in *Riccia fluitans* Biochim Biophys. Acta 848 (1986), 176-182.

Frachisse, J.M., Johannes, E. and Felle, H. - The use of weak acids as physiological tools. a study of the effects of fatty acids on intracellular pH and electrical plasmalemma properties of *Riccia fluitans* rhizoid cells. Biochim. Biophys. Acta 937 (1988), 199-210.

Gendraud, M. and Lafleuriel, J. - Charactéristiques de l'absorption de saccharose et de tétraphénylphosphonium par les parenchymes de tubercules de Topinambour dormants et non dormants cultivés in vitro. Physiol Vég. 21 (1983), 1125-1133.

Gillies, R.J. and Deamer, D.W. - Intracellular pH changes during the cell cycle in *Tetrahymena*. J. Cell Physiol. 100 (1979), 23-32.

Green, P.B. - Pattern formation in shoots a likely role for minimal energy configurations of the tunica. Int J. Plant Sci. 153 (1992), S59-S75

Haworth, R.S. and Fliegel, L - Intracellular pH in *Schizosaccharomyces pombe* - comparison with *Saccharomyces cerevisiae*. Mol Cell. Biochem 124 (1993), 131-140

Herrmann, A. and Felle, H. - Tip growth in root hair cells of *Sinapis alba* L.. Significance of internal and external Ca^{2+} and pH. New Phytol. 129 (1995), 523-533.

King, R.W. - Multiple circadian rhythms regulate photoperiodic flowering responses in *Chenopodium rubrum* Can J. Bot. 53 (1975), 2631-2638.

Kurkdjian, A. and Guern, J. - Intracellular pH: Measurement and importance in cell activity Ann. Rev. Plant Physiol. Plant Mol Biol 40 (1989), 271-303

Lee, H.C., Johnson, C., and Epel, D - Changes in internal pH associated with initiation of motility and acrosome reaction of sea urchin sperm. Dev Biol. 95 (1983), 31-45

Loyall, L. - Untersuchungen über die Verwendung des Fluoreszenzfarbstoffs Primulin zur Messung der Protonkonzentration im Apoplast pflanzlicher Zellen Master Thesis, Albert-Ludwigs-Universität, Freiburg, 1995.

Lyndon, R.F. - Control of organogenesis at the shoot apex. New Phytol. 128 (1994), 1-18.

Lyndon, R.F. and Battey, N.H - The growth of the shoot apical meristem during flower initiation. Biol Plant. 27 (1985), 339-349

Millet, B., Melin, D., and Badot, P. - Circumnutation: a model for signal transduction from cell to cell. In: E. Wagner, H. Greppin, and B Millet, eds., The Cell Surface in Signal Transduction, pp.169-188. NATO ASI Series, Springer-Verlag-Berlin, 1987.

Nucitelli, R. and Deamer, D.W. - Intracellular pH: Its Measurement, Regulation, and Utilization in Cellular Functions. Liss-New York, 1982.

Pönitz, J and Roos, W. - A glucose-activated electron transfer system in the plasma membrane stimulates the H^+-ATPase in *Penicilium cyclopium*. J Bact. 176 (1994), 5429-5438.

Rayle, D.L., and Cleland, R. - Enhancement of wall loosening and elongation by acid solutions. Plant Physiol. 46 (1970), 250-253

Roberts,A.W., and Haigler, C.H. - Cell expansion and tracheary element differentiation are regulated by extracellular pH in mesophyl cultures of *Zinnia elegans* L. Plant Physiol. 105 (1994), 699-706.

Roos, W. - Confocal pH-topography in plant cells - acidic layers in the peripheral cytoplasm and the apoplast. Bot. Acta 105 (1992), 253-259.

Roos, W. and Slavík, J. - Intracellular pH topography of *Penicillium cyclopium* protoplasts. Maintenance of both passive and active mechanisms. Biochim Biophys. Acta 67 (1987), 67-75.

Slavík, J. - Intracellular pH topography: determination by a fluorescent probe. FEBS Letters 156 (1983), 227-230.

Streiblová, E., and Beran, K - Demonstration of yeast scars by fluoescence microscopy. Exp. Cell Res 30 (1963), 603-605.

Taiz, L. - Plant cell expansion regulation of cell wall mechanical properties Annu. Rev. Plant Physiol. 35 (1984), 585-657

Tort, M. and Gendraud, M. - Contribution à l'étude des pH cytoplasmique et vacuolaire en rapport avec la croissance et l'accumulation des réserves chez le Crosne du Japon. C. R. Acad. Sci. Paris 299 (1984), 431-434.

Ullmann, J., Seidlová, F., Krekule, J., and Pavlová, L. - *Chenopodium rubrum* as a model plant for testing the flowering effects of PGRs Biol Plant 27 (1985), 367-372.

Vian, B., Roland, J.-C., and Reis, D. - Primary cell wall texture and its relation to surface expansion. Int.J. Plant Sci. 154 (1993), 1-9.

Walker, D.J., Smith, S.J., and Miller, A.J. - Simultaneous measurement of intracellular pH and K^+ or NO_3^- in barley root cells using triple-barreled, ion-selective microelectrodes. Plant Physiol. 108 (1995), 743-751.

Webb, D.J. and Nuccitelli, R. - Intracellular pH changes accompanying the activation of development in frog eggs: Comparison of pH microelectrodes and ^{31}P-NMR measurements In: R. Nucitelli, and D.W. Deamer, eds., Intracellular pH: Its Measurement, Regulation, and Utilization in Cellular Functions, pp 293-324. Liss-New York, 1982.

SPATIAL RESOLUTION OF CORTICAL CEREBRAL BLOOD FLOW AND BRAIN INTRACELLULAR pH AS MEASURED BY *IN VIVO* FLUORESCENCE IMAGING

Robert E. Anderson, Frederic B. Meyer, and T. M. Sundt Jr.

Neurosurgical Research Laboratory
Mayo Clinic, Rochester, Minnesota U.S.A.

INTRODUCTION

The anatomic distribution of cerebral blood flow has been investigated extensively over the past several decades and has been found to be homogenetic or heterogenetic depending upon the methodology used. Blood flow, as determined by autoradiography, has shown a heterogenous distribution in various animal studies. This technique is limited to only one measurement per study since the brain tissue has to be sampled, sliced, mounted on a microscope slide, and exposed to photographic film. Blood flow studies utilizing ^{133}Xe and ^{85}Kr with multiple probes and/or Anger-type cameras have demonstrated both linear and nonlinear distributions. PET scanning techniques have demonstrated heterogeneous distributions in normal brain in humans. If the volume as seen by the scintillation detector is decreased in size, then the flow distribution tends to be less homogeneous and more heterogeneous. This is probably due to the fact that large field probes will yield an average value within the field of view of the measurement.

The distribution of intracellular hydrogen ion concentration in the cerebral cortex has not been well defined. Measurements using pH micro-electrodes have been limited to one area of cortex because of the physical constraints of the electrode. Several investigators have used umbelliferone or neutral red as pH_i indicators to measure intracellular pH of cortical brain slices *in situ* using fluorescent photographic techniques or spectrometery techniques, respectively. These techniques are limited to only one physiologic intervention. Radiolabeled DMO ([^{14}C]5,5-dimethyloxazolidine-2,4-dione-2) and triple tracer techniques have been used to measure intracellular brain pH *in vivo*, but are also limited to one physiologic intervention as the brain has to be sampled for tissue analysis. Alternatively, umbelliferone has been used to measure *in vivo* cortical intracellular pH in both ischemic and seizing brain in a serial fashion using a fluorescent microphotometer.

The purpose of this experiment was to design and construct instrumentation to perform *in vivo* computerized digital imaging of umbelliferone fluorescence for the purpose of mapping CBF and brain pH_i over a large region of cortex.

MATERIALS AND METHODS

Animal Preparation

Six white New Zealand rabbits weighing between 3.5 and 4.5 kg were anesthetized with Halothane. They were induced, operated, and studied under 4.0%, 2.5%, and 1.5% Halothane, respectively. After the tracheostomy was performed, the animals were placed on a Harvard respirator and were given 0.15mg/kg pancuronium bromide to abolish respiratory efforts.

Catheters were inserted into the right femoral artery and vein for monitoring blood pressure, sampling arterial blood gases, and administration of drugs. A PE-50 catheter was inserted into the right lingual artery for the retrograde delivery of the indicator umbelliferone into the internal carotid artery.

The skin, subcutaneous tissue, and muscle were excised over the right supraorbital ridge and parietal area. A craniectomy was performed utilizing a high speed air drill with the aid of an Olympus operating microscope. The dura was carefully removed and the cerebrum covered with a single sheet of Saran Wrap™ which kept the brain moist and prevented surface oxygenation.

Following the surgical preparation, the animal was moved from the operating table and placed on an intravital type microscope stand. The microscope was focused on an area centered about the suprasylvian gyrus. Arterial blood pressure was measured by a Statham strain gauge attached to the femoral artery catheter and recorded on a Grass model 78 polygraph. The animals were kept normothermic by the use of a small heating blanket and monitoring of body temperature with a rectal digital thermometer. Arterial P_aCO_2, P_aO_2, and pH_a measurements were performed on a London Radiometer blood gas analyzer.

In Vivo Video Fluorescent Instrumentation

The system, illustrated in Figure 1, consists of a honeycomb optical bench onto which is mounted a light source, intravital type microscope, animal stage, and camera assembly. Illumination for excitation of umbelliferone is provided by a modular system consisting of a 500W xenon arc light source and exciting filter changer. The excitation energy is passed through a 3 in. diameter, 100 mm liquid optical filter which is filled with distilled water and cooled by an external water chiller to reduce infrared wavelengths. A custom built filter changer was used for selecting the excitation of either 340±5nm or 370±5nm under computer control. The mirror is driven by a scanner controller which is addressed by a computer.

A Nikon SMZ-10 stereoscopic zoom microscope body is attached to a vertical stand mounted on an optical bench. A Ploem™ type illuminator is attached to this stand below the objective lens. The filter changer is mounted behind the stand at the same optical axis as the illuminator. The image field size can be varied from 2.3 mm to 14 mm diameter, while the

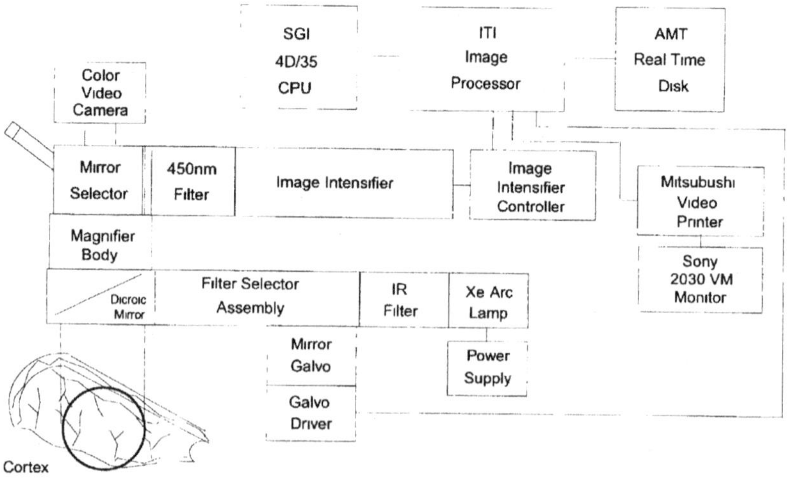

Figure 1. Block diagram of real time computer video serial fluorescent system.

illumination field size is always 18 mm in diameter. The working distance of the system is 50 mm, allowing room for use of recording electrodes. The excitation energy at 340 nm and 370 nm as measured by a digital power meter was 200 $\mu W/cm^2$ and 790 $\mu W/cm^2$, respectively. The animal platform and head micromanipulator is mounted on the optical bench below and lateral to the microscope unit.

A Nikon SMZ-10 trinocular body is attached to the top of the microscope body. This body has a beam splitting arrangement whereby 100% of the image is forwarded either to the observer for visual inspection or to an image-intensified camera.

The image-intensified camera utilizes a single stage microchannel plate intensifier with relay optics for coupling to a monochrome CCD chip camera. A 450±5 nm interference filter is inserted between the CCTV adapter attached to the trinocular body tube and the camera assembly. This is the selected wavelength for recording of the umbelliferone fluorescent emission.

The fluorescent video image is processed by an image analyzer and a Personal Iris 4D/35 computer. The acquired images were stored on a 10 MByte/sec real time disc system. The processed image is displayed on a Sony 13' color video monitor. Processed images are printed along with the calibration bars utilizing a color video printer.

Custom software was written to acquire images, perform image processing, display processed images, and store on magnetic tape. Paired images, one for each excited wavelength, were acquired with 16-frame averaging at 5 second intervals for 180 seconds. Acquired images were then corrected for background NADH fluorescence prior to processing.

The images from the 340 nm excitation were processed to compute CBF using the one-minute initial slope index. The CBF image was then displayed and stored on tape for final analysis. Paired images from the 340 nm and 370 nm excitation were ratioed and the resultant pH_i image was then displayed and stored on tape for final analysis.

Because of anatomical variation of the microvasculature from animal to animal, images cannot be averaged frame by frame from different animals at the same physiological time frame according to the experimental protocol. The mean and standard error from image sites were the result of using an xy coordinate value measurement system. The frames corresponding to the same time measurements were anatomically selected according to the xy coordinates and the value taken at that site. CBF and brain pH_i profiles were done by picking six pixel points along a line between arterial and venous surface conducting vessels. The first and sixth pixel points were located centrally over the vessel while the second and fifth pixel points were 10μ from the wall external to the vessel. The third and fourth pixel points were located equidistant from the second and fifth points. In this study, the imaging system allows the measurement of either regional (rCBF) or focal cortical blood flow (fCBF) with umbelliferone. Regional cortical blood flow can be obtained by averaging over "n" pixels ($>4.5\mu^2$), while focal CBF measurement is made with one pixel ($4.5\mu^2$).

Characteristics of Umbelliferone

Umbelliferone is the generic term for 7-hydroxycoumarin. It is non-toxic and freely diffusible across the blood-brain barrier. It is prepared for injection by dissolving 0.2 g of indicator in 200 ml of 5% glucose-saline at 90^0C for 30 min. The solution is then filtered with a 0.22-μm mesh filter prior to injection. The volume of injectate was 0.5 ml in this study.

This pH-sensitive indicator has two fluorophors; anion at 370 nm excitation and isobestic at 340 nm excitation with a common emission at 450 nm. The fluorescence of the anion varies directly with pH, and with indicator concentration, while fluorescence of the isobestic form varies directly only with indicator concentration. Therefore, it is possible to create a nomogram from the ratio of 340 nm and 370 nm to determine brain pH_i. The pK_a which is not altered by brain tissue is 7.5, has an overall measured range of pH 5.8 to 8.0. The scheme for calibrating the video fluorometer for the pH measurements is to set a ratio vs fluorescence curve based upon six different pH buffer values and calculate a fitted "n^5" degree polynomial curve to set up the calibration bar for the display.

RESULTS

Systemic Variables and Video Acquisition

Measurements of P_aCO_2, P_aO_2, pH_a, MAPB, body temperature, glucose, and hematocrit were measured. There were no significant differences between animals. Depicted in Fig. 2 are video pictures of a typical animal during a P_aCO_2 and P_aO_2 reactivity experiment.

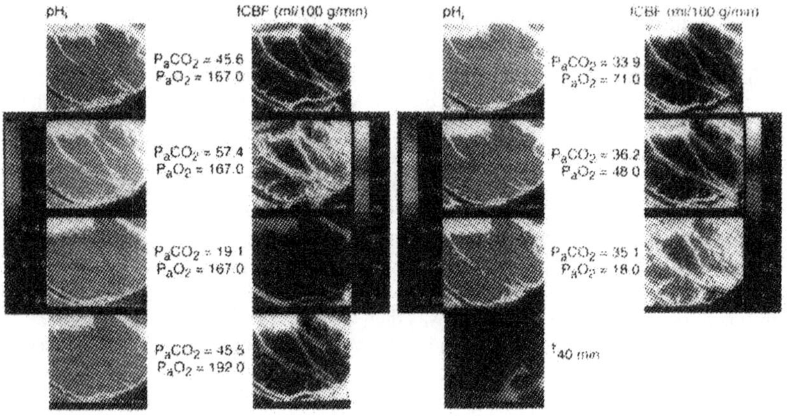

Figure 2. Video pictures of a typical animal during: A) P_aCO_2 reactivity at 40, 60, 20, 40 mmHg, B) P_aO_2 reactivity at 80, 40, 20 mmHg and at death. The left half of the picture is brain pH_i, the right half is cortical blood flow with their respective calibration bars. Each video frame is 1.5 x 1.5 cm.

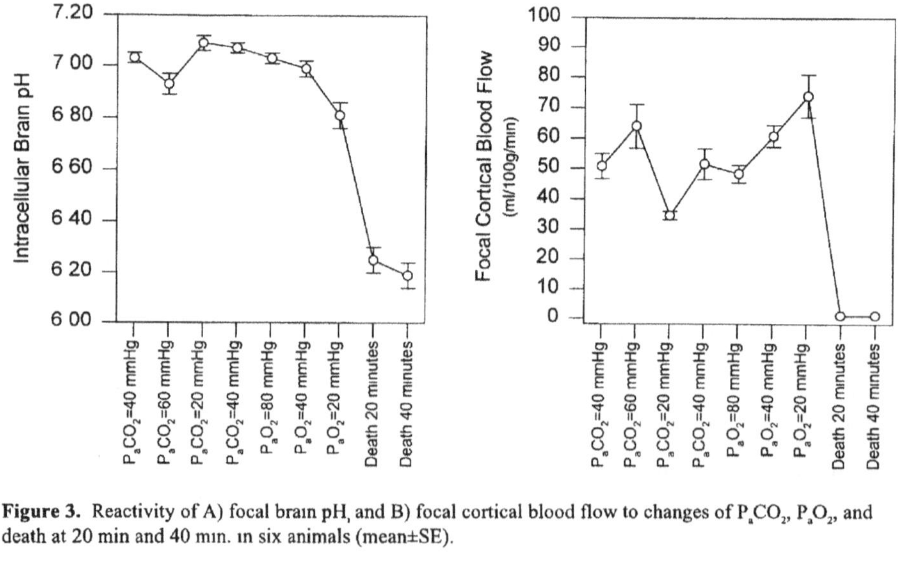

Figure 3. Reactivity of A) focal brain pH_i and B) focal cortical blood flow to changes of P_aCO_2, P_aO_2, and death at 20 min and 40 min. in six animals (mean±SE).

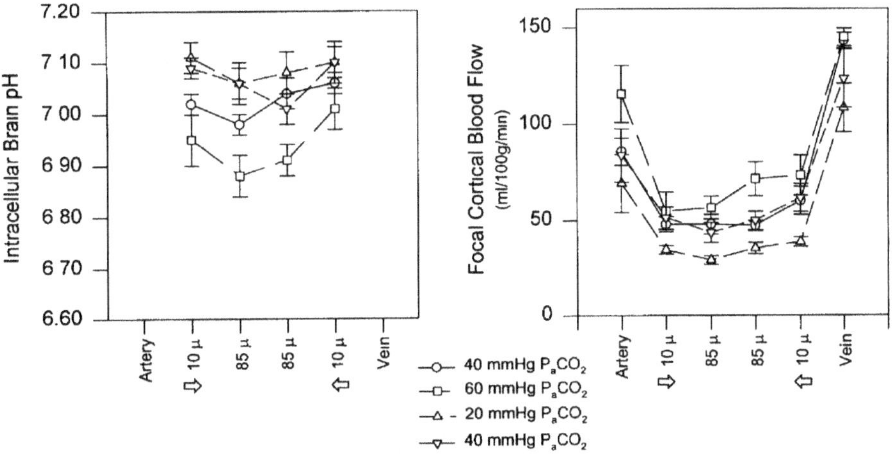

Figure 4. Profile of focal brain pH_i and focal cortical blood flow durng changes of P_aCO_2 in relationship to surface conducting arterial and vein (mean±SE).

Brain focal Cortical Blood Flow vs. P_aCO_2

Average focal cortical CBF during normocapnia was 48.5±10.5 ml/100g/min with a P_aCO_2 reactivity of 0.74 ml/100g/min/mmHg-P_aCO_2 (Fig. 3B). The fCBF profile as depicted in Fig. 4B shows that there is a heterogeneity in the flow values between the surface conducting vessels with a variation of 2-50% and that flow is higher on the venous side compared to the arterial side.

Brain focal Cortical Blood Flow vs. P_aO_2

Focal CBF showed a stepwise change from 51.5±4.9 ml/100g/min to 74.0±7.0 ml/100g/min when P_aO_2 decreased from 173.2 mmHg to 22.0 mmHg (Fig. 3A). The fCBF profile also showed heterogeneity of flow values between surface conducting vessels and that flow was

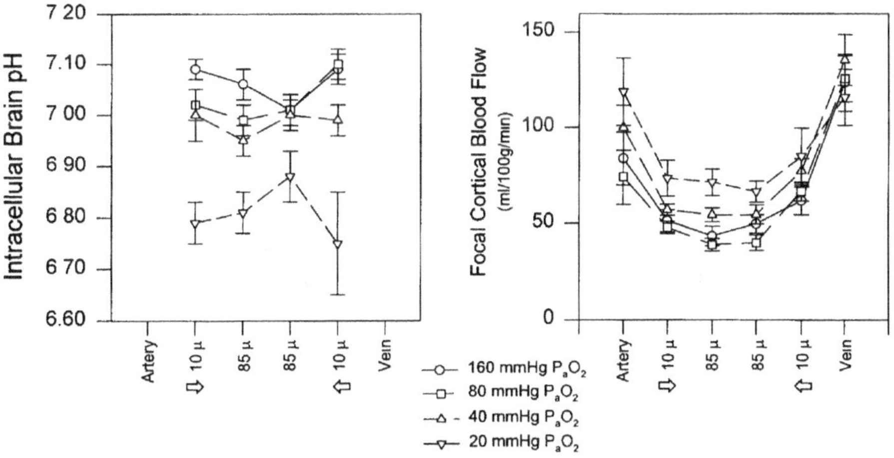

Figure 5. Profile of focal brain pH_i and focal cortical blood flow during changes of P_aO_2 in relationship to surface conducting arterial and vein (mean±SE).

higher on the venous side compared to the arterial side. However, at a P_aO_2 of 22.0 mmHg, the flow on the venous side was similar to the arterial side (Fig. 5B).

Focal brain Intracellular pH vs. P_aCO_2

Average focal brain pH_i during normocapnia was 7.05±0.02 and varied maximally 0.06 pH units during changes in P_aCO_2 from 22.9±1.4 mmHg to 61.6±2.8 mmHg (Fig. 3A). Focal brain pH_i profile showed variations from 0.005 to 0.040 pH units in the parenchymal area and as much as 0.070 pH units 10μ to the adjacent vessels, the arterial side being slightly more acidic than the venous side (Fig. 4A).

Focal brain Intracellular pH vs. P_aO_2

Focal brain pH_i remained stable when P_aO_2 was decreased from 173.2±5.5 mmHg to 22.0±1.1 mmHg (Fig. 3A). Focal brain pH_i profile varied similarly as in the P_aCO_2 reactivity with the arterial side being more acidic than the venous side by 0.080 pH units 10μ to the adjacent vessels (Fig. 5A).

CONCLUSION

The application of *in vivo* computerized digital image processing presented here facilitated the acquisition of the data in real time to interactively construct pH_i and focal cortical blood flow measurements from the acquired images. Although image processing has been in use over the past decade, advances in computer and image processing technology have resulted in increased speed and resolution. The use of frame averaging to increase signal-to-noise ratio and *pseudo-coloring* allow analysis of images in greater detail. These advances significantly enhance the utility of real time fluorescence microscopy.

The development of computerized video microfluorometry allowed the visualization of changes in brain pH_i and focal CBF *in vivo* within discrete areas of cortical tissue. These results demonstrate both the heterogeneity of cortical blood flow in relationship to the topical location from surface conducting vessels and the tight regulation of focal brain pH_i. The spatial resolution of 4.5 μm with this technique vs. 100-500 μ for autoradiogaphic and film techniques, and 7.0-8.5 mm for PET, should facilitate *in vivo* analysis of focal and regional brain pH_i and CBF regulation in both normal and pathologic brain.

ACKNOWLEDGMENTS

The authors thank Ms. Patricia Friedrich and Mr. Robert Carlson for their technical assistance and Ms. Mary Soper for preparation of the manuscript. This project was supported by National Institutes of Health Grant RO1-25374.

IS A POTENTIAL-SENSITIVE PROBE DIS-C_3(3) A NERNSTIAN DYE?: TIME-RESOLVED FLUORESCENCE STUDY WITH LIPOSOMES AS A MODEL SYSTEM

Petr Heřman, Jaroslav Večeř, and Aleš Holoubek

Institute of Physics of the Charles University
Ke Karlovu 5, 12116 Prague, Czech Republic

INTRODUCTION

Fluorescence probe diS-C_3(3) (3,3'-dipropylthiacarbocyanine iodide) is a redistribution dye used for the assessment of diffusion membrane potential in living cells. Despite of a wide usage of the dye, detailed mechanisms of its fluorescence response on changes of the membrane potential have not been well documented yet. Studies, where diS-C_3(3) was used as an indicator of the membrane potential, involve an implicit assumption that diS-C_3(3) is a 'Nernstian dye', eg. that the dye redistributes between the extracellular medium and cytoplasm according to a diffusion membrane potential and its equilibrium concentrations inside and outside the cell follow the Nernst equation. Validity of this assumption is critical when diS-C_3(3) is used for a quantitative assessment of the membrane potential.

After penetration of the dye inside the cell a fraction of diS-C_3(3) binds to intracellular structures (proteins, nucleic acids, etc.). In experiments with cell suspensions one cannot easily distinguish between fluorescence of the dye originating from the extracellular medium and the emission of the free dye which is coming from an interior of the cells. The situation is complicated even more by the fact that the cells contribute very little to the total fluorescence intensity because of their small fractional volume. Fortunately the bound form of the dye exhibits different spectral and decay characteristics from the free dye[1-4]. This property provides a handle for tracking of changes of an intracellular free dye concentration in suspensions.

In order to show that diS-C_3(3) is really a 'Nernstian dye' and to establish a reliable procedure for measurement of diffusion membrane potential in suspensions we studied large unilamellar vesicles (LUV) with encapsulated bovine serum albumine (BSA) to simulate cytoplasm. Different diffusion membrane potentials were set by K^+ ions.

MATERIALS AND METHODS

Valinomycin was purchased from Boeringer-Mannheim, diS-C_3(3) from Molecular

Probes (USA), Sepharose 6B from Pharmacia, egg yolk lecithin, bovine serum albumine pH 7.0 (BSA), and Tris-base from Sigma. All other reagents (analytical grade) were obtained from Fluka.

Preparation of Liposomes with Encapsulated Protein

Lipid dissolved in chloroform was dried under nitrogen and kept in a high vacuum at least 15 min. Then the lipid was hydrated in 50 mM Tris buffer at pH 7.5 in a presence of 150 mM KCl and 4% (w/w) BSA and hand-shaked multilamellar vesicles (MLV) were formed. The final concentration of lipid in the suspension ranged from 20 to 30 mM. Large unilamellar vesicles were extruded through a polycarbonate filter with a pore size of 400 nm using LiposoFast extrusion apparatus from Avestin (Canada) described by MacDonald et al.[5] The extraliposomal protein was removed on a 40 cm long Sepharose 6B column equilibrated with the Tris buffer containing 150 mM KCl. The eluent was monitored by tryptophane fluorescence at 340 nm and two well-separated peaks, one from LUV-associated BSA and the other from free BSA, were detected. An efficiency of separation was checked by a second passage of the eluent through the column that resulted to a single peak corresponding to LUV. No traces of free BSA were detected.

Adjustment of Potential and Labeling

The suspension of LUV was diluted in Tris buffer containing concentration of KCl ranging from 0 - 150 mM. Osmotic effects due to a difference in extraliposomal and intra-liposomal KCl concentration were compensated by a presence of an appropriate amount of choline chloride in the dilution buffer. Transmembrane diffusion potential $\Delta\varphi$ was set by an addition of 1 μM K^+-selective ionophore valinomycin. Its magnitude was determined from the Nernst equation: $\varphi_{in} - \varphi_{out} = \Delta\varphi = RT/F \cdot \ln(K^+_{out}/K^+_{in})$, where K^+_{out} is the extraliposomal concentration of KCl and $K^+_{in} = 150$ mM is the concentration of KCl trapped inside liposomes. By this procedure we adjusted diffusion membrane potential in the range between 0 mV and -150 mV.

The suspension of LUV was labeled by an injection of an aqueous stock solution of $1 \cdot 10^{-4}$M diS-C_3(3) into the sample. Final concentration of the probe in the sample was adjusted to $1 \cdot 10^{-7}$M.

Fluorescence Measurements

An apparatus for time-resolved fluorescence experiments was based on a laser excitation source and a time-correlated single photon counting as a detection system. The excitation source consisted of a cavity dumped dye laser synchronously pumped by an argon ion laser. The excitation pulses with duration of about 10 ps were generated at 550 nm with Rhodamine 540 as a laser dye. The emission wavelength of 575 nm was selected by a monochromator with an interference filter in front of the input slit. Samples were thermostated during measurements and fluorescence decays were measured under a 'magic angle' conditions in 1024 points. An apparatus function was determined by a REF procedure of Večeř et al.[6] with diS-C_3(3) and Rhodamine 610 as reference compounds. All experiments were repeated with an unlabeled sample in order to check for background fluorescence and light scattering.

Spectral characteristics were measured using an optical multichannel analyzer OMA2 and the excitation wavelength of 514.5 nm.

RESULTS AND DISCUSSION

Sensitivity of spectral and time characteristics of diS-C_3(3) to different diffusion potentials on liposomal membranes is demonstrated in Figure 1. Together with a red spectral shift (Figure 1A) there was recorded a pronounced increase of an apparent lifetime of diS-C_3(3) in the suspension (Figure 1B). The effects are more pronounced when membrane potential increases. The fluorescence decays can be satisfactorily analyzed by a triple-exponential model. The shortest decay component always corresponds to a lifetime of the free dye in the buffer which is (130±10) ps irrespective of KCl and choline chloride concentration. Therefore this component was fixed to 130 ps during the data analysis. Two long-lived components (~0.6 ns and ~1.1 ns) are almost independent of membrane potential and stay constant across the emission spectrum.

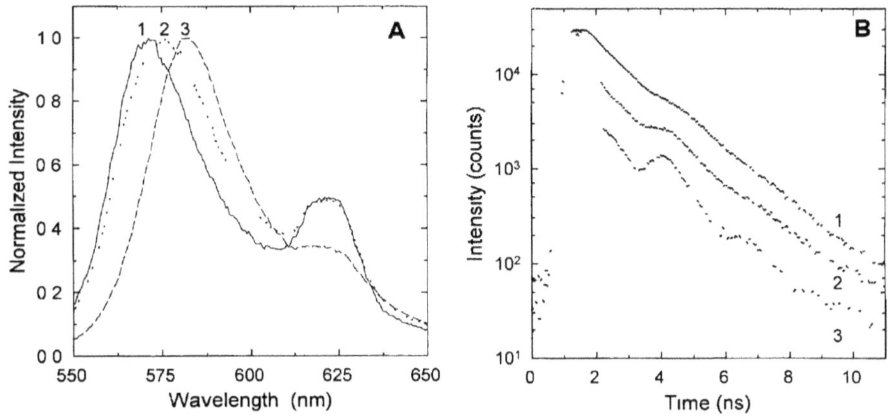

Figure 1. (A) Spectral and (B) fluorescence decay responses of diS-C_3(3) on a change of diffusion membrane potential in a suspension of liposomes filled with BSA: (1) free dye in a buffer, (2) liposomes without membrane potential, (3) liposomes with membrane potential of -100 mV. The peak near 620 nm corresponds to Raman scattering of the excitation light in water.

In order to evaluate an origin of the long decay components spectral and time characteristics of diS-C_3(3) were measured in the solution of BSA and MLV. It was found that an .interaction of the dye with lipids, BSA, and liposomes filled with BSA leads essentially to the same changes of fluorescence characteristics, eg. increase of fluorescence intensity, red spectral shift, and triple-exponential fluorescence decay was observed. It strongly indicates that the long lifetimes are associated with the bound dye and do not depend on the nature of a binding site. This conclusion was also supported by measurements of fluorescence decays in 3N KOH where a charge interaction of the cationic dye with OH^- takes place and essentially the same three lifetimes were observed.

Due to the double-exponential decay of diS-C_3(3) bound to lipids we tried to distinguish whether the fluorescence signal originates from the surface of the liposomes or comes from the hydrophobic core of the membranes. We prepared 1:1 (v/v) mixture of the buffer with cyclohexan, stained the solution by diS-C_3(3) and vigorously mixed everything in the cuvette. When the two phases completely separated fluorescence signals from both the polar and unpolar phases were measured. The intensity from the hydrophobic phase was found to be about 40 times lower then the one from the buffer. Considering small relative volume of the lipidic phase (in our experiments it was typically 10^{-8}) one can conclude that the signal from the membrane core is undetectable.

Time of equilibration of the dye in the sample was also evaluated. Figure 2A demonstrates a shift of the fluorescence maximum after an injection of diS-C_3(3) into a gently stirred suspension of liposomes. It is seen that the position of the fluorescence maximum quickly reaches a plateau. Fluorescence intensity behaves the same way (data not shown). It confirms that the redistribution of the dye in the suspension of liposomes is finished in about one minute after the injection of the dye.

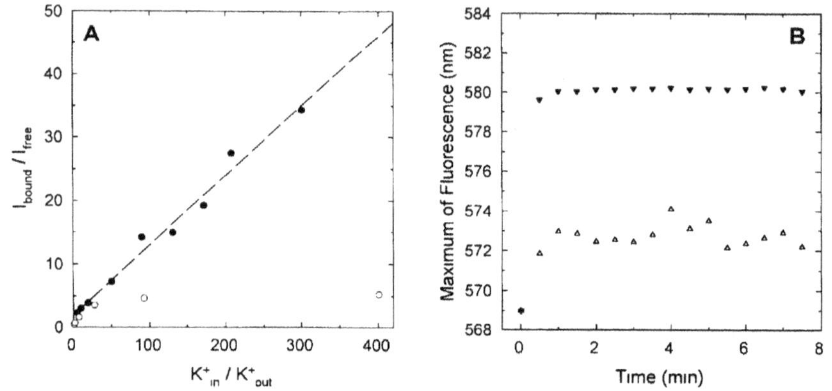

Figure 2. (A) - Equilibration of the sample after injection of diS-C_3(3) into a suspension of liposomes filled with BSA, (△) - liposomes without membrane potential, (▲) - liposomes with membrane potential of -100 mV; (B) - Calibration curve for fluorescence response of diS-C_3(3) in a suspension of liposomes; (●) liposomes filled with BSA, (○) empty liposomes. In all cases the concentration of lipid was approx. 6 μg/ml.

Because diS-C_3(3) interacts with both the membrane lipids and intraliposomal BSA the fluorescence signal of the bound dye has two components. One originates from both the inner and the outer surfaces of the liposomal membrane, the other is coming from the inner volume of the liposomes. When a loading concentration of the dye is low enough to assure proportionality between the free and the bound dye, one can write for I_{bound}: $I_{bound} = P \cdot (c_{in} + c_{out}) + C \cdot V_{in} c_{prot} c_{in}$, where c_{in} and c_{out} are free concentrations of diS-C_3(3) inside and outside liposomes, respectively, c_{prot} is a concentration of BSA enclosed in the liposomes, and V_{in} is an intraliposomal volume. The constants P and C characterize a binding of the dye to lipids and BSA, respectively. Fluorescence intensity of the free dye can be expressed as: $I_{free} = K \cdot (V_{out} c_{out} + V_{in} c_{in})$, where V_{out} is an extraliposomal volume and K is a constant. Then we obtain:

$$\frac{I_{bound}}{I_{free}} = \frac{1}{KV_{out}} \cdot \frac{P + (P + R) \cdot (c_{in} / c_{out})}{1 + (V_{in} / V_{out}) \cdot (c_{in} / c_{out})} \tag{1}$$

where $R = C \cdot V_{in} c_{prot}$. When the overall V_{in} is much smaller than the total volume of the sample ($V_{in} / V_{out} \rightarrow 0$) we can relate the experimental quantity I_{bound} / I_{free} to the ratio of the free dyes c_{in} / c_{out} by a straight line:

$$\frac{I_{bound}}{I_{free}} = A \cdot \left(\frac{P}{P + R} + \frac{c_{in}}{c_{out}} \right) \tag{2}$$

where the A is a scaling constant. The ratio of fluorescence intensities for the bound and the free dye can be calculated from the lifetimes as: $I_{bound}/I_{free} = (\sum_{bound} \alpha_{bound}\tau_{bound})/\alpha_{free}\tau_{free}$, where α_{bound} and α_{free} are preexponential factors associated with lifetimes τ_{bound} and τ_{free}.

Figure 2B demonstrates dependence of the I_{bound}/I_{free} on the ratio of K^+_{in}/K^+_{out}, eg. on diffusion membrane potential which is a straight line for liposomes filled with BSA. In empty liposomes the binding sites for diS-C_3(3) become quickly saturated and the assumption for validity of eq.(2) is violated. Then the observed deviation of the data from the expected dependence is not surprising.

Because I_{bound}/I_{free} is directly proportional to the dye concentration ratio c_{in}/c_{out}, eq.(2), and also to the potassium concentration ratio K^+_{in}/K^+_{out}, see data in Figure.2, there must be also a direct proportionality between c_{in}/c_{out} and K^+_{in}/K^+_{out} which according the Nernst equation determines the diffusion membrane potential. Otherwise we should observe a deviation of the measured I_{bound}/I_{free} from the straight line. Moreover, the proportionality constant must be a unity because at zero membrane potential when $K^+_{in}=K^+_{out}$ (eg., $\Delta\varphi$ = 0 mV) c_{in} must be equal to c_{out}. This proves that redistribution of diS-C_3(3) obeys the Nernst equation and gives us a credit for usage of the dye for quantitative assessment of the membrane potential in suspensions of liposomes or living cells. Diffusion membrane potential for a calibrated system can be estimated from the Nernst equation where the ratio of concentrations is calculated from eq.(2):

$$\Delta\varphi = \frac{RT}{F}\cdot\ln\left(\frac{I_{bound}/I_{free}}{A} - \frac{P}{P+R}\right) \tag{3}$$

For concentrated suspensions eq.(1) instead of eq.(2) should be used.

CONCLUSION

The fluorescence indicator diS-C_3(3) can be safely used as a 'Nernstian dye' and a simple experimental protocol for quantitative assessment of diffusion membrane potential in suspensions was developed.

REFERENCES

1. J. Plášek, B. Denksteinová, F. Surreau, Assessment of membrane potential using confocal microspectrofluorimetry, *J. Fluorescence* 3:157 (1993)
2. J. Plášek, R.E. Dale, K. Sigler, G. Laskay, Transmembrane potentials in cells: a diS-C_3(3) assay for relative potentials as an indicator of real changes, *Biochim. Biophys.Acta* 1196:181 (1994)
3. P. Heřman, J. Večeř, B. Denksteinová, D. Gášková, J. Plášek, H. Kurzweilová, K. Sigler, F. Surreau, Monitoring of membrane potential by means of fluorescent dyes and time-resolved fluorescence spectroscopy, *Folia Microbiol.* 39:521 (1994)
4. B. Denksteinová, D. Gášková, P. Heřman, J. Večeř, K. Sigler, J. Plášek, J. Malínský, Speed of accumulation of the membrane potential indicator diS-C_3(3) in yeast cells, in: *this volume*
5. R.C. MacDonald, R.I. MacDonald, B.P.M. Menco, K Takeshita, N.K. Subbarao, and Hu Lan-rong, Small-volume extrusion apparatus for preparation of large unilamellar vesicles, *Biochim. Biophys. Acta* 1061:297 (1991)
6. J. Večeř, A.A. Kowalczyk, L. Davenport, R.E. Dale, Reconvolution analysis in time-resolved experiments - an alternative approach: reference-to-excitation-to-fluorescence (REF) reconvolution., *Rev.Sci.Instr.* 64:3413 (1993)

KINETIC BEHAVIOR OF POTENTIAL-SENSITIVE FLUORESCENT REDISTRIBUTION PROBES: MODELLING OF THE TIME COURSE OF CELL STAINING

Jaroslav Večeř and Petr Heřman

Institute of Physics of the Charles University
Ke Karlovu 5, 12116 Prague, Czech Republic

INTRODUCTION

According to our experimental data[1] the staining of yeast cells by the redistribution probe diS-C_3(3) is a slow process which occurs in a time scale of tens of minutes. During this time interval molecules of the dye redistribute between an extracellular medium and a cell interior to reach the Nernst equilibrium. As the staining curves are not easy to interpret we decided to model this process on a computer. Among many questions to answer we try to explain: (i) whether there is a change of a fluorescence spectrum when cells with zero membrane potential are stained, (ii) how the staining curves would change if mitochondria are present in cells, or (iii) what is caused by a heterogeneity in cell sizes.

Taking into the account a binding of the dye to intracellular structures the time evolution of the staining process is modelled. Together with dye concentrations and fluorescence intensities of both free and bound dye a spectral position of the total emission maximum is calculated as a parameter which can be easily measured.

Our experiments with yeasts, protoplasts and liposomes show that the cell wall substantially limits a rate of the staining process[1,2]. The staining of mitochondria or a binding of dye to intracellular structures are much faster processes. Having simulated behavior of a suspension of identical cells with nonzero membrane potential an evolution of other systems in modified conditions is forecasted.

SIMULATIONS

The Model System

A schema of the system under study is displayed in Fig. 1. In the simplest situation we have a suspension of identical cells without mitochondria. When dye is added to the suspension the staining process starts and the extracellular concentration of the free dye c_o decreases as probe molecules enter the cell interior with the rate k_{oi}. Simultaneously the intracellular concentration of a free dye c_{if} increases. Free molecules of dye either leave cytoplasm with the rate k_{io} or bind to the intracellular structures (mainly proteins) with the

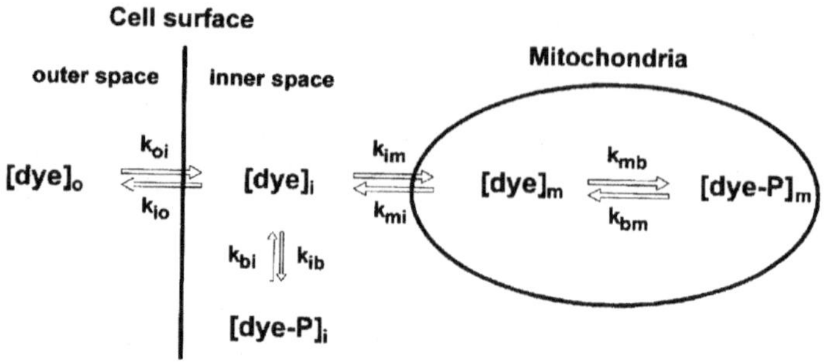

Figure 1. A kinetic schema describing processes which occur when redistribution fluorescent dye is added to a suspension of cells containing mitochondria. Indices *o*, *i* and *m* refer to extracellular, cytoplasmatic and mitochondrial environments, respectively. For other details see text.

rate k_{ib}. The bound dye with a concentration c_{ib} dissociates with the rate k_{bi}. Low concentrations of the dye are expected to avoid a saturation of binding sites. The in-rate k_{oi} is given bye the out-rate k_{io} modified by the Nernst factor: $k_{oi}=k_{io}\exp(-\Delta\varphi\cdot F/RT)$, where $\Delta\varphi$ is a difference between intracellular and extracellular diffusion potentials, T is absolute temperature, F is the Faraday constant, and R is the universal gas constant. The rate of binding is given by: $k_{ib}=K_{prot}k_{bi}$, where K_{prot} is an equilibrium constant for binding of dye to proteins.

When mitochondria are present in cells, free dye from the cytoplasm enters a mitochondrial interior with the rate k_{im}. Free mitochondrial dye with concentration c_{mf} either returns to the cell cytoplasm with the rate k_{mi} or binds to proteins with the rate k_{mb}. The bound dye dissociates with a rate constant k_{bm}. Analogically $k_{mb}=K_{prot}k_{bm}$ and $k_{im}=k_{mi}\exp(-\Delta\varphi_m F/RT)$, where $\Delta\varphi_m$ is a diffusion potential on the mitochondrial membrane.

Fluorescence Spectra of Stained Cell Suspensions

Binding of dye to proteins leads to a red spectral shift of about 10 nm which gives possibility to calculate a ratio of intensities for free and bound dye, I_f/I_b, in the range of

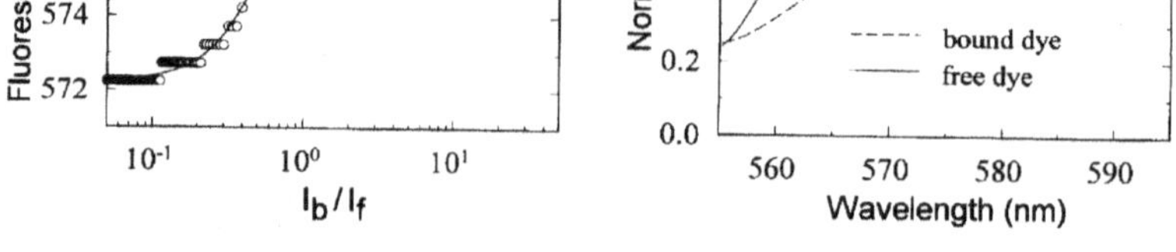

Figure 2. Position of a fluorescence maximum for suspension of cells stained by the fluorescent probe diS-C_3(3) as a function of a ratio of bound and free dye intensities (left). The solid line was calculated using experimentally measured spectra of free and bound dye (right).

about two orders of magnitude. This procedure provides the same information as a decomposition of the total emission spectrum into free and bound dye spectral components, which is often used although it is much more complicated.

The position of a fluorescence maximum as a function of I_f/I_b was calculated from individual spectra of free dye in buffer and dye bound to 5% (w/w) solution of BSA in a buffer. The result is displayed in Fig.2. As the original spectra were measured with a resolution of 0.5 nm, discrete values of the spectral maximum are displayed simultaneously with the best analytical fit of the data.

Cell Staining Equations

The staining process of the model system from Fig.1 can be described by the following system of ordinary differential equations:

$$\frac{d}{dt}(c_o V_o) = -k_{oi}\, c_o + k_{io}\, c_{if}$$

$$\frac{d}{dt}(c_{if} V_i) = k_{oi}\, c_o - (k_{io} + k_{im} + k_{ib})\, c_{if} + k_{mi}\, c_{mf} + k_{bi}\, c_{ib}$$

$$\frac{d}{dt}(c_{mf} V_m) = k_{im}\, c_{if} - (k_{mi} + k_{mb})\, c_{mf} + k_{bm}\, c_{mb}$$

$$\frac{d}{dt}(c_{ib} V_i) = -k_{bi}\, c_{ib} + k_{ib}\, c_{if}$$

$$\frac{d}{dt}(c_{mb} V_m) = -k_{bm}\, c_{mb} + k_{mb}\, c_{mf}$$

where V_o, V_i, and V_m are volumes of the extracellular medium, cytoplasm and mitochondria, respectively. The set of equations is written for numbers of dye molecules (quantities in brackets on the left hand side of the equations) which helps to understand a kinetic behavior of the system. If every equation is divided by the corresponding volume, new rate constants can be defined to remove a volume dependence of the rates with the subscript *b*.

When mitochondria are not present in the cells, there remain only three equations for c_o, c_{if} and c_{ib} because all "mitochondrial" rate constants are set to zero. In the case of a heterogeneous system, where two sizes of cells with identical mitochondria are present, four more equations are needed because all concentrations, volumes and rate constants (except of c_o and V_o) must be additionally indexed as (1) and (2). Only the first equation is changed to describe an interaction between both systems.

The rate constants must be adjusted to overlay reasonably simulated and experimental curves. The rate k_{io} directly depends on the total surface of the cells: $k_{io}=k\cdot 4\pi r^2\cdot N$. In a total sample volume $V=1$ml there was typically $N=5\cdot 10^6$ cells with an average radius $r=5\mu$m. One percent of a cell volume was supposed to be occupied by mitochondria. It implies that $V_o=0.997382$ ml, $V_i=0.002592$ ml and $V_m=0.000026$ ml. The constant *k* was set to the value of $3\cdot 10^{-4}$ cm$\cdot$min^{-1} to have simulated curves in a proper time range. The rate k_{oi} can then be calculated from k_{io} for any given membrane potential $\Delta\varphi$.

The rates k_{ib} and k_{mb} depends on the total volume of cytoplasm: $k_{ib}=k_b V_i$ and $k_{mb}=k_b V_m$. In order to have $k_b \gg k_{io}/V_i=3k/r$, the value of k_b was set to $3\cdot 10^3$ min^{-1}. The constant K_{prot} was adjusted to the value of 5 to obtain a fraction of the bound dye expected form the experimental data.

As we expect the binding of dye to proteins to be faster then a penetration through the mitochondrial membrane, the rate k_{mi} was set to the value of $1.3 \cdot 10^{-3}$ $cm^3 \cdot min^{-1}$. Then conditions $k_{mi}/V_m \gg k_{io}/V_i$ and $k_{mi}/V_m < k_b$ are fulfilled. For lower values of k_{mi} the concentration $c_{mf}(t)$ starts to deviate from $c_{if}(t)$ at $\Delta\varphi_m=0$. The rate k_{im} can be again calculated from k_{mi} for any mitochondrial potential $\Delta\varphi_m$. As a cell wall limits the rate of staining, higher values of k_b and k_{mi} than those selected do not change simulation results.

Fluorescence intensities of free and bound dye were calculated using a ratio of fluorescence quantum yields $q_b/q_f=10$. The value was taken from fluorescence lifetime measurements[2]. A difference in extinction coefficients of the free and the bound dye at the excitation wavelength was neglected.

RESULTS

A typical result of simulation for a suspension of identical cells with both mitochondrial and cell-membrane potentials is given in Fig.3. At the time $t=0$ all concentrations are equal to zero except $c_o(0)=1 \cdot 10^{-7}$ M. For $t \rightarrow \infty$ an equilibrium is reached and the ratios c_{if}/c_o and c_{mf}/c_{if} are given by membrane potentials $\Delta\varphi$ and $\Delta\varphi_m$. Also concentrations of bound dye should be determined by the equilibrium constant K_{prot}: $K_{prot}=c_{mb}/c_{mf}=c_{ib}/c_{if}$. In reality the system is very close to the equilibrium in about 20 minutes[1] and the maximum of the total emission spectrum λ_{max} is positioned at 580.9 nm. It corresponds to $I_b/I_f=3.9$. When the mitochondrial potential is set to zero or cells do not contain mitochondria then $\lambda_{max}=580.0$ nm and $I_b/I_f=2.0$, but the equilibration is slightly faster. If the cells have zero membrane potential, $\Delta\varphi=0$, the change of λ_{max} is negligible (from 572.3 to 572.5 nm).

An influence of a heterogeneity in cell sizes was also simulated by adding a small fraction of cells with different size to the suspension of "standard" cells ($r=5$ μm). Simulations revealed that smaller cells are stained faster than the "standard" ones in terms of changes of intracellular concentrations of dye. Such behavior can be expected because the rate k_{oi} depends on a cell surface but the concentration c_i depends on a cell volume. Under some circumstances the dye concentration in the smaller cells does not increase monotonically but reaches its maximum and then decreases as a staining of larger cells continues and the system is not in the equilibrium yet. To change a total fluorescence spectrum of the "standard" cells, the overall volume of cells with a different radius has to be high enough. For example, there is practically no change of λ_{max} if less then 10% of

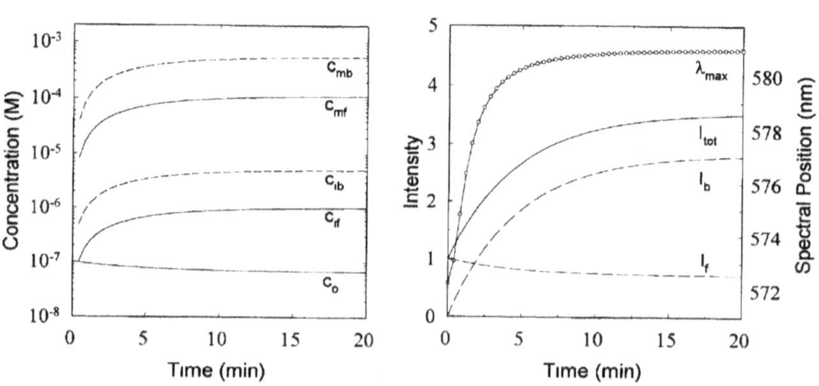

Figure 3. Simulation of a staining process; (left) - time dependencies of concentrations, (right) - fluorescence characteristics; $\Delta\varphi$=-70mV and $\Delta\varphi_m$=-120mV, T=300K

small cells (r=2 μm) is present in the suspension of "standard" cells (r=5 μm). However, the same fraction of large cells (r=10 μm) can change λ_{max} substantially.

CONCLUSIONS

To obtain staining curves which resemble experimental data only two parameters must be adjusted in the set of differential equations. The first one is the constant k of dye penetration through a walled cell membrane, which determines a time scale of the staining processes, the second one is the equilibrium constant K_{prot} which adjusts a portion of the bound dye and consequently λ_{max}. Both parameters can be precisely tuned up to overlay experimental data by the simulated curves. All other parameters can be either calculated or measured experimentally.

The following behavior of a stained cell suspension can be expected:

(i) The contribution of a bound dye to the total fluorescence spectrum is very small for cells with zero membrane and mitochondrial potentials and consequently no shift of the emission maximum can be expected during the staining process.

(ii) If mitochondria with additional negative potential are present in cells, longer time of staining can be expected to reach thermodynamical equilibrium. There will be also larger red shift of the fluorescence maximum.

(iii) Small cells in heterogeneous suspensions are stained faster than the large ones. As fluorescence intensity of a bound dye depends on a cell volume, a large fraction of small cells is needed to change fluorescence spectrum compared to a homogeneous suspension. The opposite is true for cells with a large radius.

REFERENCES

1. B. Denksteinová, D. Gášková, P. Heřman, J. Večeř, K. Sigler, J. Plášek, and J. Malínský, Speed of accumulation of the membrane potential indicator diS-C_3(3) in yeast cells, In: *this volume*
2. P. Heřman, J. Večeř, and A. Holoubek, Is a potential-sensitive probe diS-C_3(3) a Nernstian dye?: time-resolved fluorescence study with liposomes as a model system, In: *this volume*

SPEED OF ACCUMULATION OF THE MEMBRANE POTENTIAL INDICATOR DIS-C_3(3) IN YEAST CELLS

B. Denksteinová,[1] D. Gášková,[1] P. Heřman,[1] J. Večeř,[1]
K. Sigler,[2] J. Plášek[1] and J. Malínský[1]

[1]Institute of Physics, Charles University, Ke Karlovu 5,
121 16 Prague
[2]Institute of Microbiology, Czech Academy of Science,
Vídeňská 1083, 142 20 Prague, Czech Republic

INTRODUCTION

The carbocyanine dye diS-C_3(3) (3,3'- dipropylthiacarbocyanine iodide), whose the steady-state fluorescence spectra were measured in yeast cell suspensions, belongs to the group of slow (Nernstian, or redistribution) dyes which report on membrane potential by their voltage-sensitive partition between the extracellular medium and the cytosol.[1-3] Since the emission spectrum shifts and the quantum yield of fluorescence increases upon binding of the dye in the cell, two fluorescence parameters,[4-5] the wavelength of emission maximum and the intensity of fluorescence at this wavelength, were used to monitor the redistribution of the dye inside/outside the cells. To demonstrate that the dye accumulation in cells, as revealed by observed fluorescence changes, is actually membrane-potential-driven we used the uncoupler CCCP (carbonyl cyanide 3-chlorophenylhydrazone) which drastically increases membrane permeability for protons and depolarizes the cell membrane.[6]

The use of yeast cells, especially the presence of the cell wall, causes significant problems if one wants to measure the membrane potential by a redistribution dye. The uptake of the dye into yeasts is much slower in comparison to other cells and depends on many factors. Both fluorescence parameters at equilibrium probe distribution and the initial speed of cell staining, were found to depend on the following cellular characteristics: (i) concentration of cells in suspension (ii) cell respiratory activity (iii) cell growth phase (iv) properties of the cell wall.

MATERIALS AND METHODS

Yeast cells. Yeast cells were obtained by preculturing *Saccharomyces cerevisiae* S6/1, RXII and RXII rho$^-$ mutants in a YEPG growth medium (0.5 % yeast extract, 1 % bactopeptone; 2 % glucose for "high-glucose" medium and 0.2 % glucose for "low-glucose" medium) on a reciprocal shaker for 6 to 8 h at 28 °C and then performing a main

Fluorescence Microscopy and Fluorescent Probes
Edited by J. Slavík, Plenum Press, New York, 1996

culture in fresh YEPG for 16h at 28 °C. The cells were grown to the exponential or the stationary phase, washed twice in distilled water and suspended in citrate-phosphate buffer (pH 4.6) to the desired cell concentration. Protoplasts were prepared by a standard method.[7]

Chemicals. DiS-C_3(3) was purchased from Molecular Probes (USA). It was prepared in 1 mmol/L stock solution in EtOH, which was added to the yeast suspension to 0.1 μM final concentration. CCCP (SERVA) was prepared in a 10 mM stock solution in dimethylformamide (DMF) and added to the cells to a final concentration of 10 μM.

Procedure. A 10 ml volume of agitated yeast suspension of the given concentration was supplied with an equal volume of diS-C_3(3) solution in citrate-phosphate buffer (pH 4.6) to a final probe concentration of 0.1 μM. Fluorescence spectra of the sample were recorded every 30 s from the moment of probe addition (t=0) using an optical multichannel analyzer (OMA). The fluorescence was excited by a argon-ion laser at 514.5 nm. The suspension was kept at 25 °C.

RESULTS

Dye Response To Membrane Potential Changes

Fig. 1 shows the effect of adding the proton gradient uncoupler CCCP to the cell suspension. When cells are depolarized by the CCCP, the final wavelength of emission maximum and the intensity of fluorescence at this wavelength reach the same value irrespective of whether the uncoupler was added simultaneously with the dye or with a delay. No fraction of the dye thus remains irreversibly bound to cell constituents and it can be safely assumed, that both parameters indeed report on changes in membrane potential.

Effect of Yeast Cell Concentration and Respiratory Activity on Staining Characteristics.

The two main points illustrated in Fig. 2 are that the magnitude of spectral shift

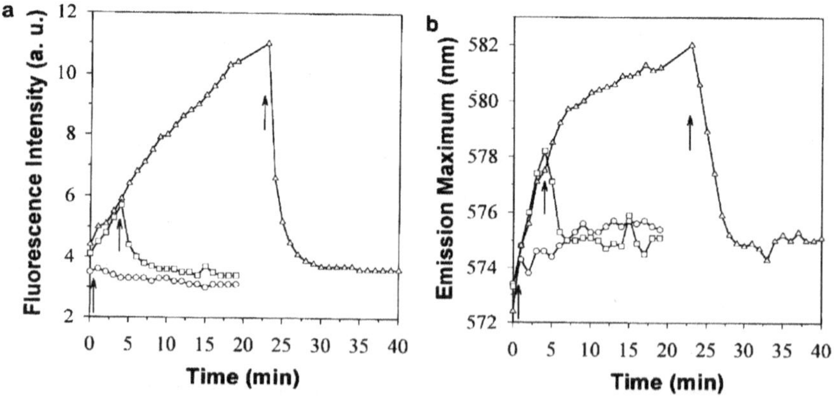

Figure 1. Time dependence of fluorescence intensity (a) and the wavelength of emission maximum (b) of diS-C_3(3) fluorescence in a cell suspension. 10 μM CCCP added after 20 min (△), after 5 min (□) or immediately (○).

corresponds to the volume fraction occupied by the cells, i.e. to the cell concentration in the suspension, and that the changes in the dye fluorescence are considerably slower in rho⁻ mutants defective in mitochondrial function.

Effect of Cell Growth Phase and Cell Wall Properties or Its Absence on Staining Characteristics.

The staining pattern of exponential-phase cells considerably differs from that of their stationary-phase counterparts in that the dye accumulates more in stationary cells (Fig. 3). In addition, cells grown in a medium with lower glucose concentration, in which the cell wall appears to represent a lower barrier for probe penetration, exhibit higher and faster dye accumulation.

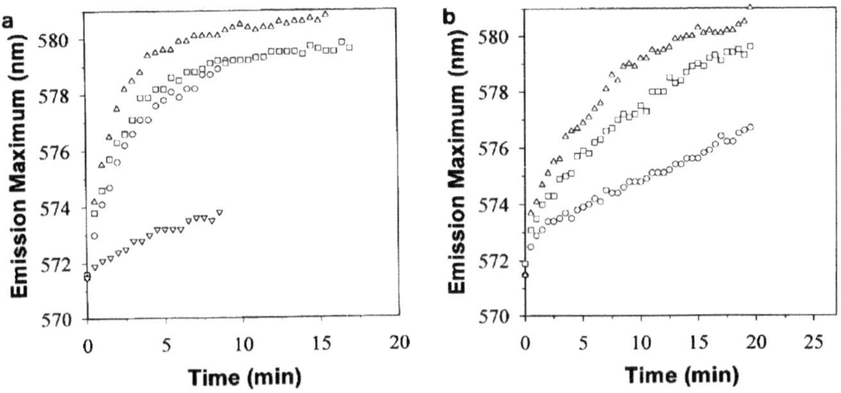

Figure 2. Time course of the change in position of fluorescence maximum of diS-$C_3(3)$ in a cell suspension: (a) wild-type R XII cells (b) respiration-deficient (rho⁻) mutants of R XII cells; cell concentrations in suspensions were 2×10^7 cells/ml (△), 1×10^7 cells/ml (□), 5×10^6 cells/ml (○) and 2×10^6 cells/ml (▽).

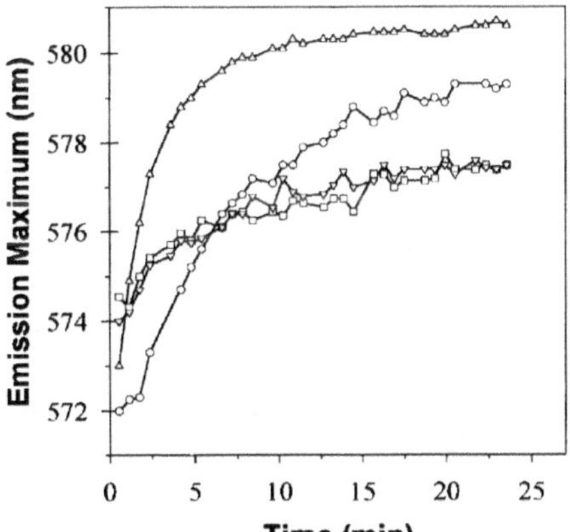

Figure 3. Time course of the change in the wavelenght of emission maximum of diS-$C_3(3)$ fluorescence in a cell suspension (10^7 cell/ml): exponential-phase cells cultured in a "high-glucose" medium (□), exponential-phase cells cultured in a "low-glucose" medium, (▽) stationary cells cultured in a "high-glucose" medium (○), stationary cells cultured in a "low-glucose" medium (△).

When the cell wall is completely removed, the resulting protoplasts behave similarly to other wall-less cells (hemapoietic cells, T47d, COS cells) stained with diS-C_3(3): the dynamic equilibrium in dye redistribution across the membrane is reached within 3 - 5 min (Fig. 4).

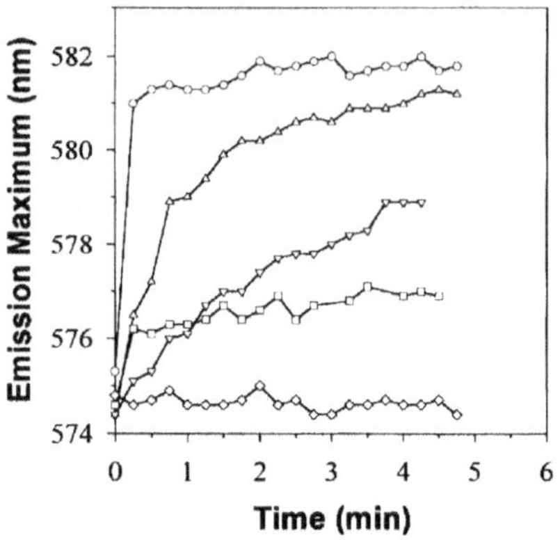

Figure 4. Time dependence of the wavelength of emission maximum of diS-C_3(3) fluorescence in a cell suspension in sorbitol: protoplasts 2.10^6 cells/ml (□), protoplasts 5.10^6 cells/ml (○), stationary intact cells 5.10^6 cells/ml (▽), stationary intact cells 2.10^7 cells/ml (△), diS-C_3(3) in sorbitol (◇).

DISCUSSION

The first question asked in our study was whether diS-C_3(3) fluorescence could report on fast transients in yeast membrane potential. The data on the effect of the proton gradient uncoupler CCCP show that both fluorescence intensity and the position of fluorescence maximum can be used as good markers of these transients since membrane depolarization by CCCP caused a fast (< 3 min) drop in both parameters to a fixed minimum value. This ultimate value was independent of whether the uncoupler was added together with the probe, after probe equilibration, or some time during the equilibration. The measurement can thus serve for qualitative detection of potential transients even for short experimental protocols when the probe has not yet reached equilibrium distribution.

Another question concerned the effect of the cell wall. In walled cells such as yeast, equilibration of reporter probes (e. g. lipophilic cations) is considerably slowed down (40 - 60 min) by the barrier presented by cell wall. We assessed this effect by determining the rate of probe accumulation in "thick-walled"and "thin-walled" and in protoplasts (wall-less cells). "Thick-walled" cells grown at a high glucose concentration took 20-30 min to equilibrate the probe, while "thin-walled" ones grown at low glucose concentration required 10-15 min and wall-less protoplasts 3-5 min. The equilibration rate therefore depends crucially on cell history and simple measurements of probe fluorescence at a fixed interval after the beginning of cell staining may give spurious results.

The third question asked was whether mitochondrial potential contributed to the overall potential reported by the probe. Stationary cells, which use respiration as the main metabolic pathway, took up more probe than the largely fermentative exponential cells. However, respiration-deficient rho$^-$ mutants behave a different way.

REFERENCES

1. A.S. Waggoner, Dye indicators of membrane potential, *Ann.Rev. Biophys.Bioeng.* 8:47(1979)
2. J.C. Smith, Potential-sensitive molecular probes in membranes of bioenergetic relevance, *Biochim. Biophys. Acta* 1016:1(1990)
3. L.M. Loew, Potentiometric membrane dyes, in: *Fluorescent and Luminiscent probes for Biological Activity*, W. T. Mason, ed., Acad. Press, San Diego (1993)
4. A. Peña, S. Uribe, J.P.Pardo and M. Borbolla, The use of a cyanine dye in measuring membrane potential in yeast, *Arch.Biochem.Biophys.* 231:217(1984)
5. J. Plášek, R. E. Dale, K. Sigler and G. Laskay, Transmembrane potentials in cells: a DiS-C_3(3) assay for relative potentials as an indicator of real changes. *Biochim. Biophys. Acta* 1196:181(1994)
6. H. M. Shapiro, Cell membrane potential analysis, in: *Methods in Cell Biology* (41), Acad. Press, Inc. (1994)
7. L. Kováč, H. Bednářová, M. Greksák, Oxidative phosphorylation in yeast. I. Isolation and properties of phosphorylating mitochondria from stationary phase cells. *Biochimica et Biophysica Acta* 153:32 (1968)

SPECTRAL EFFECTS OF SLOW DYE BINDING TO CELLS AND THEIR ROLE IN MEMBRANE POTENTIAL MEASUREMENTS

Jaromír Plášek[1] and Karel Sigler[2]

[1] Institute of Physics of the Charles University,
Ke Karlovu 5, 121 16 Prague
[2] Institute of Microbiology, Czech Academy of Sciences,
Vídeňská 1083, 142 20 Prague, Czech Republic.

INTRODUCTION

Slow (Nernstian, or redistribution) dyes monitor membrane potential by their voltage-sensitive partition between the extracellular medium and cytosol, which brings about changes in probe fluorescence intensity.[1-3] Two different effects are generally responsible for these changes: (i) fluorescence quenching due to the aggregation of fluorochromes upon their accumulation in cells, and (ii) the appearance of a new fluorescent component which is typical of a dye fraction bound to cytosolic macromolecules and intracellular membranes.

Here we show how an experimental protocol for a membrane potential assay can influence the proportion between the free dye fluorescence from the medium and that of the bound intracellular dye, and thus an apparent voltage-sensitive response of a trimethine dye. We decided to distil the main features of slow dye fluorescence changes in cell suspensions, as available in the current literature, and to present them as schematic charts which may help the readers to interpret their particular results of assays on living cells. We drew the charts for cationic dyes only (especially carbocyanines) since their modification for anionic dyes should involve a simple reversal of the direction of dye accumulation changes upon cell hyperpolarization.

RESULTS AND DISCUSSION

The Effect of Dye Association with Cells on the Fluorescence Emission Spectra of Carbocyanines

Upon accumulation in cells, virtually all slow potentiometric dyes exhibit some spectral changes in their fluorescence which are related to the occurrence of a new emission component.[4] The fluorescence emission spectra of 3, 3'-dipropylthiadicarbocyanine iodide (diS-C_3(5)) and 3,3'-dippropylthiacarbocyanine iodide (diS-C_3(3)) in aqueous media and in cells are shown as examples in Fig.1. The intensity of such new spectral bands depends

Fluorescence Microscopy and Fluorescent Probes
Edited by J. Slavík, Plenum Press, New York, 1996

considerably on the dye. For diS-C_3(5), for example, this cell-bound fluorescence is relatively week while for diS-C_3(3) it may considerably exceed the intensity of the probe fluorescence from the medium. It was demonstrated that binding of probe molecules to intracellular proteins contributes considerably to the existence of the cell-bound fluorescence component.[5]

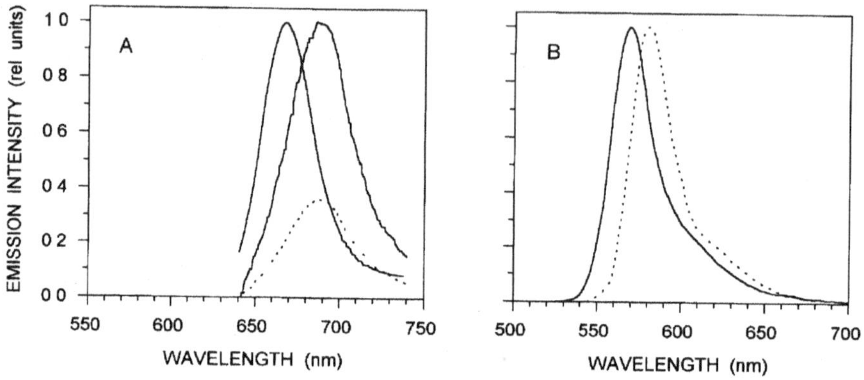

Figure 1. The difference between the fluorescence emission spectra of selected carbocyanine dyes in aqueous media and in cells. (A) diS-C_3(5) fluorescence (0.6 μM diS-C_3(5) was added to 0.17% suspension of erythrocytes in NaCl-Tris (pH 7.4)); full line (left) - fluorescence from the cell medium, full line (right) - peak-height normalized fluorescence component from cells, dotted line - fluorescence from cells (to scale), according to Sims et al.[6] (B) diS-C_3(3) fluorescence; full line - 1 μM diS-C_3(3) in Hepes, dotted line - diS-C_3(3) bound to T47D cells.[7]

The Contribution of Bound Dye to Overall Emission Spectra in Cell Suspensions

The contribution of bound dye fluorescence to overall emission spectra will obviously depend on the relative dye-to-cell concentration, as schematically demonstrated in Fig.2. At a very low cell concentration and/or when the dye concentration is so high that intracellular dye binding is saturated (see also next charts), the free dye fluorescence in the medium dominates the overall spectra. With a low dye and high cell concentration, the intracellular bound dye will dominate. In a general case, the contributions of both dye forms, i.e. free (in medium) and boud (to cytosolic proteins and intracellular membranes) can be comparable.

If it is assumed that two distinct dye forms coexist in individual cells and/or cell suspensions, i.e. the free (aqueous) and the bound form, then experimental emission spectra, $F(\lambda)$, in such samples can be expressed in terms of the linear combination:

$$F(\lambda) = aF_a(\lambda) + bF_b(\lambda)$$

where $F_a(\lambda)$ and $F_b(\lambda)$ are the peak-height- or area-normalized emission spectra of free and bound forms of the dye, respectively, and a and b are the corresponding fractions of these spectra in the overall spectrum $F(\lambda)$.[5]

The Dependence of an Apparent Probe Response on the Emission Wavelength Selection

For obvious reasons, the intensity of bound-dye fluorescence components should increse upon the dye accumulation in cells (for cationic dyes, such as carbocyanines this corresponds to cell hyperpolarization). On the other hand, the dye accumulation in cells leads to the decrease in its concentration in the medium and thus to the decrease of the free dye component. Considering the existence of the spectral difference between these two fluorescence components we can explain why the apparent voltage sensitive response of slow dyes can be influenced by the actual selection of emission wavelength. This is schematically illustrated in Fig. 3.

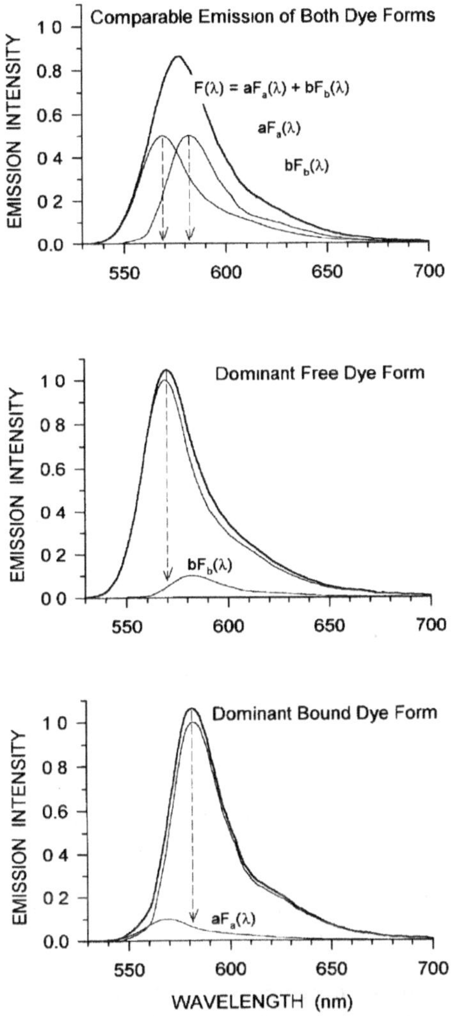

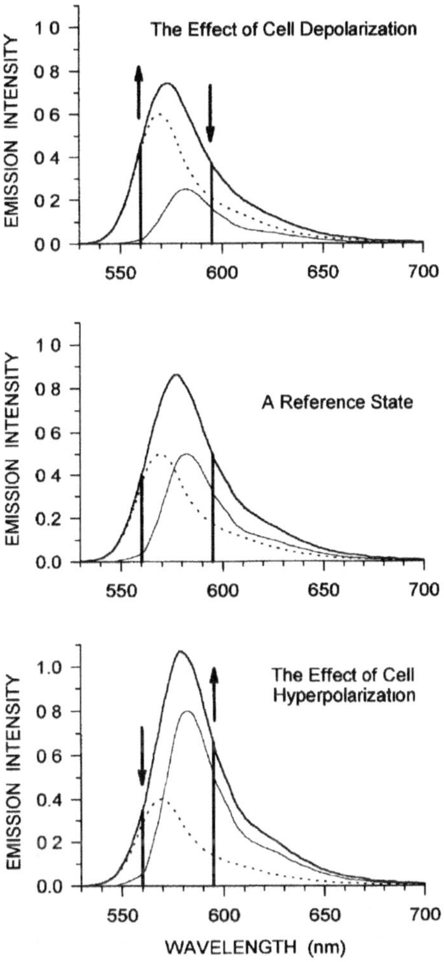

Figure 2. (left column) Effect of relative dye-to-cell concentration on probe fluorescence spectra: thick lines - overall fluorescence spectra, thin lines - spectral components corresponding to free (marked as F_a) and bound (marked as F_b) dye forms. The spectra used in this schematic diagram correspond to diS-C_3(3) fluorescence (according to Plášek et al.[7]).

Figure 3. (right column) Effect of the spectral range selection on observed fluorescence intensity changes in membrane potential assays performed with cationic carbocyanine dyes: thick lines - overall fluorescence spectra, dotted and thin lines - spectral components corresponding to free and bound dye forms, respectively; vertical bars represent the intensity of overall fluorescence at selected wavelengths (arrows indicate the expected direction of the apparent change in the probe fluorescence intensity).

The Binding of Slow Dyes to Cytoplasm is Saturable

The voltage-sensitive response of slow dyes can be modulated by the actual dye concentration in the medium owing to the fact that the intensity of the bound-dye fluorescence component cannot rise infinitely. When the average dye concentration in cells exceeds a critical aggregate concentration the free dye concentration in the cytosol must remain unchanged upon further accumulation of the dye in cells (a critical aggregate concentration is a characteristic parameter analogous to the critical micellar concentration in lipid-water systems;[8] for dicarbocyanines, for example,[4] this critical concentration is less than 5 μM). Then the extent of dye association with cytosolic macromolecules and intracellular membranes, and hence the bound dye fluorescence intensity, must also remain constant.[9] The intensity of the bound-dye fluorescence component will exhibit a plateau when the total intracellular dye concentration exceeds a certain limiting value. Such a concentration dependence of bound dye fluorescence intensity is schematically shown in Fig. 4A. The extracellular dye concentrations used in this schematic diagram follows roughly the situation typical for the fluorescence of bound diS-C_3(3) in murine hemopoietic cells which exhibited a pronounced saturation when the extracellular dye concentration increased from 5×10^{-8} M to 5×10^{-7} M.[5]

For cationic dyes their intracellular concentration in a cell with plasma membrane potential V is higher than that in a depolarized cell (V = 0), the difference being characterized by the multiplicative factor *exp(FV/RT)*, as predicted by Nernst equation. The saturation level of the dye accumulation in hyperpolarized cells will therefore be reached at lower extracellular dye concentrations than in depolarized cells. The relationship between bound-dye fluorescence intensity *vs* extracellular dye concentration plots and the cell membrane potential, as schematically depicted in Fig. 4A, provides a background for the interpretation of the dependence of voltage-sensitive probe response on the dye concentration used.

Let's start with the analysis of the bound dye fluoresccence properties. When fluorescence intensity *vs* membrane potential curves are measured at low dye concentrations ($c = c_L$), the bound dye fluorescence (measured in logarithmic units) will increase linearly with -V. For $c = c_M$, the calibration curves will exhibit a saturation-like behaviour. For $c = c_H$, the bound dye fluorescence intensity will be constant, independent of the cell membrane potential, Fig. 4B.

Usual experiments, however, are performed without discriminating between the free- and bound-dye fluorescence components. When the extracellular dye concentration is so high that the bound dye fluorescence can reach a saturation level even in depolarized cells (which is typical for cells stained with 1 μM diS-C_3(5)), only the fluorescence of the free dye in medium can report on membrane potential chages. Upon cell hyperpolarization, additional accumulation of the dye in cells will be observed which is followed by formation of non-fluorescent aggregates, and thus by the well-known fluorescence quenching, Fig. 4C. On the other hand, when the bound dye fluorescence dominates the overall probe signal (e.g. at low dye and high cell concentrations), cell hyperpolarization will be revealed by an increase in fluorescence intensity. This scheme explains why such dyes as diO-C_5(3) and diI-C_5(3) revealed cell hyperpolarization by quenching of fluorescence intensity when high (about 1 μM) dye concentrations were used, while their fluorescence increased at low dye concentrations (about 0.05 μM).

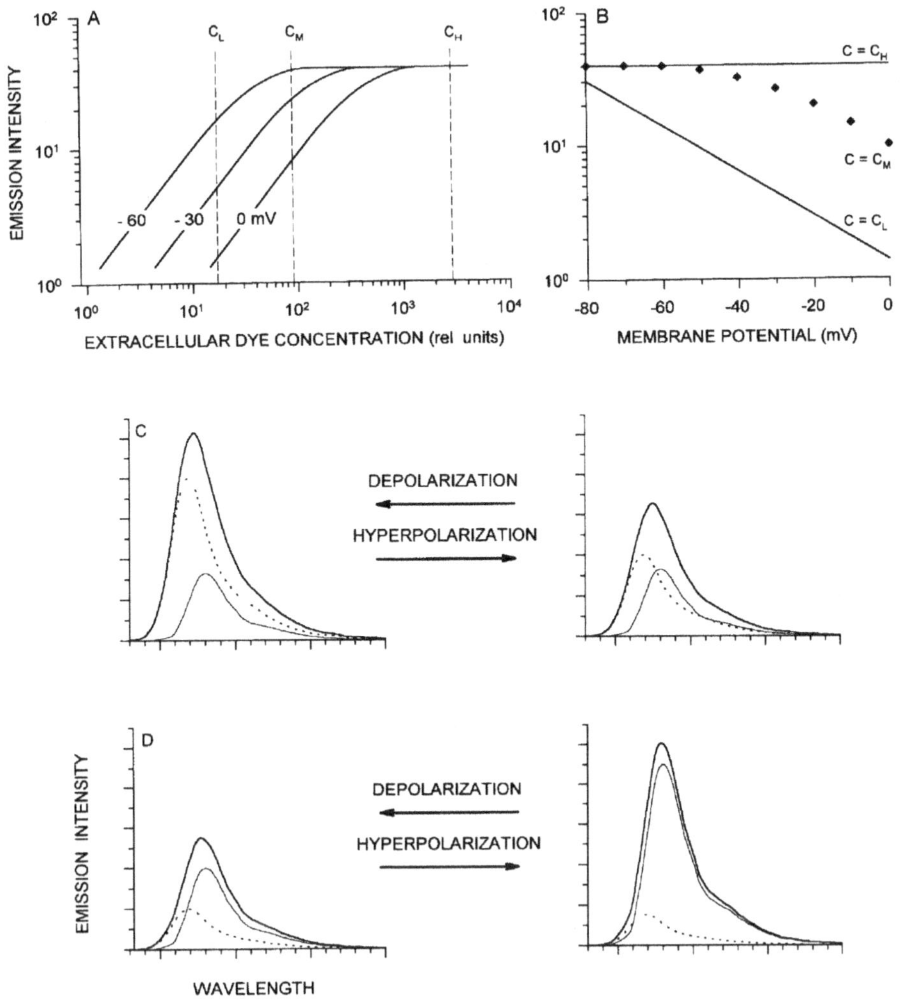

Figure 4. Effect of extracellular dye concentration on the voltage-sensitive response of carbocyanine dyes: (A) Schematic representation of dye binding saturation effects. The extracellular dye concentrations used in this schematic diagram derive roughly from the finding that the fluorescence of bound diS-C_3(3) in murine hemopoietic cells exhibited a pronounced saturation when the extracellular dye concentration increased from 50 nM to 0.5 μM, i.e. by one order of magnitude, see Plášek and Sigler.[5] (B) Effect of extracellular probe concentration on elicited voltage-sensitive responses. The saturation curves similar to those shown in (A) were calculated for the set of indicated membrane potential values. The calculated data were then used to construct schematic voltage-sensitive responses of a probe for its extracellular concentrations equal to c_L, c_M and c_H, respectively. (C) Voltage-sensitive responses of carbocyanine dyes, expected in samples where the extracellular dye concentration is so high that the bound dye fluorescence can reach a saturation level even in depolarized cells. (D) Voltage-sensitive responses of carbocyanine dyes, as expected in samples with low dye and high cell concentration, i.e. when the intracellular bound dye dominates the overall fluorescence spectra. In (C) and (D) thick lines indicate overall fluorescence spectra, dotted and thin lines represent spectral components corresponding to free and bound dye forms, respectively.

CONCLUSIONS

The binding of slow dyes to cytosolic macromolecules and intracellular membranes can considerably contribute to the overall probe signal. This effect should be taken into account when interpreting the results of various membrane potential studies. It offers a plausible explanation for known discrepancies in observed voltage-sensitive responses of such carbocyanine dyes as diO-C_5(3) and diI-C_5(3). For more details on this subject see our forthcoming review Slow Fluorescent Indicators of Membrane Potential: a Survey of Different Approaches to Probe Response Analysis.[4]

REFERENCES

1. A.S. Waggoner, Dye indicators of membrane potential, *Ann. Rev. Biophys. Bioeng.* 8:47 (1979).
2. J.C. Smith, Potential-sensitive molecular probes in membranes of bioenergetic relevance, *Biochim. Biophys. Acta* 1016:1 (1990).
3. L.M. Loew, Potentiometric membrane dyes, in: *Fluorescent and Luminiscent probes for Biological Activity*, W. T. Mason ed., Academic Press, San Diego (1993).
4. J. Plášek and K. Sigler, Slow Fluorescent Indicators of Membrane Potential: a Survey of Different Approaches to Probe Response Analysis, *J. Photochem. Photobiol. B: Biology* (in press)
5. J. Plášek, R.E. Dale, K. Sigler and G. Laskay, Transmembrane potentials in cells: a DiS-C_3(3) assay for relative potentials as an indicator of real changes. *Biochim. Biophys. Acta* 1196:181 (1994).
6. J. Sims, A.S. Waggoner, C.-H. Wang, and J.F. Hoffman, Studies on the mechanism by which cyanine dye measure membrane potential in red blood cells and phosphatidylcholine vesicles, *Biochemistry* 13 3315 (1974).
7. J. Plášek, B. Denksteinová and F. Sureau, Assessment of membrane potential using confocal microspectrofluorimetry, *J. Fluorescence* 3:157 (1993).
8. M. Reers, T.W. Smith and L.B. Chen, J-Aggregate formation of a carbocyanine as a qualitative fluorescent indicator of membrane potential, *Biochemistry* 30:4480 (1991).
9. J. Plášek and V. Hrouda, Assessment of membrane potential changes using the carbocyanine dye diS-C_3(5): synchronous excitation spectroscopy studies, *Eur. Biophys. J.* 19:183 (1991).

EXPLOITATION OF RHODAMINE B IN THE KILLER TOXIN RESEARCH

Valášek L., Vondrejs V., Bartůněk M., Janderová B.

Department of Genetics and Microbiology
Charles University, Faculty of Natural Sciences
Vinicna 5
Prague 2, 128 44
Czech Republic

INTRODUCTION

Killer yeasts secrete proteins called killer toxins which are lethal to certain sensitive yeast strains. The killing ability (K^+) and sensitivity (R^-) to various killers can be routinely screened on the formation of clear zones around heavily streaked killer strains on background lawns of sensitive strains (Woods and Bevan 1968). The research of killer phenomenon and particularly the applications of killer toxins in selection techniques (Vondrejs et al. 1991) call for rapid assays for estimating the killer toxin activity and evaluating the susceptibility of sensitive cells and protoplasts. As the well assay (Woods and Bevan, 1968) and assay based on plating out treated cultures and counting survivals (e.g. Bussey and Sherman, 1973) are time consuming, various fluorescent dyes were examined (Spacek and Vondrejs 1986, Kurzweilova and Siegler 1993) for developing rapid procedures based on staining the killed cells or protoplasts.

In this communication several modified versions of a Rhodamine B assay (Spacek and Vondrejs, 1986) are described, and their exploitation is demonstrated.

MATHERIALS AND METHODS

Yeast strains in this work were obtained from dr. Janderova or dr. Palkova (Charles University, Prague). Rhodamine B was a Sigma product No R1879. Culture supernatant fluids K1J containing the killer toxin K1 produced by Saccharomyces cerevisiae X3 or GOSMH and all cultivation media were prepared according to Vondrejs et al. (1983).

Rhodamine B assay I (RBAI): Stationary phase cultures of selected strains of yeasts were diluted with YEPG medium to the final O.D (540 nm) = 0.15 and cultivated aerobically at 28°C for 2 hrs. Aliquots containing 10^7 cells were centrifuged, pelets were resuspended each in 1 ml K1J, properly diluted with medium J (pH 4.7) and 0.1 ml of 2M glucose was added. After 2 hrs at 22°C 0.1 ml 0.002 M Rhodamine B was added to each sample. After 1 hr incubation the samples were centrifuged and washed with distilled water. Fractions of stained cells were determined using fluorescence microscopy (Spacek and Vondrejs, 1986).

Rhodamine B assay II (RBAII): Aliquots of exponentially growing culture in YEG medium at 28°C containing 10^6 cells were harvested by centrifugation. Pelets were resuspended each in 1ml properly diluted K1J, and 0.1 ml 2M glucose and 0.1 ml 0.002M Rhodamine B was added. After 3 hrs incubation at 22°C the samples were centrifuged, washed with distilled water and resuspended each in 0.1 ml water for determining the fractions of stained cells.

Rhodamine B assay III (RBAIII): Aliquots containing 10^7 protoplasts which were prepared according to Vondrejs et al. 1983 were centrifuged and pelets were resuspended in 1 ml properly diluted osmotically stabilized K1J (with 1M sorbitol). 0.1 ml 2M glucose was added. After 2 hrs incubation at 22°C each sample was divided to two parts. 0.5 ml was used for determining the frequency of protoplast reversion. 0.1 ml 0.001M Rhodamine B was added to the second part. After 1hr incubation at 22°C stained samples were centrifuged, pelets were washed with 1M, and resuspended each in 0.1 ml 1M sorbitol for determining the fractions of stained protoplasts.

Protoplast reversion assay (PRA): Aliquots of protoplast samples treated with K1 toxin (see RBAIII) were properly diluted, resuspended each in 5 ml of osmotically stabilized agar with 1M sorbitol, and poured onto plates containing osmotically stabilized medium J, pH 4.7. The frequency of protoplast reversion was determined by counting colonies after several days of incubation at room temperature.

RESULTS AND DISCUSSION

Estimating the K1 toxin activity in K1J by RBAI or RBAII was based on determining the fraction of stained cells F of supersensitive strain Saccharomyces cerevisiae S6/1 as a function of volume V of K1J added to individual samples. The volume V(0.5) of K1J which causes staining 50% of cells under standard conditions of the assay can be easily determined (Fig. 1) to evaluate approximately the K1 toxin activity in relative units.

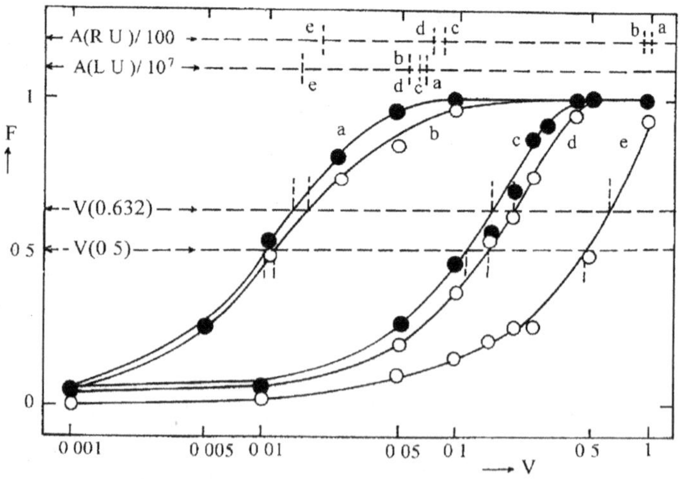

Figure 1.

$$A[R.U.]=1/V(0.5) \quad (1)$$

It is not surprising, that R.U. for RBAI (R.U.I) and RBAII (R.U.II) are different, because more toxin molecules are required for killing 10times more molecules used in RBAI. The difference in the average susceptibility of conditioned stationary phase cells used in RBAI and exponential phase cells used in RBAII causes very likely an additional increase in sensitivity of RBAII in comparison with RBAI. It was shown that R.U.I/R.U.II is about 12 for both killer toxins used in this work, and the activity of the K1 toxin in the GOSMH supernatant fluid is about 1.2 fold higher than the activity in the X3 supernatant fluid. The main advantage of RBAII is that it can be exploited even for estimating the activity in supernatant fluids with relatively low K1 toxin concentrations, and also for evaluating the susceptibility of less sensitive strains. It is likely, that the sensitivity of RBA could be further increased by decreasing the number of cells in individual samples, however, the exploitation of less concentrated cell suspensions is not practical when F is determined by counting the cells using fluorescence microscopy. On the other hand, the exploitation of stationary phase cells in RBAI makes this assay more convenient for estimating high activities resp. susceptibilities, because the stationary cultures can be more easily prepared by overnight cultivations than the exponential ones.

More sophisticated evaluating the K1 toxin activity in lethal units (L.U.) was suggested by Bussey and Sherman (1973) and Kurzweilova and Siegler (1993). L.U. represents the average amount of toxin needed to kill a single cell. At multiplicity $m=\ln 1/(1-F)=1$, where $F=0.633$, the amount of toxin in L.U. is equal to the total number of cells in the sample N. Activity A in L.U./ml can be calculated using the equation (2).

$$A[L.U]=N/V(0.632) \quad (2)$$

When the activities A[R.U.] resp A[L.U] in GOSMH supernatant fluids were related to the activities in X3 supernatant fluids, almost the same values were obtained independently on the variant of RBA(I,II) used for determining "the dose response curves".

It is obvious that the assays described above can also be employed for evaluating the relative susceptibility (S) to select toxin of different strainsccc (Y) in relation to the standard prototrophic supersensitive strain S. cerevisiae S6/1 (M).

$$S =V(0.5,M) / V(0.5,Y) \quad (3)$$

$$S' =V(0.632,M) / V (0.632, Y) \quad (4)$$

The dose response curves obtained on the basis of RBAIII or PRA (Fig. 2) can be exploited for evaluating the susceptibility of protoplasts using the equation 3 or 4. It should be stressed that PRA gives primary data on survived revertants and reverted protoplasts represent very small fractions of total protoplast populations even in untreated samples. This fact may explain why the S values obtained on the basis of PRA and RBAIII sometimes remarcably differ (Fig 2a,b), while the dose response curves and their characteristics obtained by plating cells and counting survivals are very similar to those obtained by RBAI resp.RBAII (data not shown). Two sensitive strains S. cerevisiae S6/1 and P95 and four resistant strains S. diastaticus CC164, S. kluyveri SPK15, S. oviformis SPK19 and Candida utilis C7 were selected for evaluating the susceptibility of their protoplasts. Three of the resistant strains, C. utilis C7, S. kluyveri SPK15 and S. oviformis SPK19 were shown to be resistant even as protoplasts using both PRA and RBAIII. Susceptibility of S. diastaticus CC164 to the K1 toxin treatment, however, was very similar to the sensitive strain P95. In both cases the reverting protoplasts were more sensitive than

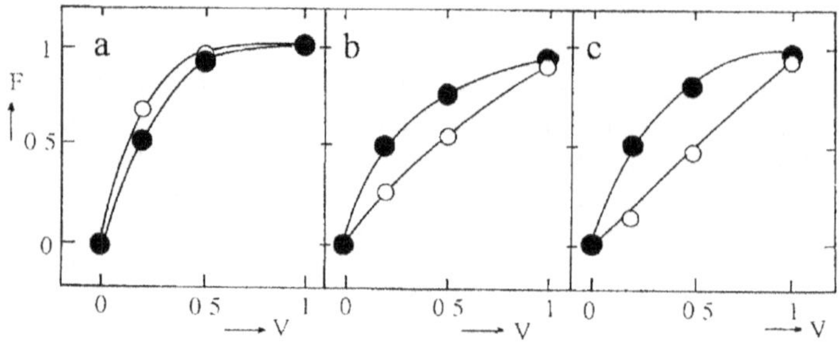

Figure 2.

the total population of protoplasts assayed by RBAIII (Fig. 2b,c). While in case of S6/1 the difference among the dose response curves obtained using RBAI, RBAIII and PRA was very small (Fig. 1 and 2a), in case P95 the susceptibility of intact cells (RBAI) was remarcably lower than the susceptibility of total protoplast population (RBAIII) and particularly than the susceptibility of the reverting protoplasts (PRA) (Fig 1, 2b).

The observations mentioned above clearly show that the cell wall is playing an important role in setting up the susceptibility of sensitive yeasts to the K1 toxin treatment, and in some cases it may even protect the sensitive protoplast against attack of killer toxins. Several explanations of the protective role of the cell wall can be offered at present. All of them are based on the fact that they are the specific receptors in the cell wall of sensitive yeasts (Bussey et al., 1979), which are required for the effective transfer of K1 toxin molecules through the cell wall to hypothetic receptors in the cytoplasmatic membrane. In general, when the number of the specific receptors in the cell wall is decreased or if the receptors are occupied e.g. by other killer toxin produced by the sensitive strain the susceptibility of intact cells of this strain will be decreased but the susceptibilty of protoplasts will stay unchanged. The second explanation fits only when the killer toxin produced by the sensitive strain has different membrane receptors. Also the increased number of nonspecific binding sites in cell walls may cause some decrease in the susceptibility of sensitive strains. The results of experiments with Candida utilis C7 did not confirm the observations of Zhu and Bussey (1989) that spheroplasts of some strains of Candida sp. are sensitive to the K1 toxin treatment. It should be noted that different strains of Candida sp. and techniques for evaluating the susceptibility of protoplasts were used by these authors.

ACKNOWLEDGEMENTS

This work was supported by the grant 204/0053 from the Grant agency of Czech Republic.

LITERATURE

Bussey, H., Saville, D., Hutchins, K., and Palfree, R.G.E., 1979, Binding of yeast killer toxin to a cell wall receptor on sensitive Saccharomyces cerevisiae, *J. Bact.* **140**, 888-892.

Bussey, H., and Sherman, D., 1973, Yeast killer factor ATP leakage and coordinate inhibition of macromolecular synthesis in sensitive cells, *Biochem. Biophys. Acta* **298**, 868-887.

Kurzweilova, H., and Siegler, K., 1993, Factors affecting the susceptibility of sensitive yeast cells to killer toxin K1, *Folia Microbiolica* **38**, 424-526.

Spacek, R., and Vondrejs, V., 1983, Rapid method for estimation of killer toxin activity in yeasts, *Biotechnol. Lett.* **8**, 701-706.

Vondrejs, V., Palkova, Z., and Sulo, P., 1991, Application of killer system in yeasts in selection techniques, *Biotechnol. Curr. Prog.* **1**, 227-247.

Vondrejs, V., Psenicka, I., Kupcova, L., Dostalova, R., Janderova, B., and Bendova, O., 1983, The use of killer factor in the selection of hybrid yeast strains, *Folia Biologica* **29**, 372-384.

Woods, D.R., and Bevan, E.A., 1968, Studies on the nature of the killer factor produced by Saccromyces cerevisiae, *J. Gen. Microbiol.* **51**, 115-126.

Zhu, H., and Bussey, H., 1989, The K1 toxin of Saccromyces cerevisiae kills spheroplasts of many yeast species, *Appl. Enviro. Microbiol.* **55**, 2105-2107.

"IN SITU" ESTIMATES OF THE SPATIAL RESOLUTION FOR "PRACTICAL" FLUORESCENCE MICROSCOPY OF CELL NUCLEI

Bernd Rinke[1,2], Joachim Bradl[1], Bernhard Schneider[1], Markus Durm[1,4], Michael Hausmann[1], Horst Ludwig[4], and Christoph Cremer[1,3]

[1] Institute of Applied Physics, University of Heidelberg, Albert-Ueberle-Str. 3-5, D-69120 Heidelberg, Germany
[2] Graduate College "Tumor diagnosis and therapy based on 3D radiologic and laser methods"
[3] Interdisciplinary Centre for Scientific Computing (IWR), University of Heidelberg
[4] Institute of Physical Chemistry, University of Heidelberg

ABSTRACT

Axial and lateral responses obtained from ideal point objects through a fluorescence light microscope can be used to calculate the spatial resolution of the system from the point spread function. In practice, however, the experimental conditions given by a biological object can have an additional, considerable influence on the final resolution. Therefore, it is important to understand how parts of the setup contribute to the optimal response, e. g. cover glasses of different thickness, immersion and mounting media, or optical inhomogeneities of the biological, not point like objects themselves. Here, imaging properties of a confocal laser scanning fluorescence microscope are studied in situ in female lymphocyte cell nuclei. They were stained with a red fluorochrome (propidium iodide). Inside the nuclei, the centromeric regions of the two X chromosomes were specifically labelled by a green fluorochrome (FITC). Lateral and axial responses through the object and the labelling site were investigated. The increase or decrease (15%-85%) of the fluorescence intensities were used as an estimate for the spatial resolution of the system. This estimate was two times larger in the axial direction than in the lateral direction. The results suggest that such measurements can also be used in fluorescent biological objects as an internal standard to estimate the quality of resolution in "practical" quantitative fluorescence microscopy.

INTRODUCTION

Fluorescence light microscopy in combination with DNA-DNA in situ hybridization techniques [Lichter and Cremer, 1992] is the most often applied tool in cytogenetics to visualize chromosome territories in cell nuclei. Especially confocal laser scanning microscopy (CLSM) [Shotton, 1989] with the possibility to obtain optical sections of the investigated specimen without distortion by cutting them physically offers possibilities to quantify the overall threedimensional organization and internal structures of the cell nucleus. To investigate the correlation between structure and function of chromosome territories in cell nuclei, i.e. the functional compartmentalization of chromatin, the requirements for measurements in the submicron range increase [Krystosek and Puck, 1990; Cremer et al., 1993; Zirbel et al., 1993; Janevski et al., 1995; Kamei, 1995].

The most important feature of a CLSM [Wilson and Sheppard, 1984; Stelzer et al., 1986; Pawley, 1995] is its ability to obtain a defined depth discrimination by confinement both, the illuminated focal region and the detected area of the emitted light. Normally this is done by means of a pinhole in front of the light source and a high numerical aperture objective resulting in a diffraction limited light spot in the focal plane. Additionally, a pinhole in front of the detector is confocally arranged. This has the advantage to exclude out of focus contributions to the image leading to a "nice sharp" picture. However, for physical reasons the lateral resolution is better than the spatial resolution in the direction of the optical axis [Sheppard, 1988; Nakamura, 1993] Moreover, the spatial resolution can be considerably influenced by the practical conditions of biological measurements. Therefore, it is important to know the imaging properties of the CLSM under the real practical conditions in a particular experiment. This means: how do biological buffer systems, on the shelf cover glasses, or the specimens themselves influence the resolution. As a measure of the resolution of a CLSM, usually the full width half maximum (FWHM) of the point spread function (PSF) of the complete imaging system is used. For the axial resolution of dry objectives this measure is compatible with the gradient of a reflection signal of short wavelength light reflected by a high reflective mirror. For the PSF of immersion objectives, fluorescent beads of subwavelength size are applied to measure the FWHM under highly defined experimental conditions. In practice, these definitions are far away from many routine conditions in cytogenetics. For instance, since an objective is constructed and corrected for a given thickness of cover glasses and a special immersion fluid, care has to be taken in preparation of the microscope ready slide. Furthermore, using embedding or mounting media of different refractive indices may result in optical distortions. The focal shift increases with the refractive index mismatch and the depth of the investigated region [Carlsson, 1991; Visser et.al., 1992; Hell et.al., 1993; Bradl et.al., 1994; Bradl et.al., 1995; Jacobsen and Hell, 1995; Wilson and Juskaitis, 1995]. Here, we present a direct method to estimate the resolution from the registered cell nuclei themselves.

MATERIALS AND METHODS

Confocal images were recorded with a Leica TCS 4D with a 100x / 0.7-1.4 NA PL APO objective. The distances between consecutive optical sections was 100 nm, the lateral pixel size was 100 nm in an image frame of 256x256 pixels. With this objective, theoretically, an optimum lateral resolution of about 200 nm and an axial resolution of about 400 nm can be expected for an excitation wavelength of 488 nm (Ar^+-laser).

Cell nuclei of female human lymphocytes were hybridized in situ with a X-chromosome α-satellite probe according to the Fast-FISH protocol [Celeda et.al., 1994; Durm et.al., 1995]. The labelling sites were stained with fluorescein-isothiocyanate (FITC) and the complete cell nuclei were counterstained with propidium iodide (PI) (0.2 μg / ml). After air-drying in a prewarmed dry chamber (40 °C), the slides were mounted with Vectashield and covered with a standard cover glass of 160 μm thickness.

For microscopy, an immersion oil of n=1.518 was used. Choosing appropriate filter combinations, the PI and FITC images were acquired simultaneously in two channels without crosstalk. The images of 10 female cell nuclei with detectable FITC labelled X-centromeric regions were recorded. Vertical XZ-sections and horizontal XY-Sections were taken through the centre of each of the X-centromeric regions.

IMAGE ANALYSIS AND RESULTS

In figure 1, two optical sections (XZ and XY) through a cell nucleus and one hybridization site are shown. Both images represent combined grey value images from the FITC and PI detection channels. For the quantitative analysis, the grey value distributions of each channel were examined. First the maximum grey value (=maximum of the fluorescence intensity) in the FITC-image was localized, and eight parallel neighbouring horizontal X- and vertical Z-lines were extracted. This was also done for the PI-image representing the surrounding nuclear area at the same coordinates. Each set of eight lines was averaged. From these data, a grey value distribution normalized to the maximum intensity was calculated (figure 2).

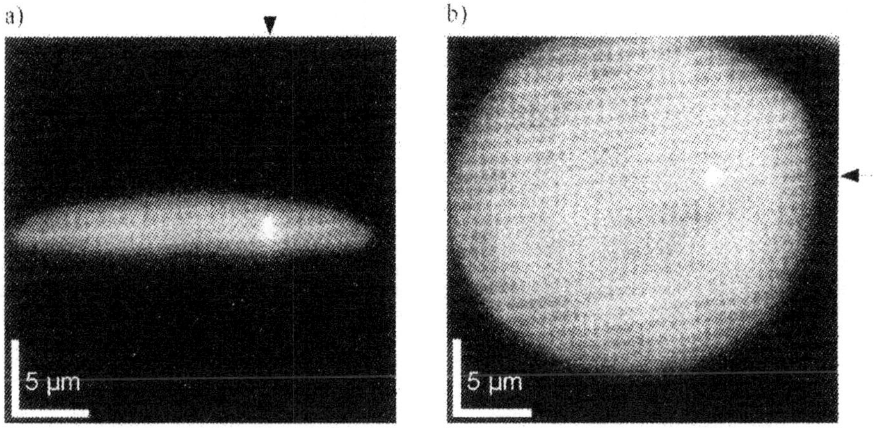

Figure 1. Confocal XZ-section a) and XY-section b) through a PI-counterstained female lymphocyte nucleus with FITC-labelled X-centromeric regions. Overlay of grey value images of the two separately registered fluorescence channels. The second X-centromeric region was positioned in a different optical layer and therefore was not visible in these sections. The black arrows indicate the image lines of the resulting grey value distributions in figures 2a,2b.

Under the assumption that the hybridization site as well as the cell nucleus are large compared to subwavelength objects ("point-like" objects), the surface of both fluorescent objects were regarded to be formed of "infinite" plane horizontal and vertical fluorescent layers. In this case, the distances (Δ) between the 15% and 85% intensity values can be regarded as a measure compatible to the full width half maximum of the PSF. Since cell nuclei and hybridization sites are "thick" fluorescent layers, the two borderlines (left/right or top/bottom, respectively) have to be evaluated independently. In the example of figure 2a (vertical section), the PI-fluorescence intensity distribution resulted in a Δ_{PI}-value of 1200 nm at the top (=left profile increase) (minimum distance of the PI-stained nuclear surface to the microscope lens in this section); on the bottom (contact of the cell nucleus with the glass slide), the Δ_{PI}-value was 1100 nm (=right profile decrease). For the FITC-image, Δ_{FITC} was 800 nm at the top and 900 nm at the bottom. As figure 2b shows, the lateral FITC distribution was sharper. Averaging the left and right increase resulted in 540 nm. The lateral increases of the PI images (=cell nucleus) were not evaluated because the assumptions were not fulfilled for preparative reasons.

In table 1 the results obtained from 10 nuclei with 19 registered labelling sites are summarized. For the axial PI-fluorescence increases and decreases, the Δ_{PI}-values were calculated after fitting (least square) the experimental grey value distributions by a polynom of 7th order. For the axial sections, $\Delta_{increase}$ gives the Δ-values for the top (i.e. minimum distance to the microscope lens) whereas $\Delta_{decrease}$ gives the Δ-values for the bottom. For the lateral sections, the left and right increases for the FITC signals are shown. To estimate the mean resolution of the system for the different cell nuclei, the complete data sets were averaged. For the FITC signals, $\Delta_{lateral}$ was 432 nm and Δ_{axial} was 814 nm and 864 nm, respectively. Although both values were worse than the FWHM of the ideal PSF, the principle difference between lateral and axial FWHM was maintained. For the PI images, the mean value $\Delta_{increase,mean(top)}$ was 943 ± 200 nm, and the mean value $\Delta_{decrease,mean(bottom)}$ was 1095 ± 255 nm. These values may be regarded as an upper experimental estimate for the axial PSF's.

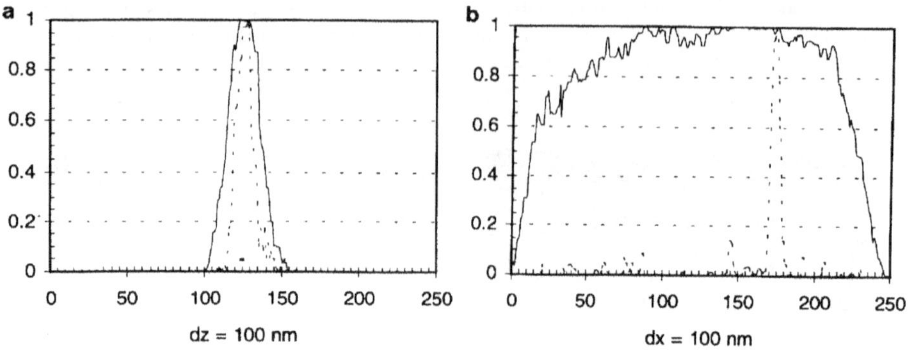

Figure 2. Grey value distributions of the two detection channels independently normalized to the maximum of its fluorescence signal. Solid line represents the PI-signal and the dotted line the FITC-signal. Ordinate: normalized intensity, Abcissa: 1 unit = 100 nm. Distributions were obtained from a) XZ-section and b) XY-section indicated by arrows in figure 1.

signal no.	$\Delta_{PI,increase}$ axial	$\Delta_{PI,decrease}$ axial	$\Delta_{FITC,increase}$ axial	$\Delta_{FITC,decrease}$ axial	$\Delta_{FITC,left}$ lateral	$\Delta_{FITC,right}$ lateral
1	1114	1574	734	692	471	459
2	997	1497	999	681	355	264
3	1124	1707	--	--	--	--
4	1197	1196	--	--	--	--
5	812	1253	1153	1117	350	268
6	1099	884	1001	887	576	458
7	691	663	752	1098	330	508
8	943	1020	983	732	435	723
9	657	1040	864	774	401	348
10	732	999	912	1082	584	222
11	759	992	683	718	549	319
12	776	1015	801	752	325	288
13	954	980	724	569	378	260
14	1350	873	806	664	315	462
15	785	1040	965	674	317	434
16	1216	1108	807	892	457	618
17	1001	1069	1045	860	332	725
18	944	926	681	678	469	479
19	773	970	773	970	594	619
stand.dev.	200.0	255.7	140.1	169.0	132.8	
mean	943.4	1095.1	863.7	814.1	432.1	

Table 1. Δ-values 1-19 for the 10 individual cell nuclei after evaluation of the PI and FITC images (-- = values excluded from evaluation due to high signal noise).

DISCUSSION

The influence of specimen and mounting media on the spatial resolution of a CLSM was investigated for "practical" fluorescence light microscopy. Cell nuclei and hybridization spots were considered to be "thick" fluorescent layers, and the 15% to 85% increase in the grey value intensity was taken as a an upper estimate of the FWHM of the PSF. In practice, the resolution, i.e. the minimum detectable distance between two given biological objects is not determined by the FWHM of an ideal PSF but by the FWHM of a measured PSF under the real experimental conditions used. Therefore, in practical applications it may be useful to estimate the resolution directly from the object images. The applicability of the 15%-85% signal increase as an useful estimate was shown.

The principle effect of different lateral and axial resolution (factor of 2 to 3) was also found in the data obtained inside the cell nuclei (FITC data). Comparing the data obtained inside the cell nuclei (FITC data) and at the borderline inside the mounting medium (PI data), the data suggest a medium dependent difference (about 850 nm to about 1000 nm). These data, however, coincided within one standard deviation. Due to the sample size of 10 nuclei, this difference could not be confirmed with statistical significance. The approach presented here provides an easy method for online control of the system resolution in the cellular objects used.

ACKNOWLEDGMENTS

B.Rinke receives a scholarship from the Deutsche Forschungsgemeinschaft (DFG) as a member of a DFG graduate college. We thank Dr. T. Cremer, Institute of Human Genetics, Heidelberg, for stimulating discussions and the access to the Leica TCS 4D. The financial support of the Deutsche Forschungsgemeinschaft is gratefully acknowledged.

REFERENCES

Bradl, J., Hausmann, M., Schneider, B., Rinke, B , Cremer, C., 1994, A versatile 2π-tilting device for fluorescence microscopes, *J. Microsc.* 176:211-221.

Bradl, J., Rinke, B., Schneider, B., Hausmann, M , Cremer, C., 1995, Improved resolution in "practical" light microscopy by means of a glass fibre 2π-tilting device, *Proceedings of BIOS 95*, submitted

Carlsson ,K., 1991, The influence of specimen refractive index, detector signal integration, and non-uniform scan speed on the imaging properties in confocal microscopy, *J. Microsc.* 163:167-178

Celeda, D , Aldinger, K., Haar, F.-M., Hausmann, M , Durm, M., Ludwig, H , and Cremer, C., 1994, Rapid Fluorescence In Situ Hybridization With Repetitive DNA Probes: Quantification by Digital Image Analysis, *Cytometry* 17:13-25.

Cremer, T., Kurz, A., Zirbel, R., Dietzel, S , Rinke, B., Schrock, E., Speicher, M R , Mathieu, U., Jauch, A , Emmerich, P., Scherthan, H., Ried, T., Cremer, C., Lichter, P , 1993, Role of Chromosome Territories in the Functional Compartmentalization of the Cell Nucleus, *Cold Spring Harbor Symp on Quant Biology* 58:777-792

Durm, M., Haar, F -M., Hausmann, M , Ludwig, H , Cremer, C , 1995, Quantitative Studies for Optimization of Fast-FISH with Repetitive α-Satellite Probes, *Biotechniques*, submitted.

Hell, S., Reiner, G., Cremer, C., Stelzer, E.H.K., 1993, Aberrations in confocal fluorescence microscopy induced by mismatches in refractive index, *J Microsc.* 169:391-405.

Jacobsen, H., and Hell, S.W., 1995, Effect of the specimen refractive index on the imaging of a confocal fluorescence microscope employing high aperture oil immersion lenses, *Bioimaging* 3:39-47.

Janevski, J., Park, P.C., and de Boni, U , 1995, Organization of Centromeric Domains in Hepatocyte Nuclei: Rearrangement Associated with De Novo Activation of the Vittellogenin Gene Family in Xenopus laevis, *Exp. Cell Res.* 217.227-239.

Kamei, H., 1995 A Nuclear Dot-like Structure That Has a Relationship with Perinuclear Intermediate Filaments, *Exp. Cell Res.* 218:155-165.

Krystosek, A., and Puck, T.T , 1990, The spatial distribution of exposed nuclear DNA in normal, cancer, and reverse-transformed cells, *Proc. Natl. Acad Sci. USA* 87.6560-6564.

Lichter, P , and Cremer, T., 1992, Chromosome analysis by non-isotopic in situ hybridization, *in.* "Human Cytogenetics - a practical approach Vol. I. ", pp 157-192, D.E. Rooney, and B.H Czepulkowski, eds., IRL Press, Oxford.

Nakamura, O., 1993, Three-dimensional imaging characteristics of laser scan fluorescence microscopy: Two-photon excitation vs. single-photon excitation, *Optik* 93:39-42.

Pawley, J B , 1995, "Handbook of Biological Confocal Microscopy", 2nd edition, Plenum Press, New York

Sheppard, C.J R , 1988, Depth of field in optical microscopy, *J Microsc* 149:73-75.

Shotton, D M., 1989, Confocal scanning optical microscopy and its applications for biological specimens, *J. Cell Science* 94:175-206.

Stelzer, E H.K , Marsman, H.J.B., Wijnaendts van Resandt, R.W , 1986, A setup for a confocal scanning laser interference microscope, *Optik* 73:30-33.

Visser, T. D., Oud, J L , Brakenhoff, G J , 1992, Refractive index and axial distance measurements in 3-D microscopy, *Optik* 90:17-19.

Wilson, T., and Sheppard, C.J.R., 1984, "Theory and Practice of Scanning Optical Microscopy", Acad. Press., London.

Wilson, T, and Juskaitis, R., 1995, The axial response of confocal microscopes with high numerical aperture objective lenses, *Bioimaging* 3·35-38.

Zirbel, R.M., Mathieu, U R., Kurz, A., Cremer, T , and Lichter, P., 1993, Evidence for a nuclear compartment of transcription and splicing located at chromosome domain boundaries, *Chromosome Res.* 1:93-106

REQUIREMENTS FOR A COMPUTER-BASED SYSTEM FOR FISH APPLICATIONS

Wolf Malkusch

Image Analysis Division
KONTRON ELEKTRONIK GmbH
D-85385 Eching, FRG

INTRODUCTION

In situ hybridization techniques allow the detection of specific nucleic acid sequences in chromosomes, cells, or tissue sections (Cremer et al., 1995). The specimen preparation is followed by a denaturation of the DNA, the hybridization with the desired probes, and the labelling with the fluorochromes (Brecevic et al., 1994). Mostly three, but sometimes up to seven differently marked DNA probes are used in fluorescent in situ hybridization experiments (Ried et al., 1992). As a counter staining either red fluorescent propidium iodide is used, or blue fluorescent DAPI (Polak and McGee, 1990).

Using the methods of fluorescent microscopy the different signals can be collected separately from each other, when suitable filters are used. Clear signal improvement can be achieved using electronic image processing systems. They offer numerous correction and enhancement possibilities for each single step, starting from the image acquisition down to the final result image. Furthermore they offer all possibilities of image archiving for the original input images as well as for the result images or all intermediate steps.

Using the KS 400 image processing system as an example, it will be shown which processing steps a modern system should offer, when used in FISH applications.

FLUORESCENCE MICROSCOPY

Today three methods are used for the production of FISH images. In the first case all different filters for all probes are used in one single set. With these double or triple band filter sets certainly the biggest compromises are necessary regarding the separation of the signals. Clearly better results are obtained, when the different exciting filters (FITC, TRITC, Rhodamine, Texas-red, DAPI) are mounted in different positions of a filter slider. But in this case the handling is more tedious. Following each image capture the filter change has to be performed manually. As a consequence of the different optical filter properties inside the illumination path there is a resulting geometrical shift between the intermediate images that has to be corrected electronically. Nevertheless is the big advantage of this possibility to sample the different signals separated from each other in an optimum way using integrating cameras.

With the third method the exciting filters are mounted outside the imaging path in an automatic filter wheel. Now there is no necessity to move the filter slider with the beam splitter any more, resulting in no shift between the intermediate images. An automatic control of the filter wheel is simplifying the whole handling during the image acquisition. The use of an integrating camera furthermore solves the intensity problems occurring with very smooth signals.

Having this in mind, an electronic image processing system that is to be used in this field should include a number of essential functions. First of all there have to be enough free control channels for the supervision of the microscope and camera. The integration time of the camera has to be adjustable in a variable way, the filter wheel has to be controlled, and the shutter has to be opened and closed to keep damage away from the specimen.

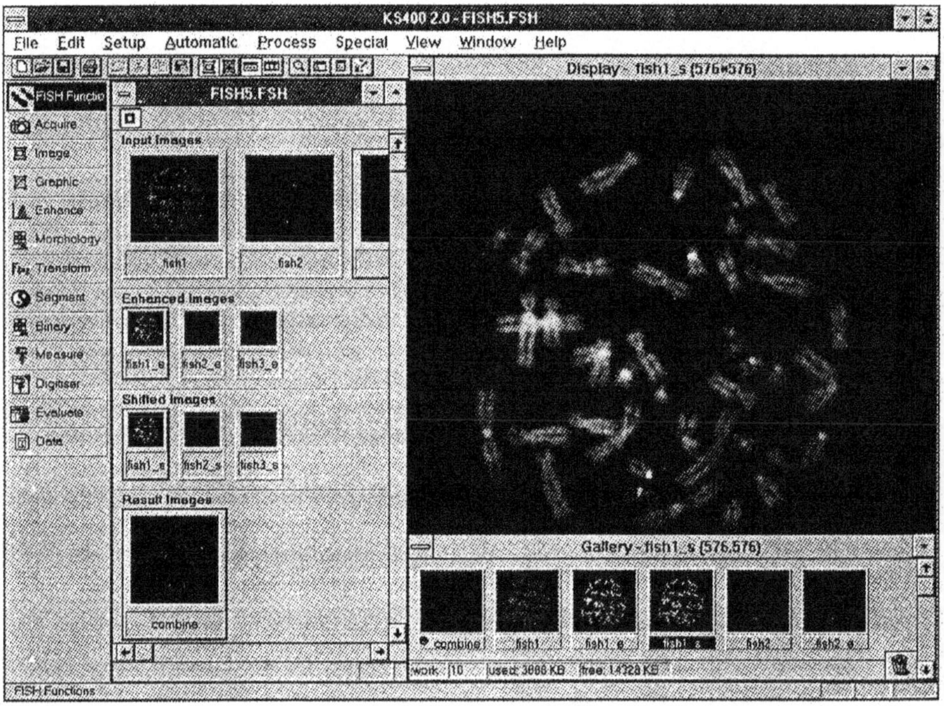

Figure 1. Special FISH software (user interface) embedded in the generic KS 400 image processing system, enabling a further evaluation of the resulting FISH images.

SEPARATE SOFTWARE FOR FISH APPLICATIONS

As Windows-based systems are nowadays widely spread and well-known, a Windows-based special FISH imaging software is especially to be recommended. There is a series of generic image processing systems available, all offering more or less functions suitable for the processing of FISH images. But for the routine work this is very difficult and time consuming. This is the reason why a specific software is necessary, including on the bases of a general image processing system all essential functions combined in an user interface that is easy to understand and to handle. Furthermore all functions have to be adaptable to the users needs.

The requirements on an imaging system for FISH applications is firstly split in the image acquisition of all signals, the separate processing of each intermediate image, an eventually necessary position correction of the single images, and finally the combination to a resulting colour image. A good user interface should keep this path always visible for the user. As an example the KS FISH software should be presented here. The user interface is shown in figure 1.

Each intermediate step is presented in this software with its results as a reduced image. The combined result of all functions is always and immediately visible in the large image window. Behind each reduced image of all single processing steps a menu is hidden with all control variables for this function. The menu can be selected by simply clicking into the image with the computer mouse.

If desired, the complete user interface can be integrated in the general purpose KS image processing software, as shown in figure 1. In this case all functions of a modern image analysis are available for further processing and measurements.

COMPOSITION OF FISH IMAGES

Some signals in FISH applications may be extremely weak. In these cases it is worth to use a cooled integrating black-and-white camera. During the image processing phase the specimen should be protected by the opening or closing of a shutter. The exposition times and gain values have to be adjusted and tested in a very simple way. As an example for an easy to understand adjustment window the dialogue box for the camera live adjustment is shown in figure 2.

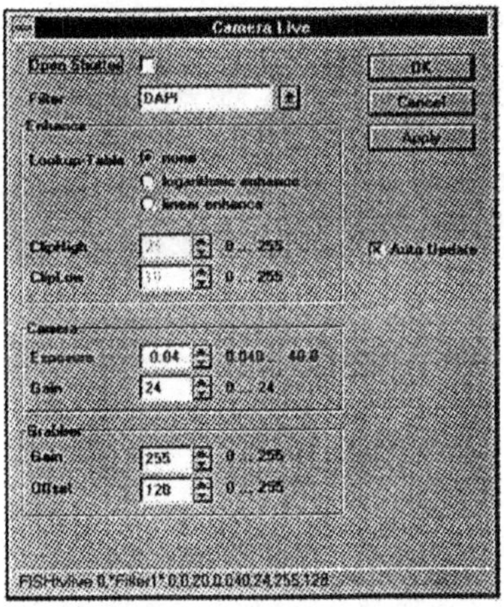

Figure 2. Camera Live Dialogue Box.

In this dialogue box there is furthermore the possibility to use different lookup tables together with the camera live image (logarithmic or linear). As a result even very low intensities will produce a visible live image. This is especially of advantage when looking for suitable positions in the specimen on the computer screen.

The enhance step should offer different possibilities of contrast manipulation (manual and automatic). Functions for electronic noise reduction (median filter, sigma filter) using variable pixel matrices are also essential. Simply usable correction algorithms are necessary to eliminate the image shift, especially when using a manual filter slider. This can be achieved either by a simple value input for the X- and Y-amount, or better by clicking to the correction position on a pixel matrix. Finally all produced intermediate images have to be combined to one resulting colour image. Very good results can be achieved using the maximum method. Sometimes it is of advantage when the available intermediate images can be combined to different resulting colour images at the same time. In this way different signals can be presented more clearly.

An optimum system for use in FISH applications should certainly include simple ways for archiving all input and result images. Additionally all standard output devices like video printers, laser printers, and slide machines should be supported.

As a result of such a system figure 3 shows a metaphase composed with the KS FISH software. For this paper an example with only one marker was chosen, as this can also be presented in black-and-white.

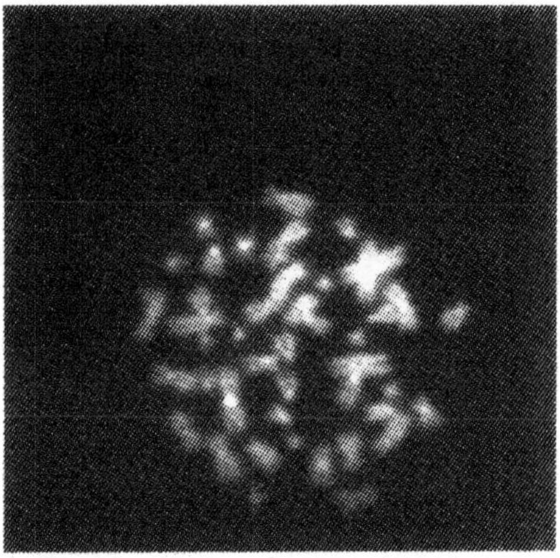

Figure 3. DNA probe detected with FITC marked antibodies; karyotype 45,X/46,X,r(x); counter stain propidium iodide.

REFERENCES

Brecevic, L., Binhert, F., Schinzel, A., 1994, Fluoreszierende in-situ Hybridisierung (FISH) - Ein Bericht vom Institut für Medizinische Genetik der Universität Zürich, Zeiss Aktuell 42:2.

Cremer, T., Jauch, A., Ried, T., Schröck, E., Lengauer, C., Cremer, M., Speicher, M., 1995, Fluoreszenz-in-situ-Hybridisierung (FISH), Dt. Ärztebl. 92(22):31.

Polak, J.M., McGee, J.O., 1990, "In-Situ-Hybridization: Principle and Practice," Oxford University Press, Oxford.

Ried, T., Baldini, A., Rand, T.C., Ward, D.C., 1992, Simultaneous visualization of seven different DNA probes by in situ hybridization using combinatorial fluorescence and digital image microscopy, Proc Natl Acad Sci USA 89:1388.

FLUORESCENT DYES AND DYE LABELLED PROBES FOR DETECTION OF NUCLEIC ACID SEQUENCES IN BIOLOGICAL MATERIAL

Ian Durrant[1] and Leslie Gubba[2]

[1]Research and Development
Amersham International
Amersham, UK
[2]Amersham Life Science Inc.
Pittsburgh, USA

INTRODUCTION

The traditional method for the detection of nucleic acid sequences on membranes and *in situ* utilises radiolabelled compounds. These systems are still widely used in many applications, particularly where sensitivity and robustness are required[1]. However, in many situations there is less need for sensitivity and a more overriding need for resolution and speed of detection. To this end, effort has gone into the production and design of good, reliable, non-radioactive alternatives. Originally, these systems utilised hapten molecules, attached to nucleotides, to label the probes that were used to detect the presence of specific nucleic acid sequences. These hapten molecules were then detected through antibody conjugates. These too are now widely used in a number of applications[2]. In the field of chromosome *in situ* hybridization, detection through fluorescence labelled antibodies found great favour due to the ease of detection with modern microscopes and camera systems and, particularly with chromosome analysis, due to the low level of associated background fluorescence.

As a result a number of fluorescent dyes have become established in molecular biology applications; notably fluorescein and rhodamine derivatives have been used in fluorescent *in situ* hybridization (FISH) and primer extension *in situ* reactions (PRINS)[3]. These dyes were initially used as labels of antibodies which, in turn, were specific for hapten labels attached to nucleotides incorporated into the probes used in these systems. Recently, we have utilised suitable derivatives of these dyes to attach directly to nucleotides so that the antibody incubations are no longer required. This has eased the ability to perform multiple hybridizations on the same slide[4] and has enabled initial investigations into quantification of results to be performed.

Following the availability of these dyes it became evident that there was a need for more colours that could be used in combination. These new dyes would need to have discrete, and therefore narrow, emission spectra and be brighter than those currently available. Such criteria are fulfilled by cyanine dye based compounds; they have narrow excitation and emission spectra and are significantly more photo stable than the those currently in use [5]. The basic cyanine dye structure can be modified in a variety of ways so that they fluoresce at distinct points across the visible range in a set of 5-6 dyes, thus increasing the range of discernible colours which can be applied to labelled nucleotides, probes and antibodies for a wide variety of applications.

MATERIALS AND METHODS

Fluor labelled compounds

All the fluor-labelled compounds were obtained from Amersham International (Amersham, UK). The original set of compounds consisted of labelled nucleotides only and these were labelled with fluorescein (FluoroGreen; RPN2121), rhodamine (FluoroRed; RPN2122) and coumarin (FluoroBlue; RPN2123); see Table 1.

Table 1. Spectral properties of fluor labelled nucleotides

Fluor-labelled dUTP	Excitation max (nm)	Emission max (nm)	E ($M^{-1}cm^{-1}$)
FluoroGreen	495	518	66000
FluoroRed	548	572	60000
FluoroBlue	342	442	15000

The new set of fluors (Cy Dyes™; see Table 2) have been used to label nucleotides, chromosome specific probes and whole chromosome paint probes.

Probe labelling

Probe labelling reactions, leading to the incorporation of fluor-labelled nucleotides, were performed by a modified version of the nick translation reaction. The nick translation reaction mix consisted of 60μM each of dATP, dGTP, dCTP; 12μM dTTP; 20μM fluor-labelled dUTP; 1μg template DNA. The reaction was initiated by the addition of reaction buffer and DNA Pol 1 / Dnase I mix and incubated at 15°C for 4 hours.

Applications

Metaphase chromosome spread, interphase nuclei preparation and hybridizations were performed using standard techniques [4,6]. Briefly, blood samples were cultured and arrested at the interphase or metaphase. Cells were collected, lysed, fixed and spread onto glass microscope slides. Hybridization was performed in 50% formamide; 10% dextran sulphate; 2x SSC at 42°C overnight with 1μg/ml labelled probe. Detection of fluor-labelled hybrids was achieved using fluorescence microscopy on a fluorescent microscope with results being collected through a standard 35mm camera or through a CCD camera and image capture onto a computer.

Table 2. Excitation and emission properties of the Cy dye range

Fluor	Colour	Excitation max (nm)	Emission max (nm)	E ($M^{-1}cm^{-1}$)
Cy2	Green	489	505	~150000
FluorX	Green	494	520	68000
Cy3	Orange	552	565	150000
Cy3.5	Scarlet	581	596	150000
Cy5	Far red	650	667	250000
Cy5.5	Near IR	678	703	250000
Cy7	Near IR	743	767	~250000

Apoptosis detection was performed by the TUNEL technique using Cy3 labelled nucleotide (terminal transferase labelling of genomic DNA strand breaks). The method was derived from published techniques[7] and briefly consisted of sample collection and fixation, cell permeation with saponin and elongation of free 3' ends at 37°C for 1 hour using 1unit of terminal transferase; 6μM Cy3-dCTP; 0.6μM dCTP. The detection of the fluor-labelled fragments was achieved as described for fluorescent *in situ* hybridization probes.

Fluor-labelled DNA ladders were prepared from restriction enzyme digests of DNA by filling in the overhanging ends with either fluorescein or Cy3 labelled nucleotides. The ladder DNA was separated on a standard agarose electrophoresis gel system and the fluor-labelled samples were detected directly on the Vistra FluorImager 575 (Amersham International). Ladders can be viewed separately or combined in the same lane and distinguished by the individual colour associated with the fluorescent label.

DISCUSSION

Dyes

Fluorescence is the most commonly used end point in chromosome analysis. The success of this technique has been built on the historically available dyes, such as fluorescein and rhodamine. The new set of dyes attempts to bring some science to the selection and design of more suitable dye ranges. The new dyes are characterised, in particular, by being significantly brighter than the traditional ones and therefore offering significantly increased sensitivity thus making the technique of direct detection much more feasible; direct detection was possible with the original dyes but the whole system is now more robust due to the extra sensitivity available. The extra light output is derived from two factors; the first is the extremely high absorption coefficient for the new range and the second is the increase in photostability offered by the Cy dyes.

Probe labelling

A method is described for the labelling of probes with these dye labelled compounds using the traditional nick translation reaction. However labelling can be achieved by any standard non-radioactive polymerase catalysed DNA or oligonucleotide labelling reaction conditions including; random priming, amplification, primer extension *in situ* and oligonucleotide tailing. In amplification reactions it is recommended that the extension reaction time is increased, compared to standard conditions, to allow for the slower incorporation of fluor-labelled nucleotides. In addition, the dTTP/fluor-dNTP ratio may be altered; higher levels of fluor-dNTP (for example 1:1 ratio) may lead to lower yields of heavily labelled probe where as lower amounts (for example 4:1) give good reaction yields with lower incorporation of label.

Applications

The traditional dyes and the Cy dyes have been used in various applications, such as chromosome painting, chromosome copy number counting and chromosome identification, using both fluorescence microscopes and confocal microscopes. In these applications the new dye systems work extremely well and are very easy to use. There is no requirement to amplify the signal thus saving time and the use of antibody conjugates. The signal should be visible to the dark adapted eye in a good fluorescence microscope fitted with the appropriate filter sets, a number of which are commercially available. The systems can be used in the aid of mapping of genes, in looking for translocations and ploidy status. The nucleotides can be used to label individual probes including single copy gene specific sequences. These probes will benefit in interpretation if the microscope is fitted with a suitable photon capturing device which can add in the region of 5 fold extra sensitivity to the detection process. Indeed such systems are strongly recommended for all fluorescence microscopy work as the signal is output to a computer system that can be used for image processing.

Fluor-labelled nucleotides can also be used in other molecular biology applications such as mRNA *in situ* hybridization, detection of apoptosis *in situ* and for direct detection in gels and on blots in conjunction with a suitable fluorescence imaging machine. Messenger RNA can be detected directly by use of fluor-labelled probes. However, the level of signal may be obscured in some cases by the level of autofluorescence seen in biological tissue. This is not as significant a problem in isolated cells and the use of longer wavelength dyes may overcome this problem as the level of autofluorescence is known to decrease towards the red end of the visible spectrum.

For apoptosis a commonly used assay method is the terminal transferase catalysed labelling of new 3'-ends formed during the process. The fluor-labelled nucleotides can be used in this reaction to yield a fluorescent end-point to the assay that can be viewed directly by fluorescence microscopy, and potentially by flow cytometry.

Fluorescence has become well established in fields associated with microscopy and cytometry. However, until recently there was little possibility of extending fluorescence into blot and gel based detection systems. The recent availablity of the FluorImager machine, which captures the light produced from fluor-labelled probes on blots, gels and microtitre plates, has changed that position. Some of the dyes, both traditionally and newly available, can be used successfully in conjunction with this machine. Like fluorescence microscopy the advantages of using the machine and fluorescence are the ability to view a number of probes simultaneously and to be able to accurately quantify the results.

CONCLUSION

Fluorescence is a powerful detection system which can offer resolution, speed and sensitivity. New dyes, new optics and imaging systems will enable fluorescence to be used in a ever increasing number of applications in nucleic acid detection.

REFERENCES

1. I. Durrant, B. Dacre, and M. Cunningham, Evaluation of novel formulations of 35S- and 33P-labelled nucleotides for in situ hybridization, *Histochem. J.* 27: 89 (1995).
2. I. Durrant, S. Brunning, L. Eccleston, P. Chadwick, and M. Cunningham, Fluorescein as a label for non-radioactive in situ hybridization, *Histochem. J.* 27: 94 (1995).
3. J. Koch, Probe labeling and hybridization in one step, *in* "Nonradioactive labeling and detection of biomolecules," C. Kessler, ed., Springer-Verlag, Berlin (1992).
4. J. Wiegant, C.C. Wiesmeijer, J.M.N. Hoovers, E. Schuuring, A. d'Azzo, J. Vrolijk, H.J. Tanke, and A.K. Raap, Multiple and sensitive fluorescence in situ hybridization with rhodamine-, fluorescein-, and coumarin-labeled DNAs, *Cytogenet Cell Genet.* 63: 73 (1993).
5. H. Yu, J. Chao, D. Patek, R. Mujumdar, S. Mujumdar, and A.S. Waggoner, Cyanine dye dUTP analogs for enzymatic labeling of DNA probes, *Nucleic Acids Res* 15: 3226 (1994).
6. N. Arnold, M. Bhatt, T. Ried, J. Wienberg, and D.C. Ward, Probe labeling and hybridization in one step, *in* "Nonradioactive labeling and detection of biomolecules," C. Kessler, ed., Springer-Verlag, Berlin (1992).
7. M. Thiry, Highly sensitive immunodetection of DNA on sections with exogenous terminal deoxynucleotidyl tranferase and non-isotopic nucleotide analogues, *J. Histochem. Cytochem.* 40: 411 (1992).

FLUORESCENCE IN SITU HYBRIDIZATION (FISH) IN CYTOGENETICS OF LEUKEMIA

Kyra Michalová,[1,2] Zuzana Zemanová,[1] Jana Březinová,[2] Věra Michalová,[1]

[1] 3rd Medical Department, General Faculty Hospital
[2] Institute of Hematology and Blood Transfusion
128 08 Prague 2
Czech Republic

INTRODUCTION

Fluorescence in situ hybridization (FISH) has become a powerful tool for the analysis of chromosomal abnormalities in metaphase and interphase cells. The principle of the method is in the base pairing of the DNA probe to complementary sequences in the studied specimen. The hybridization of specific DNA or RNA probes to the cellular targets attached to the microscopic slides is widely used for identification of chromosomal translocations, deletions, amplifications of specific genes and changes in chromosome number[1,2,3]. Advances in the use of FISH with all the innovation and the new modifications of the basic method made one of the biggest impacts and meant a breakthrough in detection and diagnosis of human malignancies. During the last two years FISH was used in our laboratory for:

- identification of constitutive and acquired numerical and structural chromosomal abnormalities
- detection of minimal residual disease or early relapse in patients treated for leukemia by bone marrow transplantation (BMT) and/or chemotherapy
- determination of cytogenetic pattern of nondividing or terminally differentiated cells.

Several types of DNA probes can be employed for these purposes. For structural rearrangements whole chromosome painting probes (WCP) which hybridize to multiple chromosomal sequences are commercially available. To determine numerical and sex chromosome changes alpha-satellite DNA probes which detect centromeric repetitive sequences of each autosome and gonosome of the human chromosomal complement are established. Specific unique chromosomal sequences which can confirm all chromosomal rearrangements i.e. deletions, translocations or inversions with corresponding breakpoints could be also utilized[4,5,6] and are available for many different rearrangements.

Recently, every case of translocation, deletion and any other structural or numerical change found by conventional cytogenetic analysis in bone marrow cells of patients with leukemia has been verified in our laboratory by FISH. Results of this study showed that

FISH is more efficient than conventional cytogenetics in detecting residual malignant cells. For chromosomal rearrangements (translocations, inversions, deletions, amplifications) FISH is an extremely sensitive method which not only verifies but also interprets with more precision the findings of meticulous classical cytogenetic examinations.

MATERIAL AND METHODS

Chromosome Preparations

Bone marrow cells were obtained by sternal puncture of patients suffering from different subtypes of preleukemia or leukemia. Cells were cultivated for 24 and/or 48 hours without stimulation. Mitoses were harvested after hypotonic treatment with 0.075M KCl and slides were prepared by conventional technique. Cytogenetic examination was performed on G-banded slides by method described by Yunis[7].

Fluorescence In Situ Hybridization

FISH was prepared according to the manual and protocols recommended by different manufacturers of DNA probes with slight modifications. Washing the slides in 0,2xSSC allowed repeated examinations and applications of several labeled probes on one chromosomal slide. Double-color fluorescence was possible using DNA probes labeled both by digoxigenin and biotin. Microscope Zeiss Axiophot equipped with epifluorescence, 50W halogen lamp and single filters of different wave length or triplebandpass filter (Dapi, FITC, Texas Red - AF-Analysentechnik, Tubingen) was used for signal determination. Following DNA probes were used to detect sex chromosomes complement: DXZ1 Biotin (ONCOR) and pY3.4 labeled by digoxigenin, in some experiments CEP® X Spectrum Orange™ and CEP® Y Spectrum Green™ (VYSIS) were introduced. At least 200 interphase nuclei were screened in each patient. The number of mitoses examined depended on their availability on the chromosomal preparations.

Whole chromosome painting probes (WCP) manufactured by CAMBIO and AGS were differently labeled for single-color or double-color fluorescence. Samples for this type of FISH examination were selected on the basis of the result of the G-banded chromosomes showing simple or complex translocations and in some cases undetectable marker chromosomes of unknown origin.

Translocation breakpoints t(9;22)(q34;q22) in mitoses and interphase nuclei of patients with chronic myeloid leukemia (CML) and Ph positive acute lymphocytic leukemia (ALL) were followed by cosmid probes for BCR/ABL fusion gene prepared by ONCOR. Translocation t(15;17)(q22;q21) was determined in chromosomal preparations of bone marrow cells of patients with acute promyelocytic leukemia (APL) in dividing and non-dividing cells by double-color fluorescence with cosmid probe manufactured by ONCOR. In every case at least 200 non-dividing nuclei were screened by FISH.

Statistical Evaluations

Non-parametric Friedman's matched test and non-parametric Wilcoxon's test for paired observations with BMDP statistical software were used.

RESULTS

More than 220 examinations with WCP helped to define precisely the origin of 120 marker chromosomes and 43 numerical changes. Chromosome painting always followed conventional cytogenetic analysis of G-banded chromosomes. In many cases complex translocations involving small chromosomal regions undetected by classical analysis were found. The most interesting single cases were quoted in Michalová et al.[8,9] and will be published as case reports elsewhere. The color pictures with the diagrammatic representation of reciprocal translocations and explanation of the origin of marker chromosomes found in patients with preleukemia and acute myeloid leukemia were presented in the poster.

Table 1a,b. Percentage of recipient's cells found in bone marrow with classical and FISH methods.

a) Males

Pat.	A	B	C
1.	*	20	15,5
2.	*	*	7
3.	N	37,5	26,5
4.	N	*	12
5.	*	*	2,5
6.	N	*	1
7.	*	*	8
8.	*	*	6
9.	*	*	11
10.	N	*	1
11.	9	*	2,5
12.	N	*	0,5
13.	*	*	0
14.	N	50	31,5
15.	N	*	0,5

b) Females

Pat.	A	B	C
1.	*	N	5
2.	*	*	1,5
3.	*	*	3,3
4.	*	*	6,
5.	*	*	2,
6.	*	*	2
7.	*	*	1,5
8.	8,3	*	2
9.	*	*	3
10.	*	*	2
11.	*	*	11,5
12.	N	*	3,5

A - classical cytogenetics
B - Fish mitoses
C - FISH nuclei
* - no recipient's cells were found, only donor's cells were present
N - no mitoses present

The results of study with alpha-satellite DNA probes are shown in Table 1. Complete results of classical and FISH analyses of mitoses and interphase nuclei of patients treated by bone marrow transplantation (BMT) were carried out by double-color fluorescence with the probes DXZ1/pY3.4 after series of control studies in blood and bone marrow cells of healthy donors. (The results of control studies are not shown). All transplanted patients received bone marrow graft from sex-mismatched healthy donors and therefore the differences in sex chromosome complement after BMT could be informative when evaluated by classical and FISH methods. The change of the sex chromosomes in bone marrow cells of the recipient could be expected in two weeks after transplantation. Cytogenetic studies and FISH on mitotic and interphase nuclei were carried out concurrently at varying intervals after BMT. Classical cytogenetic and FISH examinations can monitor success of the BMT in these patients, however, its success is dependent on the quality and quantity of sample received. In this respect, the results of FISH quoted in the Table 1a,b clearly show that FISH is more sensitive method and gives answer in every examined patient.

In Table 2a are presented results of classical cytogenetics of G-banded chromosomes and FISH performed in patients with CML and in Table 2b the results of the study of patients with APL are presented. The tables again enable comparison of methods of classical and molecular cytogenetics. Statistical analysis proved significant differences of all three methods used for evaluation of the presence of donor's and recipient's cells. The most sensitive was found evaluation of interphase nuclei by FISH (on the level of statistical significance 0,05).

Table 2a,b. Percentage of residual leukemic cells after BMT in patients with CML(2a) and after treatment in patients with APL(2b).

a) BCR/ABL:

Pat.	A	B	C
1.	*	N	0
2.	N	*	2
3.	N	5	3
4.	50	N	82,5
5.	*	N	4
6.	50	N	10,5
7.	100	100	89,5
8.	5	*	0,5
9.	53	11,1	5,5
10.	N	64,7	62
11.	93,3	90	92,7
12.	3,6	*	4
13.	*	*	3,4
14.	45	55	44,5
15.	100	85	87
16.	N	*	3
17.	N	*	0

b) MYL/RARA:

Pat.	A	B	C
1.	*	2,7	0,5
2.	*	*	0
3.	49,1	53,3	49,5
4.	*	6	3

A - classical cytogenetics
B - FISH mitoses
C- FISH nuclei
N - no mitoses present
* - no residual leukemic cells were found, only normal mitoses were present

DISCUSSION

Cytogenetics has become a critical part of the clinical evaluation of many types of hematopoetic disorders. Chromosome analysis yields important informations for the diagnosis and prognosis of certain hematopoetic neoplasias, and is also used to follow the progression or regression of an abnormal cell line in bone marrow of treated patients[10]. FISH has been widely used during the last few years to study the biology and molecular biology of hematopoetic malignancies and to classify and monitor the response to therapeutic interventions. Both, structural and numerical chromosomal abnormalities are amenable to FISH. Also single copy genes which underwent rearrangement in certain leukemias could be traced by FISH. In our studies we used commercially available probes to confirm the results of the classical cytogenetic analysis by FISH. Except such confirmation we detected in some cases further structural chromosomal changes which could be easily overlooked by traditional techniques. In this respect FISH has proved to be a valuable complementary method for studies of complex chromosomal rearrangements of malignant cells.

In patients treated by chemotherapy and/or bone marrow transplantation FISH is a sensitive and quantitative new method for detecting specific chromosomal aberrations and/or sex chromosome complement not only in metaphase spreads, but especially in interphase nuclei[11]. The technique allows to screen a high number of the non-dividing cells and therefore can give the answer about the success of the treatment even in those cases where the classical method is limited by marrow hypoplasia and the inability of the cells to divide. It can also follow the proliferation capacity of residual leukemic cells in vitro and quantitatively monitor and predict possible relapse of the disease[12]. In this study we have shown that interphase cytogenetics performed with chromosome-specific DNA probes contributes to a prompt detection of the presence of residual host cells in bone marrow of patients who underwent sex-mismatched BMT. In addition to its sensitivity FISH has proved to be a very fast and reliable method. The few discrepancies found in some patients between the results of classical cytogenetic examination and FISH are rather theoretical in character, as the molecular biology and molecular cytogenetic methods open the new possibilities to follow even non-dividing cells. Studies of large series of patients using both methods for a long time after therapy should lead to the new understanding of the role of residual leukemic cells during the regression or progression of the disease.

This study was supported by grant of Czech Ministry of Health IGA MZ CR No.2309-3.

REFERENCES

1 B.J. Trask, Fluorescence in situ hybridization: applications in cytogenetics and gene mapping, *Trends in Genetics* 7:149 (1991)

2. M.M LeBeau, Fluorescence in situ hybridization in cancer diagnosis, *Important Advances in Oncology*, V.T. de Vita, S. Hellman and S.A. Rosenberg ed., J.B Lippincott Comp., Philadelphia (1993).

3. M Bentz, H. Dohner, G Cabot, P Lichter, Fluorescence in situ hybridization in leukemias: The FISH are spawning *Leukemia* 8 1447 (1994)

4. C.M. Price, Fluorescence in situ hybridization, *Reviews* 7: 127 (1993).

5 V.Vrazas, L.M. Ooms, Ch Rudduck, J. Szer, L J. Campbell, O.M. Garson, *American Journal of Hematology* 49:15 (1995)

6. Z Lian, K.S. Chang, E.H. Estey, K Hayes, A.B. Deisseroth, J C. Liang, Detection of residual leukemic cells in patients with acute promyelocytic leukemia by the fluorescence in situ hybridization method: Potential for predicting relapse, *Blood* 85:495 (1995).

7 J.J. Yunis, New chromosome techniques in the study of human neoplasia, *Human Pathology* 12:540 (1981).

8. K. Michalová, O Bartsch, J. Starý, J. Jelínek, J. Wiegant, E. Bubanská, Partial trisomy of 3q detected by chromosome painting in a case of juvenile chronic myelomonocytic leukemia. *Cancer Genet. Cytogenet.* 71.67 (1993)

9 K. Michalová, J. Musilová, Z. Zemanová, J. Vařáková, M Lukášová, Chromosomal translocation t(11;17)(q23;q21) present in bone marrow cells of two patients with myelodysplastic syndrome (MDS), *Medizinische Genetik* 2 193 (1995).

10. N. Clare, K. Hansen, Cytogenetics in the diagnosis of hematologic malignancies, *Hematology/Oncology Clinics of North America* 8:785 (1994).

11 A. Nagler, S Slavin, S. Yarkoni, M. Fejgin, A. Amiel, Detection of minimal residual disease after sex-mismatch bone marrow transplantation in chronic myelogenous leukemia by fluorescence in situ hybridization, *Cancer Genet Cytogenet.* 73:130 (1994).

12. P. Temperani, P. Vaccari, F Giacobbi, M G. Ferrari, G. Emilia, Assessment of minimal residual diseasein acute promyelocytic leukemia with t(15;17) by chromosome painting, *European J Haematology* 55:10 (1995)

ESTIMATION OF "START" IN *Saccharomyces cerevisiae* BY FLOW CYTOMETRY AND FLUORESCENT STAINING OF DNA AND CELL PROTEIN

Lars Guldfeldt, Henrik Siegumfeldt, Lotte Lammert and Mogens Jakobsen

Department of Dairy and Food Science, Food Microbiology
The Royal Veterinary and Agricultural University
1958 Frederiksberg C, Denmark

INTRODUCTION

For industrial fermentations "early events" refer to the immediate response of the culture when added to the substrate. At that time the culture is exposed to a completely new environment leading to a number of cellular events taking place before measurable changes in the substrate are detectable. These events are considered valid indicators of the physiological status of the culture and key events necessary for an optimal fermentation and final product quality. For *Saccharomyces cerevisiae* as a starter culture "START", defined as the time when the cells are committed to cell division, is regarded as an important "early event". "START" is a "point" late in the G1-phase that yeast cells need to pass before they can replicate their DNA (Hartwell, 1994) Much of the current understanding of "START" in *Sacch. cerevisiae* comes from studies of cells carrying mutations in cell division cycle (CDC) genes (Baroni *et al.*, 1994 and Tokiwa *et al.*, 1994). "START" has been shown to be a series of tightly regulated events that must occur before the cells are committed to mitotic division (Sherlock and Rosamond, 1993). The yeast cell cycle is divided into four different cell cycle phases G1, S, G2 and M-phase. In the G1-phase cells are unbudded with shapes that approximates a prolate spheroid. The cells grow in volume during this phase of the cell cycle. The initiation of the S-phase coincides with bud emergence and during this phase DNA is doubled. When DNA has doubled the cells enter the G2-phase which is the gap between DNA replication and mitosis At the end of the G2-phase the nucleus migrates to the junction of the parent cell and the bud. Nuclear division occurs in the M-phase, the nucleus elongates such that part of it lies within the parent cell and the remaining part within the bud and the cell divides into a mother and a daughter cell (Carter, 1981). The

Fluorescence Microscopy and Fluorescent Probes
Edited by J. Slavík, Plenum Press, New York, 1996

exact point of "START" has not yet been determined but it is known to be closely linked to cAMP-dependent protein kinases and cyclin-dependent kinases which are the regulating factors for growth and the cell division cycle (Hartwell, 1994) Further, it is known that in the yeast *Sacch. cerevisiae* commitment to cell division requires growth to a critical cell size mediated by cAMP (Baroni *et al*, 1992) before traversing "START" (Mitsuzawa, 1994)

It has been the purpose of the present work to estimate "START" for *Sacch. cerevisiae* in a laboratory fermentation by flow cytometric analyses of the cells stained with propidium iodide and fluorescein isothiocyanate (FITC) respectively to measure the cell content of DNA and protein, the latter seen as an expression of cell size (Porro *et al*, 1995).

MATERIALS AND METHODS

Strains and growth condition

Three different industrial strains of *Sacch. cerevisiae* coded A, B and C were used in these studies Fermentations were done in 2 5 litres tall-tubes. The tubes were inoculated with 2.5 gram centrifuged yeast (3000 g for 5 min at 4°C) pr. litre of malt extract (11%w/w), pH 5.6 aerated to 10 ppm dissolved oxygen. Fermentations were performed at a constant temperature at 12°C. Yeasts were sampled from the middle section to examine yeasts in suspension. Growth was followed by monitoring cell dry weight for cells in suspension. The rate of fermentation was determined by measuring specific gravity using a Calculating Density Meter AP PAAR DMA 55 (Anton PAAR KG, Graz, Austria). Ethanol concentrations were determined by GC (Hewlett Packard 5890 SERIES III, Hewlett Packard, 3460 Birkerød, Denmark).

Flow cytometry

Flow cytometric analyses were performed at a Partec instrument (Partec PASIIIi, Chemunex SA, 94700 Maisons Alfort, France) connected with an argon laser (40 mW) and a mercury high pressure (HBO) lamp Both light sources can be used alternatively or simultaneously for two-wavelength excitation. All optical components, lenses, light-collectors and dichoric mirrors are made out of fluorescence free synthetic quartz Excitation filters and dichoric beam splitters are mounted on mirror-filter-blocks that easily can be changed. Light detection is done by five built-in photomultipliers, three of them for colour fluorescence analysis and two for scatter signal analysis. The flow cell is a quartz cuvette with an internal diameter of 250x250 µm that allows the use of high numerical objectives for sufficient fluorescence light detection

Yeast examinations

DNA staining: The DNA staining protocol used is described by Münch *et al.* (1992). For flow cytometric analysis the PI bound to double stranded DNA within the cells is excited by the argon ion laser at 488 nm and emission (red fluorescence) is measured at ca. 600 nm.

Protein staining: Harvested cells were washed and resuspended in PBS-buffer at pH 7.4 to a cell concentration of 10^6 cells pr. ml. The cell suspension was sonicated for 10 seconds at 20 kHz After sonication the cells were centrifuged (3000 g for 5 min at 4°C) and fixed in ice cold 70 %v/v ethanol to a final cell concentration of 10^6 cells per ml. The staining procedure was performed according to Peterson and Patkar (1992). Flow cytometric analysis was performed by exciting FITC with the argon ion laser at 488 nm and measuring green fluorescence at 525 nm.

Esterase activity. Cells were harvested by centrifugation (3000 g for 5 min at 4°C) and washed and resuspended in PBS-buffer pH 7.4 to an approximate cell concentration of 10^6 cells pr. ml. The suspension was then sonicated for 10 seconds at 20 kHz to separate clumps of cells. 10 µl fluorescein diacetate solution was added from a stock solution, 0.5 %w/v FDA (Sigma, F-7378) in acetone, to a final concentration of 120 µM After mixing, the samples were incubated in the dark at 40°C for 10 min followed by placing the samples on ice until flow cytometric analysis which must be performed within and hour after staining Before running the samples at the flow cytometer they are diluted 10 times. The argon laser (488 nm) was used for exciting liberated fluorescein within the cells, emission (green fluorescence) was detected at 525 nm.

Computational evaluation of results

The flow cytometric results were analysed using an integrated software package "PartecLIST" (Chemunex SA, 94700 Maisons Alfort, France), where numerous statistical operations during or after acquisition can be performed. PartecLIST is an advanced signal processing package handling up to 7 parameters in real time specifically designed for use with Partec PASIIIi. Typical evaluations are peak analysis, channel analysis and cell cycle analysis

RESULTS AND DISCUSSION

For three strains of *Sacch. cerevisiae*, growth and the rate of carbohydrate utilisation are shown in Fig.1. The three strains of *Sacch. cerevisiae* shows only slight variations in growth and carbohydrate utilisation.

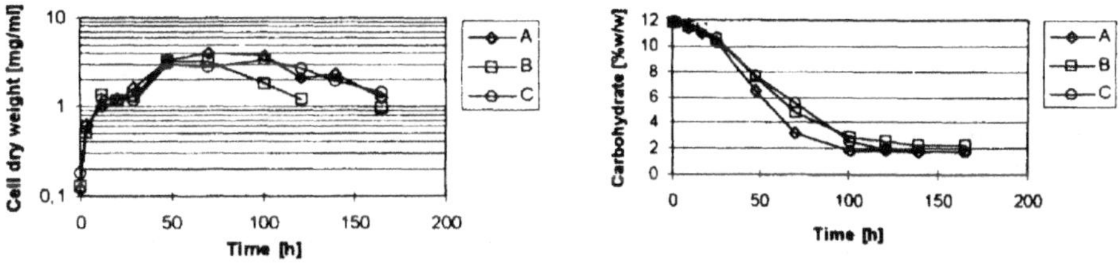

Figure 1. Left side: Growth. Right side: Carbohydrate utilisation.

The esterase activity (Fig 2), shown as a relative measure called the fluorescence index (FI), is calculated from the ratio between the peak channel number for standard beads D (Chemunex SA, 94700 Maisons Alfort, France) and the maximum peak channel number for the sample The esterase activity of the three industrial strains follows a well known pattern as described previously (Jespersen and Jakobsen, 1994) for yeasts in optimal condition. Some variation in the esterase activity in the beginning of the fermentation are seen, indicating differences between the strains in their ability to hydrolyse FDA. Similar differences have been described for other strains of *Sacch. cerevisiae* (Jespersen and Jakobsen, 1994) as well as for different species of yeasts by Haworth and Fliegel (1993). The lowest esterase activity is reached for all strains 20 - 30 hours after inoculation followed by an increase until approximately 100 hours where after a small decline can be seen.

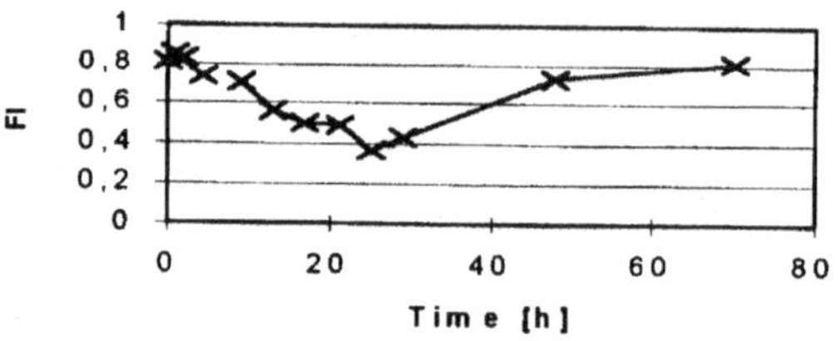

Figure 2. Esterase activity in strain A during fermentation

The protein content of strain A are illustrated (Fig 3) in the same way as the esterase activity. It has often been suggested that the cellular protein content

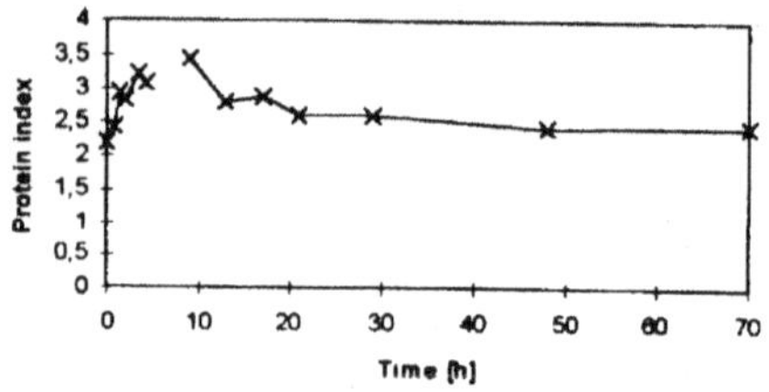

Figure 3. Relative protein content in strain A

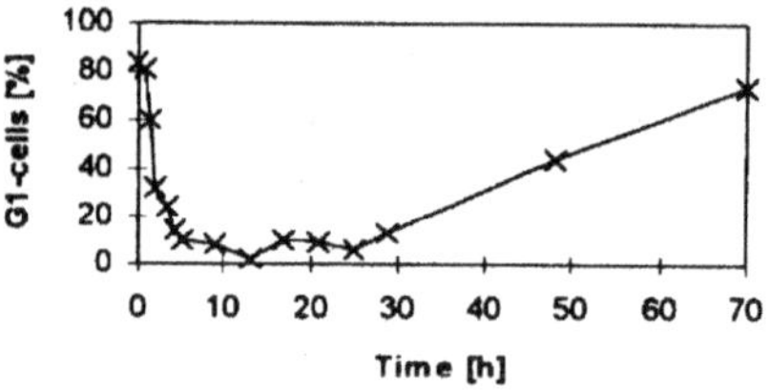

Figure 4. Number of G1-cells in strain A.

correlates well with the cell size (Porro *et al.*, 1995). We have observed (results not shown) that the protein staining with FITC is very complex depending on cell concentration, FITC concentration, incubation time and incubation temperature. Therefore staining must be performed exactly the same way each time to be able to make valid conclusions.

The number of cells in the G1-phase (Fig 4) is very similar for the strains investigated. It is seen that there is a drastic decline in the number of G1-cells during the first 2 - 4 hours followed by an increase where after the number of G1-cells remains fairly constant (data not shown). As seen from Fig.4 the decline of G1-cells begins immediately after inoculation and therefore it can be characterised as a true early event that may be used for predicting fermentations By comparison of Fig.3 and Fig.4, it appears that the maximum protein content occurs at the time of minimum number of G1-cells

CONCLUSION

The results confirm earlier findings (Jespersen and Jakobsen, 1994) that fluorescence staining and flow cytometry offers the possibility of rapid measurements of important intracellular events in *Sacch. cerevisiae*. The most interesting observation was the well defined reduction in the number of G1-phase cells 2 - 4 hours after pitching , which could prove particular useful in evaluating yeast quality and predicting yeast performance. It may also be linked with "START".

REFERENCES

Baroni, M.D., Monti, P., Marconi, G. and Alberghina, L., 1992, cAMP-mediated increase in the critical cell size required for the G1 to S transition in *Saccharomyces cerevisiae*, *Exp.Cell Res.* **201**, 299-306

Baroni, M.D., Monti, P. and Alberghina, L., 1994, Repression of growth-regulated G1 cyclin expression by cyclic AMP in budding yeast, *Nature* **371**, 339-342

Carter, B.L.A., The control of cell division in *Saccharomyces cerevisiae*, *in*: "The cell cycle", John, P.C.L., ed., Cambridge University Press, Cambridge pp. 99-117 (1981)

Hartwell, L., 1994, cAMPing out, *Nature* **371**, 286

Haworth, R.S. and Fliegel, L., 1993, Intracellular pH in *Schizosaccharomyces pombe* - comparison with *Saccharomyces cerevisiae*, *Mol.Cell.Biochem.* **124**, 131-140

Jespersen, L. and Jakobsen,M., 1994, Use of flow cytometry for rapid estimation of intracellular events in brewing yeasts, *J.Inst.Brew.* **100**, 399-403

Mitsuzawa, H., 1994, Increase in cell size at START caused by hyperactivation of the cAMP pathway in *Saccharomyces cerevisiae*, *Mol.Gen.Genet.* **243**, 158-165

Münch, T., Sonnleitner, B. and Fiechter, A., 1992, The decisive role of the *Saccharomyces cerevisiae* cell cycle behaviour for dynamic growth characterization, *J.Biotechnol.* **22**, 329-352

Peterson, M.S. and Patkar, A.Y., 1992, Flow cytometric analysis of total protein content and size distributions of recombinant *Saccharomyces cerevisiae*, *Biotechnol Techn* **6**(3), 203-206

Porro, D., Ranzi, B.M., Smeraldi, C., Martegani, E and Alberghina, L., 1995, A double flow cytometric tag allows tracking of the dynamics of cell cycle progression of newborn *Saccharomyces cerevisiae* cells during balanced exponential growth, *Yeast* **11**, 1157-1169

Sherlock, G. and Rosamond, J., 1993, Starting to cycle: G1 controls regulating cell division in budding yeast, *J.Gen.Microbiol.* **139**, 2531-2541

Tokiwa, G., Tyers, M., Volpe, T. and Futcher, B., 1994, Inhibition of G1 cyclin activity by the Ras/cAMP pathway in yeast, *Nature* **371**, 342-345

FLUORESCENCE IMAGE CYTOMETRY OF DNA CONTENT: A COMPARATIVE STUDY OF THREE FLUOROCHROMES AND FOUR FIXATION PROTOCOLS

Mehdi Benchaib, Richard Delorme, Paul-André Bryon And Catherine Souchier

Cytologie analytique
8 avenue Rockefeller
69373 LYON CEDEX 08, FRANCE

INTRODUCTION

Image cytometry is extensively applied for the determination of nuclear DNA content. Most studies are performed in transmission microscopy using Feulgen DNA stain However, fluorescent dyes may also be used DNA dyes[1] and fixative protocols[2] are known to play a critical role in the reliability of such studies.

The present study compares results obtained with three fluorescent DNA dyes and four fixative protocols.

MATERIAL AND METHODS

Reh-6 (early-early-B leukemia) cells were cytocentrifuged (Shandon Southern Product, Runcorn, England) at 450 rpm for four minutes Slides were then stored at -20°C Rat liver imprints were made directly from fresh tissue and were then frozen
Different fixative protocols were performed·
1 - first unalloyed methanol at -20 °C for 2 minutes followed by paraformaldehyde (PFA) 0 5% at +4°C for 30 minutes (Methanol/PFA cold protocol)
2 - first unalloyed methanol at room temperature(RT) for 2 minutes followed by PFA 0.5% at room temperature for 30 minutes (Methanol/PFA RT protocol).
3 - first PFA 0.5% at room temperature for 30 minutes followed by unalloyed methanol at room temperature for 2 minutes (PFA/Methanol RT protocol).
4 - ethanol 70°only at room temperature for 10 minutes (Ethanol protocol).

Cells were DNA stained after RNAse IIIA treatment (Sigma Chemical, St Louis, MO, USA.) in PBS (5mg/5 ml) for 30 minutes at room temperature. Three fluorescent dyes were used: Hoechst 33342 (Sigma Chemical) 1 50 (1mg/ml), Propidium Iodide (PI) (Sigma Chemical) at a dilution of 50 µg/ml and YOPRO-1 (Molecular probes, Eugene,OR, USA.) 1:1000. All stainings were performed for 30 minutes at room temperature. Slides were mounted in glycerol (Bio-Mérieux, Lyon, France)

FLUORESCENCE IMAGE ANALYSIS

Material

The study was done on a Quantimet 570 image analyser (QTM 570) (Leica, Cambridge, UK) fitted to an intensified black and white CCD camera (LH XX-SPU, Lhesa, Cergy Pontoise, France). The stained slides were observed under an epifluorescence microscope (Axioplan, Zeiss, Oberkochen, Germany) using a 40X pan achromat objective (NA=1.0). Three different Zeiss filter sets were used to visualize the fluorochromes (Table 1)

Table 1. Fluorochrome filter sets

Fluorochromes	Excitation Filter	Dichroic Mirror	Emission Filter
HOECHST 3342	BP 365/12	DM 395	BP 470/20
YOPRO	BP 485/20	DM 510	BP 540/25
Propidium Iodide	BP 546/12	DM 580	LP 590

Image correction

Correction for uneven epi-illumination was performed with the shading correction procedure included in QTM 570 software For this, we used a uniformly fluorescent image obtained with a Wratten 2B film (Rochester, New York, USA), one stained with PI and one with YOPRO. For Hoechst staining, a saturated solution was diluted in glycerol and a drop placed on a slide, coverslipped, allowed to spread without pressure, and sealed with clear nailpolish. For background correction, an image corresponding to a region without cell was acquired and was then subtracted from all images.

Data screening

Three hundred cells at least were analyzed for each slide. Artifacts and disrupted nuclei were rejected by setting conservative limits on the object area and shape. Touching cells were separated using a QTM 570 segmentation algorithm based on morphological filters. An interactive step was included in order to improve the automatic segmentation results.

Total nuclear DNA content was evaluated by integrated fluorescence intensity (IFI). IFI values were normalized by the G_1 peak modal value and were binned to 40 channels over the range 0.0 to 9.5. A normalized DNA content histogram was thus obtained.

The following parameters were calculated:

- Stoichiometric index (SI). ratio of IFI value at G2 peak to the value at G1 peak.
- CV_{2c}, DNA content variation coefficient for the 2c cell population, calculated as the ratio between the standard deviation and the mean, multiplied by 100.

RESULTS

Figure 1 shows DNA histograms obtained with the various fluorochromes and fixative protocols studied.

On the hepatocyte population, three DNA histogram types were obtained Type I showed a bimodal IFI distribution with one 2c mode, one 4c mode and low CVs Type II showed a bimodal distribution but larger CVs. In type III, the two modes were not well separated

Three kinds of histograms were also generated by the DNA quantitative analysis of Reh-6 cell line. Type A showed a bimodal IFI distribution, with distinct G1 phase, G2/M phase and a well visible S phase (type A). In the B type, the cell cycle phases were more difficult to separate and the CV_{4c} was more important. In the last type (type C) the different phases were not clearly distinguishable

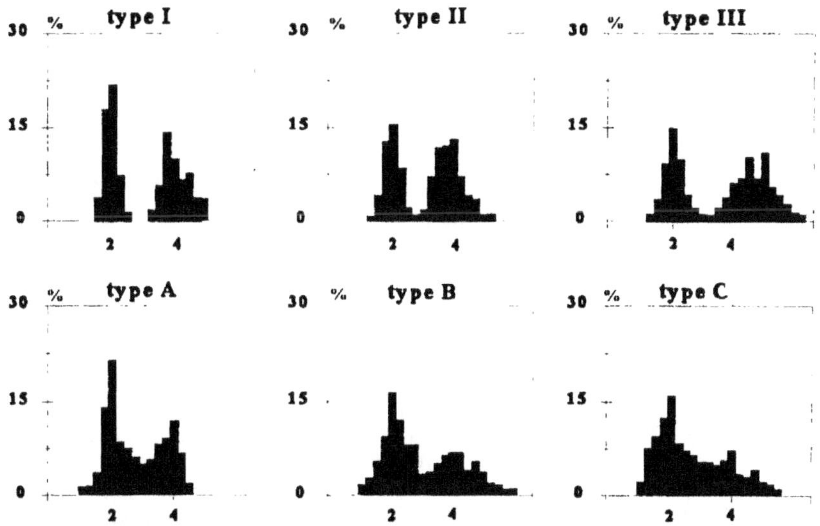

Figure 1. DNA histogram types

Histograms obtained on hepatocyte cells with I - Hoechst (PFA+Methanol, RT), II - YOPRO (Ethanol, RT), III - PI (Methanol+PFA, RT) Histograms obtained on Reh-6 cells A - Hoechst (Methanol+PFA, RT), B - PI (PFA+Methanol, RT), C - YOPRO (Methanol+PFA, 4°C)

The histogram type, the stochiometric index (SI) and the coefficient of variation of the 2c phase (CV_{2c}), are shown in Table 2

For rat hepatocytes, Hoechst provided correct results with respect to DNA histogram type, CV_{2c} (10.2%-14%) and SI (1.9-2 2) whatever the fixation protocol YOPRO and PI were more fixative dependent and the best results were obtained with the methanol/PFA (cold) protocol.

With reh-6 cells, Hoechst provided the best results whatever the fixation protocol. The CV_{2c} were equal to 8-9% with Methanol/PFA or PFA/Methanol fixation at room temperature. For YOPRO and PI, the best results were obtained with ethanol fixation.

Table 2. DNA results

	Rat hepatocytes			Reh-6 cells		
	Hist	SI	CV_{2c}	Hist	SI	CV_{2c}
Ethanol						
Hoechst	II	1 9	10 9	A	2 1	12 8
YOPRO	II	1 9	12 6	C	2 1	13 6
PI	III	2 4	17 4	B	2 0	12 5
Methanol/PFA (cold)						
Hoechst	II	2 2	13 8	B	2 0	14 9
YOPRO	I	2 2	9 3	C	2 1	20 6
PI	II	2 2	10 8	C	2 1	15 9
Methanol/PFA (RT)						
Hoechst	II	2 2	14 0	A	2 0	8 1
YOPRO	III	2 2	14 2	C	2 2	16 8
PI	III	2 2	12 6	B	2 0	19 4
PFA/Methanol (RT)						
Hoechst	I	2 1	10 2	A	2 0	9 8
YOPRO	II	2 1	15 9	C	2 1	24 6
PI	III	2 3	13 9	B	2 1	14 3

Hist Histogram type

DISCUSSION

Three fluorochromes and four fixative protocols were compared to find the combinations suitable for DNA quantification To address this question, DNA content of rat hepatocytes and Reh-6 cell line were measured by image cytometry

Different results were obtained according to the fixation protocol, the DNA dye or the cell type, even through they may be considered as stoichiometric in most cases These differences may be due to variations in dye binding according to the fixation protocol and the fluorochrome type. The fluorochromes used differ in their modes of interaction with DNA Hoechst 33342 is a classic minor groove ligand[3] with an affinity for A-T DNA sequences. PI and YOPRO are two intercalating dyes[3,4].

For the reh-6 cell line, intercalating dyes provide the most accurate DNA quantification with ethanol fixation. The fixation protocols which contained PFA might be expected to cross-link DNA[5,6], thereby rendering it less accessible to intercalating dyes such as YOPRO or PI. However, on rat hepatocytes, such differences were not observed.

DNA-topology may also significantly influence nuclei stainability[1,7,8] and might explain for the differences observed between the two cell types Indeed, chromatin condensation was more pronounced in rat hepatocytes than in Reh-6 cells Ethanol action on rat hepatocyte histones might not be sufficient for improving DNA accessibility Indeed, ethanol has been shown to be an efficient solubilizing agent for histones, thereby improving intercalating dye accessibility Hoechst 33342 seems to be less sensitive to changes in chromatin structure, and may thus be expected to result in relatively uniform stainability in cell mixtures of various types or proliferative activities[9] whereas, intercalating dyes seem to be more sensitive to the nuclear protein condensation and to the fixative protocols

In our hands, Hoechst was found to provide satisfactory quantification of nuclear DNA content whatever the cell type and the fixation protocol. YOPRO gave correct results on rat hepatocytes, except with the methanol/PFA(RT) fixative protocol, but results were poorest on Reh-6 cells. PI provided intermediate results both on rat hepatocytes and Reh-6 cells

ACKNOWLEDGMENTS

This work was supported by a grant from "la ligue contre le cancer de l'Ain et de la Saône & Loire"

REFERENCES

1. M.S Santısteban, M P Montmasson, F Gıroud, X Ronot, and G Brugal, Fluorescence ımage cytometry of nuclear DNA content versus chromatın pattern a comparatıve study of ten fluorochromes, J Hıstochem Cytochem 40 1789(1992)
2. N M. Poulın, J B Matthews, K A Skov, and B Palcıc B, Effects of fixatıon method on ımage cytometrıc measurement of DNA content and dıstrıbutıon ın cells staıned for fluorescence wıth propıdıum ıodıde, J Hıstochem Cytochem 42 1149(1994)
3 D J Arndt-Jovın and T M Jovın, Fluorescent labelıng and mıcroscopy of DNA In "Methods ın Cell Bıology, vol 30", Y L Wang, ed , Academıc Press Inc, San Dıego, (1989)
4 A Larsson, C Carlsson, and B Albınsson, Characterızatıon of the bındıng of the fluorescent dyes YO and YOYO to DNA by polarızed lıght spectroscopy J Am Chem Soc 116 8459(1994)
5 C. H Fox, F B Johnson, J Whıtıng, and P P Roller, Formaldehyde fixatıon, J Hıstochem Cytochem 33·845(1985)
6 F Traganos, Z Darzynkıewıcz, T Sharpless, M R Melamed, Denaturatıon of deoxyrıbonucleıc acıd ın sıtu Effect of formaldehyde, J Hıstochem Cytochem 23 431(1975)
7 E. Prosperı, M C Gıangaré, G Bottırolı, DNA staınabılıty wıth base-specıfic fluorochromes dependence on the DNA topology ın sıtu, Hıstochemıstry 102 123(1994)
8 A Myc, F Traganos, J Lara, M R Melamed, and Z Darzynkıewıcz, DNA staınabılıty ın aneuploıd breast tumors Comparıson of four DNA fluorochromes dıfferıng ın bındıng propertıes, Cytometry 13 389(1992)
9 Z Darzynkıewıcz, F Traganos, Unstaınable DNA ın cell nucleı Comparıson of ten dıfferent fluorochromes, Anal & Quant Cytol & Hıstol 10 462(1988)

IN VIVO TISSUE CHARACTERIZATION USING ENVIRONMENTALLY SENSITIVE FLUOROCHROMES

Georges A. Wagnières,[1*] Seiichi Iinuma,[2] Kevin T. Schomacker,[2] Tom Deutsch,[2] and Tayyaba Hasan[2]

[1]Institute of environmental Engineering, Swiss Federal Institute of Technology (EPFL), CH-1015 Lausanne, Switzerland.
[2]Wellman Laboratories of Photomedicine, Department of Dermatology Harvard Medical School, Massachusetts General Hospital, Boston, MA 02114, USA.

ABSTRACT

The purpose of the presented study is to evaluate a new approach for establishing a high and specific fluorescence contrast between a tumor and its surrounding healthy tissues. The principle consist to use fluorochromes which are sensitive to their physico-chemical environment in order to establish a "spectral" and/or "intensity" contrast. A metachromatic DNA-staining fluorescent probe showing significant changes of its fluorescence quantum yield, fluorescence life-time, excitation and emission fluorescence wavelengths has been investigated *in vivo*. An orthotopic rat bladder tumor model has been used in order to study the pharmacokinetics and biodistribution of an unsymmetric dimeric cyanine DNA-staining dye (TOTO-1). These measurements were performed *in-vivo* using a gated optical fiber based spectrofluorometer comprising a N_2 pumped dye laser as excitation source.

The preliminary results obtained suggest that the strong fluorescence contrasts which have been observed are due to two factors: 1) differences in the cell membrane integrity between the atypical and normal cells. Since this fluorescent dye is also a very good probe for evaluating the cells permeability, its uptake is probably furthered in the malignant cells. 2) Once in the cell, the binding of this dye with DNA strongly increases its fluorescence quantum yield (up to 1000X). The metachromatic properties of this dye induces therefore one kind of strong amplification of the concentration contrast. Moreover, this study suggests that a tissue discrimination can be performed *in-vivo* using fluorochrome presenting different fluorescence life-times for different physico-chemical environments.

INTRODUCTION

The probability of success for all cancer therapies is significantly improved if the lesion is detected at an early stage. Light induced fluorescence (LIF) photodetection of cancer is a

---

* To whom reprint requested should be addressed

Fluorescence Microscopy and Fluorescent Probes
Edited by J. Slavík, Plenum Press, New York, 1996

sensitive and promising technique which is potentially able to detect superficial lesions at an early stage [1-9]. As this technique can be used endoscopically, LIF cancer photodetection is of interest for detecting and delineating most of the carcinomas. This technique is based on the detection of the fluorescence emitted by endogenous or exogenous fluorochromes which are establishing an optical contrast between the lesion and its surrounding normal tissues. Contrast, and to a lesser extent, noise and spatial resolution are key parameters which govern the photodetection performance based on the commonly used exogenous markers such as porphyrin derivatives (HPD, Photofrin II). As the contrast obtained with most of these exogenous porphyrins is mainly due to the preferential accumulation of these fluorochromes in the lesion, this contrast is therefore, in a first approximation, proportional to the concentration ratio between the lesion and its surrounding normal tissues. The optical contrast observed clinically with most of these "standard" fluorochromes is unfortunately frequently not sufficient for reliably and efficiently detecting early cancers which are invisible under standard "white light" observation.

The purpose of this study is therefore to propose a new approach for inducing a fluorescence contrast. The idea consist to use fluorochromes which are sensitive to their physico-chemical environment in order to establish a "spectral" and/or intensity difference between the cancer and the normal tissues, resulting in a high and specific optical contrast. Several micro-environmental factors which are different in certain lesion in regards with the corresponding normal tissues have been reported in the literature. Among them one can mention: the pH[10], the DNA and/or RNA concentration[11], the cell membrane permeability which could modulates the access of the fluorochrome to certain micro-compartments such as the nucleus[12,13], the enzymatic activity[14,15], etc. As the changes in the cell membrane permeability are the only factors which has been reported for clinical early stage carcinoma in the upper aerodigestive tract, the esophagus and the cervix[16-18], we have focused our efforts on the study of a cell impermeant nucleic acid staining fluorochrome.

MATERIAL AND METHODS

A DNA-staining fluorescent dye (unsymmetrical dimeric cyanines) which is used as cell membrane permeability probe and which shows significant changes in its fluorescence yield, fluorescent life-time as well as in its fluorescence excitation and emission wavelengths once bound to DNA, has been investigated *in vivo* and *ex vivo*. The *in vivo* studies involved measurement of the fluorescence pharmacokinetics of this dye using a rat bladder tumor model. *Ex vivo* microspectrofluorometric studies were performed in order to obtain information about the histologic and cellular binding sites of this fluorochrome.

The Fluorochrome

The DNA-staining fluorescent unsymmetrical dimeric cyanine was provided by Molecular Probes, Inc. (P.O. Box 22010, Eugene, OR 97402-0414, USA) in units of 200 ml of a 1 mM solution in 1:4 (v/v) DMSO/H_2O. This dye was dissolved in 0.5 ml of saline before the i.v. injection in rats.

Table 1. TOTO-1 spectral properties (free or coupled to dsDNA).

λ_{EX} [nm]		λ_{EM} [nm]		$\Phi_{F(BOUND)}$ [%]	$\Phi_{F(BOUND)}$ / $\Phi_{F(FREE)}$	ε x 1000 [M^{-1} cm^{-1}]	MW
Free	Bound	Free	Bound				
481	514	630	533	34	1100	117	1303

The spectral properties in the free and bound states of thiazole orange dimer (TOTO-1) are given in Table 1. This dye exhibit a large spectral shift upon binding with DNA or RNA, but the most striking feature of this fluorochrome is the important change of fluorescence quantum yield (~10^3) between the free and bound states[19].

The Rat Bladder Tumor Model

Experiments were carried out on female Fisher CDF rats (Charles River, Wilmington, MA, USA.) bladder tumor model. The Tumor cells transplantation was performed under general anesthesia. The rats were placed in a supine position and the urinary bladder was exposed through an incision. A cell suspension containing $2x10^6$ NBT-II cells (a rat bladder carcinoma cell line provided by: ATCC, 12301 Parklawn Drive, Rockville, MD 20852-1776, USA) was injected directly into the bladder wall and the abdominal wall was closed.

Fluorescence pharmacokinetics experiments were carried out 4-6 days after the implantation of tumor cells, when the average tumor size in the bladder wall was about 4 mm in diameter and 2 mm in thickness. These measurements were performed once again under general anesthesia. The bladder was opened in order to expose the urothelium and the tumor. This surgical procedure was performed so as to spare the major blood vessels which come from the postero-lateral part of the bladder to keep the blood circulation in the tumor and normal tissue near normal conditions. The quartz fiber (600-microns core diameter) of the optical fiber based spectrofluorometer was placed in direct contact with the tissue for the fluorescence measurements. A "background" measurement was recorded on the tumor and normal tissue before the i.v. injection of fluorochrome (1 mg/kg) in a tail vein, and the fluorescence pharmacokinetics studied by measuring the dye fluorescence at different time after injection. Fifteen rats injected with cyanines were involved in this study.

All bladder tumors and normal mucous membranes were excised at the end of the pharmacokinetics study for the microspectrofluorometric investigations. These tissues samples were frozen (-40°C) and 20 μm thick tissue slabs were obtained with a conventional microtome.

The Optical Fiber Based Spectrofluorometer

A schematic of the optical fiber based spectrofluorometer is shown in Fig. 1. A pulsed nitrogen-laser-pumped dye laser system ("nitrogen": VSL-337ND from Laser Science; "dye": DLM 220 from Laser Science, Cambridge, MA, USA) containing Coumarin 500 dye (Exciton Chemical Co., Dayton, OH, USA) was used to generate the fluorescence excitation light. The 3-ns-long dye-laser excitation pulses were launched into a 600-micron core-diameter quartz optical fiber (Superguide-G; Fiberguide Industries, Stirling, NJ, USA) by a 25.4 mm focal length lens. Typical pulse energies incident on the tissue were approximately 5 μJ at 488 nm. Fluorescence from the tissue and the fluorochrome was collected by the same 600 μm fiber and optically coupled to a quartz fiber bundle which in turn goes to a f/3.8, 0.275 m polychromator (Monospec 27; Anaspec, MA, USA). The 600 μm quartz fiber bundle had a circular arrangement of fibers at the input and these fibers were arranged linearly at its output. A long-pass filter ($\lambda > 505$ nm) was inserted before this quartz fiber bundle to eliminate scattered laser light. Fluorescence was recorded using a gated and intensified 1024-diode array controlled by an optical multichannel analyzer (OMA III; Princeton Applied Research, Princeton, NJ, USA). The intensifier was gated with 100 ns pulses centered around the 3 ns laser pulse, eliminating interference from the room lights. The multichannel analyzer records a complete spectrum with each laser pulse. A typical measurement, in which 50 spectra are accumulated, requires 2 seconds. Spectra are transferred to a personal computer for subsequent signal processing, such as background subtraction and correction for the optical response of the system.

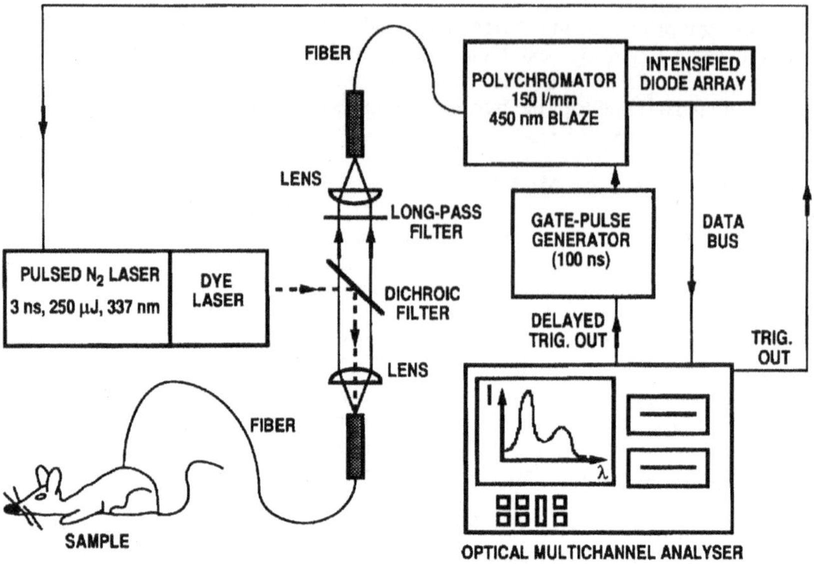

Figure 1. Schematic diagram of the LIF optical fiber-based spectrofluorometer.

Microscopic Fluorescence Imaging and Microspectrofluorometry System

We have constructed a microspectrofluorometry system in order to have the capability of performing both imaging and spectroscopy on tissue sections. The system is based on a Zeiss fluorescence microscope coupled to the OMA system described above, as well as to a monochrome scientific CCD camera (TM-765 from Pulnix, Sunnyvale, CA, USA) to which a separate image intensifier (Litton Industries Inc., Lexington, MA, USA) can be added when needed. In the imaging mode, images from the CCD camera were transferred to a frame-grabber board in an IBM-PC clone computer. A 35 mm film camera for photographic recording can also be used instead of the CCD camera. The microscope has a white illumination source which can be used for conventional "white light" observations and a mercury arc lamp for conventional fluorescence imaging. The system has all the functions of a conventional fluorescence microscope but, in addition, has the ability to perform spectral analysis on a selected microscopic region of a sample. For such fluorescence measurements with the OMA, the pulsed nitrogen-pumped dye laser described above provides fluorescence excitation light via a 600-µm diameter optical fiber (same type as described above) coupled to the microscope. The optics which is located between this fiber and the sample makes the image of the fiber output on the sample with a magnification "M" defined by: M = 20/3.G, where "G" is the magnification of the microscope objective. For example, the use of a 10X objective generates a 400-µm excitation light spot. The size and position of this homogeneous excitation light spot which illuminates the sample can be changed and the typical energy per pulse on the sample was approximately 1 µJ. The optical fiber bundle described above is located in the image plane of the microscope objective at the level of one of the microscope ocular. This fiber bundle collect light from a microscopic region of the sample and transmit it to the OMA for spectral analyses. If "d" is the diameter of the fiber bundle which guides the light to the OMA, the diameter of the probed area on the sample is given by: d/G. The use of a 600-µm fiber bundle with a 100X objective allows the collection and analysis of fluorescence from a 6-µm diameter area.

RESULTS

The fluorescence pharmacokinetics of the TOTO-1 dyes is presented on Fig. 2. The detectable fluorescence corresponded to that of the "bound" state only. No significant fluorescence contrast was observed immediately after injection, which suggests that the contrast observed later was not induced by effects related to the tissue optical properties, the geometry of the sample etc. The fluorescence intensity was maximum at about 10 hours after injection for both the tumor and the normal tissue and the maximum contrast, ~ 10, was established at longer times. Such a high contrast has never been observed with our animal-tumor model with the "conventional" fluorochromes and photosensitizers used in the field of PDT or LIF (Photofrin II, BPD-MA, ALA-PPIX).

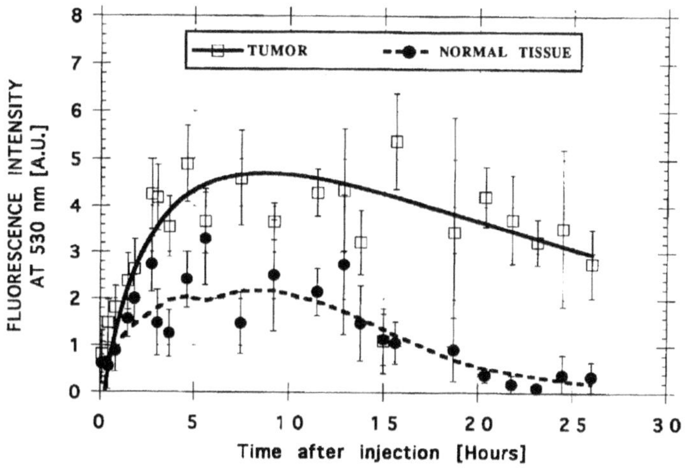

Figure 2. TOTO-1 pharmacokinetics in the tumor and normal bladder wall (1 mg/kg i.v., λ_{ex}: 488 nm).

In principle, this high contrast, observed using such environmentally sensitive fluorochromes, can be due to two mechanisms. The first one is related to dye accumulation in the lesion and the second one due to environmental effects. We have performed some microspectrofluorometric investigations of frozen tissue slices with the system described above in order to determine the origin of this contrast. The granular structure of the fluorescence images observed in the tumor tissue (not shown here) suggests that most of the fluorescence originates in the nuclei. The spectral analyses of the fluorescence emitted by one cell nucleus indicate that this fluorescence corresponds only to that of the dye in the bound state. The fluorescence pattern observed in the normal surrounding tissues was clearly more uniform, which means that the cellular nuclei fluorescence no longer play a major role in the overall fluorescence emitted by normal tissue. The spectrum of the fluorescence measured in the nuclei as well as in the cytoplasm of the normal tissues indicates once again that this emission corresponds to that of the dye in the bound state. This may be due to the binding of fluorochrome molecules with RNA.

DISCUSSION

This study demonstrated that environmentally sensitive fluorochromes can be used in order to perform a tissue differentiation *in vivo*. These preliminary results suggest that the observed fluorescence contrast is due in part to two factors: 1) differences in the cell membrane integrity between the atypical and normal cells. Since the TOTO-1 fluorescent dye is also a very good probe for evaluating the cells permeability, its uptake is probably enhanced in the malignant cells. 2) Once in the cell, the binding of this dye with DNA strongly increases its fluorescence quantum yield. Therefore, the metachromatic properties of this dye seems to induce a significant amplification of the concentration contrast. The large change in fluorescence quantum yield of this cyanine dye between its free and DNA-bound states, together with the fact that the fluorescence of the tumor is generated in the cells nuclei whereas this is not the case in the normal tissue, indicate that the fluorescence contrast is essentially due to an "environmental" effect and not to a preferential dye accumulation in the tumor.

This principle for establishing a contrast enables therefore to avoid the drawbacks related with techniques based on the use of selective fluorochromes. This point is of particular interest as a highly specific accumulation of fluorochrome is frequently very difficult to obtain in a large proportion of cancer types (excepting the use of ALA applied topically in the bladder). This is in particular the case for early invisible carcinomas which constitute the main application of interest for LIF photodetection. The principle proposed here enables also to increase the number of "discrimination factors (DF)" used for performing a differentiation between the signal of the fluorescent marker and that of the background (autofluorescence, etc.). Indeed, continuous wave LIF photodetection is based on only two of these DF (i.e. the fluorescence excitation and emission wavelength), whereas the use of environmentally sensitive fluorochrome enables moreover to exploit the change of quantum yield and the related change of fluorescence life-time. A tissue characterization based on this last parameter is of high interest as small variations of the fluorescence life-time which is sensitive to the physico-chemical properties of the environment can easily be detected. A photodetection procedure based on the use of environmentally sensitive fluorochromes is therefore potentially more specific (more than 3 DF) than the steady state fluorescence techniques (2 DF).

This study indicates therefore than time and frequency domain fluorescence spectroscopy can be used in order to perform an *in vivo* tissue characterization (cancer photodetection) based on exogenous and/or endogenous fluorochromes.

Acknowledgments

This work has been supported by: the Swiss National Fund for Scientific Research and National Priority Program in Optics I; the U.S.A. Office of Naval Research and Department of Energy.

Moreover, the authors are grateful to J. A. Parrish and M. Bamberg for their support and help.

REFERENCES

1. J.H. Kinsey and D.A. Cortese, Endoscopic system for simulatneous visual examination and electronic detection of fluorescence, *Rev. Sci. Instrum.* 51 (1):1403 (1980).
2. A.E. Profio, O.J. Balchum and F. Cartens, Digital background subtraction for fluorescence imaging, *Med. Phys.* 13:717 (1986).
3. T. Hirano, M. Ishizuka, K. Suzuki et al., Photodynamic cancer diagnosis and treatment system consisting of pulsed lasers and endoscopic spectro-image analyzer, *Lasers in the Life Sciences* 3 (2):99 (1989).
4. R. Baumgartner, H. Fisslinger, D. Jocham et al., A fluorescence imaging device for endoscopic detection of early stage cancer. Instrumental and experimental studies, *Photochem. Photobiol.* 46:759 (1987).
5. G. Wagnières, Ch. Depeursinge, Ph. Monnier et al., Photodetection of early cancer by laser induced fluorescence of a tumor-selective dye: apparatus design and realization, *Proc. of SPIE* 1203:43 (1990).
6. S. Andersson-Engels, J. Johansson and S. Svanberg, Multicolor fluorescence imaging system for tissue diagnostics, *Proc. of SPIE* 1205:179 (1990).
7. B. Palcic, S. Lam, J. Hung and C. MacAulay, Detection and localization of early lung cancer by imaging techniques, *Chest* 99:742 (1991).

8. S. Lam, C. MacAulay, J. Hung, J. LeRiche, A.E. Profio and B. Palcic, Detection of dysplasia and carcinoma *in situ* with a lung imaging fluorescence endoscope device, *J. Thoracic Cardiovascular Surg.* 105:1035 (1993).

9. G. Wagnières, D. Braichotte, A. Châtelain, Ch. Depeursinge, Ph. Monnier, M. Savary, Ch. Fontolliet, J.-M. Calmes, J.-C. Givel, G. Chapuis, S. Folli, A. Pèlegrin, F. Buchegger, J.-P. Mach and H. van den Bergh, Photodetection of early cancer in the upper aerodigestive tract and the bronchi using photofrin II and colorectal Adenocarcinoma with fluoresceinated monoclonal antibodies, *Proc. of SPIE* 1525:219 (1991).

10. S. Mordon, V. Maunoury, J.-M. Devoiselle, Y.Abbas and D. Coustaud, Characterization of Tumorous and Normal Tissue using a pH-Sensitive Fluorescence Indicator (5, 6-carboxyfluorescein) in vivo, *J. Photochem. Photobiol. B:Biol.* 13:307 (1992).

11. H.S. Rye, S. Yue, D.E. Wemmer, M.A. Quesada, R.P. Haugland, R.A. Marthies and A.N. Glazer, Stable fluorescent complexes of double-stranded DNA with bis-intercalating asymmetric cyanine dyes: properties and applications, *Nucleic Acids Res.* 20:2803 (1992).

12. P. Herlin, J. Marnay, J.H. Jacob et al., A study of the mechanism of the Toluidine blue dye test, *Endoscopy* 15:4 (1983).

13. O. Kocher, M. Amaudruz, A.M. Schindler et al., Desmosomes and gap junctions in precarcinomatous and carcinomatous conditions of squamous epithelia (an electron mocroscopic and morphometrical study), *J. Submicrosc. Cytol.* 13: 267 (1981).

14. L. Massaad, I. de Waziers, V. Ribrag, F. Janot, J. Morizet, Ph. Beaune, A. Gouyette and G.G. Chabot, Screening of principal enzymes involved in the metabolism of anticancer drugs in human and murine colonic tumors, *Bull. Cancer* 80(5):397 (1993).

15. G. Weber, "Advances in enzyme regulation", Pergamon Press, vol 15 (1976).

16. A.M. Mandard, J. Tourneux, M. Gignouxet et al., In situ carcinoma of the oesophagus : macroscopic study with particular reference to the Lugol test, *Endoscopy* 12:51 (1980).

17. R.M. Richart, A clinical staining test for the in vivo delineation of dysplasia and carcinoma in situ, *Am. J. Obstet. Gynecol.* 86:703 (1963).

18. P. Herlin, J. Marnay, J.H. Jacob et al., A study of the mechanism of the Toluidine blue dye test, *Endoscopy*15:4 (1983).

19. L.G. Lee, C.H. Chen, L.A. Chiu, Thiazole Orange: A New Dye for Reticulocyte Analysis, *Cytometry* 7:508 (1986).

SENSITIVE AND RAPID DETECTION OF ß-GALACTOSIDASE EXPRESSION IN INTACT CELLS BY MICROINJECTION OF FLUORESCENT SUBSTRATE

Odd Terje Brustugun, Gunnar Mellgren, Bjørn Tore Gjertsen, Rolf Bjerkvig and Stein Ove Døskeland

University of Bergen, Department of Anatomy and Cell Biology,
Årstadveien 19, N-5009 Bergen, Norway
Phone: +47 55 20 63 75
Fax: +47 55 20 63 60
E-mail: Stein.Doskeland@pki.uib.no

INTRODUCTION AND PRINCIPLE

Bacterial ß-galactosidase, coded by *lacZ*, has become the reporter of choice for studies of transcriptional activity on the single cell level. Unfortunately, current detection methods, like X-gal cytochemistry cannot be applied to live cells since the hydrolysis product of X-gal is toxic and the method requires fixation and permeabilization of the cells. Furthermore, a long incubation time (hours to days) is required and the method has limited sensitivity[1]. Other methods (immunocytochemical; introduction of fluorescent substrate by hypotonic shock) have also proven to be either incompatible with study of viable cells[1], or not tolerated by all cell types[2,3,4]. We will report a method for detection of ß-galactosidase in living cells without affecting morphology, viability or proliferating capacity. A more detailed description of the method is found in reference 2.

The method is based on microinjection of a substrate which is hydrolyzed by ß-galactosidase to a fluorescent product. Positive cells, staining green, are detected under UV-illumination using standard filter sets as for FITC. The sensitivity of the method was far better than other commonly used methods. The rapidity with which cells became stained and the long duration of the stained state suggested that the method could be used to fluorescently tag transfected cells and study e.g. their shape, motion and interaction with neighboring cells.

Fluorescence Microscopy and Fluorescent Probes
Edited by J. Slavík, Plenum Press, New York, 1996

PREPARATIONS

Reagents required

5-chloromethylfluorescein di-ß-D-galactopyranoside (CMFDG) and plasmid expressing *lacZ*. If plasmid is not to be transfected by microinjection, one needs reagents for e.g. lipofection.

Used in this study: CMFDG and phenylethylthio-ß-D-galactopyranoside (PETG) were from Molecular Probes, Inc., Eugene, OR, USA, 5-bromo-4-chloro-3-indolyl-ß-D-galactopyranoside (X-gal), 8-(4-chlorophenylthio)-cAMP (8-CPT-cAMP), forskolin, and 3-isobutyl-1-methylxanthine (IBMX), were from Sigma, St. Louis, MO, USA. Lipofectamine were from Gibco, Paisley, Scotland. [^{3}H] thymidine (40 - 60 Ci/mmol) was from Amersham, Little Chalfont, UK. The plasmid pCH 110 (constitutively expressing *lacZ*) was from Pharmacia, Uppsala, Sweden. Plasmid 5xCRE *lacZ* containing cAMP responsive gene elements driving the *lacZ* reporter gene[5], was generously supplied by Dr. John Scott, Vollum Institute, Portland, Oregon.

Cell lines

Cells growing in monolayer is a prerequisite for microinjection. This should be obtainable for most cell types, even for those normally grown in suspension, requiring coating with an adequate growth matrix (e.g. collagen, fibronectin)[6].

Used in this study: Kirsten *ras* transformed NIH 3T3 rat fibroblasts (R3T3) were grown in Dulbecco´s modification of Eagles minimal essential medium (DMEM) with 10% newborn bovine serum. Primary rat liver hepatocytes were isolated by perfusion, and cultured in a synthetic hepatocyte-selective medium as described by Mellgren *et al* [7]. The stable *lacZ* -positive cell line AN1 *lacZ* was derived (by lipofectin transfection and selection with G-418 [Sigma Chemical Co.]) from a human glioma cell line (D-54MG) kindly supplied by professor Darell D. Bigner at Duke University, N.C.[8] The cells were grown in DMEM supplemented with 10% newborn bovine serum, 4 times the standard concentration of non-essential amino acids, 2% L-glutamine, penicillin (100 U/ml) and streptomycin (100 µg/ml).

Microinjection, transfection and detection

Cells were seeded at 15,000 - 25,000 cells per cm^2. Fibroblasts and glioma cells were injected 16 - 30 h after seeding, and hepatocytes 50 - 72 h after seeding.

R3T3 cells were exposed for 6 h to 1 ml serum-free medium containing Lipofectamine (24 µl) and 2 µg of either 5xCRE *lacZ* or of pCH 110. After transfection the cells were trypsinized, pooled and plated in dishes suitable for microinjection. Alternatively, R3T3 cells and primary hepatocytes were transfected by microinjection of plasmid DNA (about 0.2 µg/µl of injected solution). CMFDG was microinjected at a concentration of 10 mM in 50% (v/v) DMSO/water. The specific ß-galactosidase inhibitor PETG was sometimes injected to estimate non-specific fluorescence. Alternatively, ß-galactosidase activity was detected by the formation of visible, blue precipitates through the hydrolysis of X-gal, essentially as described by MacGregor *et al.* [1].

Equipment required: (Examples used in this study) Microcapillaries and puller (model P-87 from Sutter Instrument Co., Novato, CA, USA) or pre-pulled microcapillaries (Eppendorf). Micromanipulator (Eppendorf 5170), microinjector (Eppendorf 5242) and an incubator providing 37°C and 5% CO_2 mounted on an inverted microscope with UV-lamp (Zeiss Axiovert 35M). Filters as for FITC-detection (BP 450-490 nm for excitation, barrier filter LP 520 nm, and chromatic beam splitter FT 510 nm, all from Zeiss). We used NUNC culture dishes (10 cm^2) with gridlines.

RESULTS AND DISCUSSION

Specificity and stability. Very weak non-specific autofluorescence was detected in non-transfected fibroblasts, unaffected by microinjection of either vehicle or CMFDG, and unaffected by injection of the specific ß-galactosidase inhibitor[3] PETG. This inhibitor, when injected before the injection of substrate in transfected cells, blocked completely the fluorescence development, further underscoring the specificity of CMFDG.

Studies on the stability of the fluorescent product of CMFDG (chloromethylfluorescein) under minimal illumination indicated that the fluorescent product had a long, but not indefinite, effective half life in the fibroblasts. If observing cells under constant illumination is necessary, recovery of light emission can be noted after 10-20 min in the dark, i.e. substrate survives illumination, even if product is destroyed.

Effect on cell viability and DNA replication. In order to exclude toxic effects of injected CMFDG itself or of its products the glioma cells were left for 9 - 14 h after injection and then scored for DNA replicative activity by [^{3}H]-thymidine incorporation and autoradiography as previously described[9]. There was no decline in cell number or DNA replication in cells injected with CMFDG (Fig. 1; Table 1). Similar results were obtained with fibroblasts and primary hepatocytes (not shown).

Comparison of sensitivity. 5xCRE-*lacZ* -transfected R3T3 fibroblasts were treated with cAMP-elevating agents (forskolin and IBMX), and *lacZ* expression monitored by microinjection of CMFDG. The percentage of fluorescent cells increased to about 50%

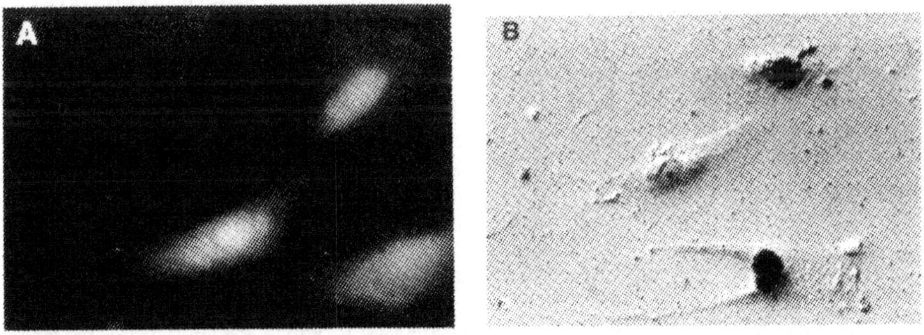

Figure 1. Panel A shows AN1 *lacZ* human glioma cells injected with CMFDG 23 h after seeding. Panel B shows an autoradiogram of the same cells 9 h later, after a 1 h pulse with [^{3}H]thymidine (3.0 μCi/dish). Note the high number of autoradiographic grains overlying the nuclei in active DNA replication. The panels show the same cells, and it can be seen that some have migrated slightly and changed shape. Magnification 600x. © Academic Press.

within 12 h after cAMP challenge. In contrast, less than 5% of the cells stained positively by the X-gal method (Fig. 2). The microinjection method proved therefore far more sensitive than the X-gal cytochemical method for detection of transfected cells.

Table 1. DNA replication and density of AN1 *lacZ* glioma cells microinjected with fluorogenic ß-galactosidase substrate

Agents injected	% S-phase cells (L.I.)	±SEM	Cells/mm^2	±SEM	n
None	22.7	2.8	220	14	12
Vehicle	22.6	2.3	221	12	9
10 mM CMFDG	21.0	3.2	209	20	8

Note. Human glioma AN1 *lacZ* cells were injected with vehicle (50% dimethylsulfoxide in water) only or with 10 mM of the fluorogenic substrate CMFDG. 12 h later the cells were exposed to [^{3}H] thymi dine, and processed for autoradiography. The percent of nuclei labeled with [^{3}H] thymidine (labeling index, L.I.) was determined as well as the number of cells per mm^2, in which all cells had been injected with vehicle or with CMFDG. Uninjected neighboring fields served as controls. The standard error of the mean (SEM) and the number of separate experiments (n) is indicated.

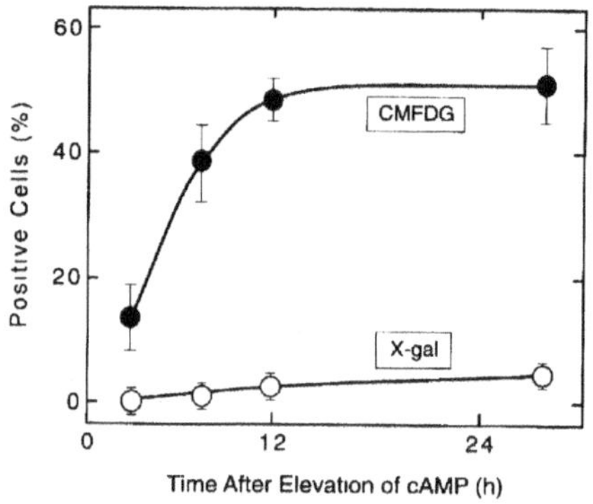

Figure 2. The time course of appearance of detectable *lacZ*-expression in transfected fibroblasts as evaluated by microinjection of fluorescent substrate and X-gal cytochemistry. R3T3 fibroblasts were lipofectin transfected with 5xCRE *lacZ* and 48 h later exposed to cAMP challenge with 10 µM forskolin and 0.1 mM IBMX. The expression of ß-galactosidase was evaluated 4, 8, 12 and 28 h after start of cAMP challenge by conventional X-gal cytochemistry (○) or by injection of the fluorescent substrate CMFDG (●). The number of fluorescent cells was scored 20 min after injection. The cell dishes were stained using the X-gal method 30 min after the cells had been injected. The bars represent standard error of the mean of three to ten separate experiments. © Academic Press.

We have also compared fibroblasts retrovirally transfected with the gene for green fluorescent protein (GFP)[10] and *lacZ* governed by the same RSV promoter. Results indicates that X-gal is slightly more sensitive than GFP, which means that the presently described method is far more sensitive than the GFP fluorescence method.

Primary rat hepatocytes microinjected with the *lacZ* reporter gene under the control of a cAMP-responsive (CRE) promoter showed about the same percentage of positive cells and similar cAMP-induction of ß-galactosidase as fibroblasts. Again, detection by CMFDG injection proved much more sensitive than X-gal cytochemistry (Fig. 3).

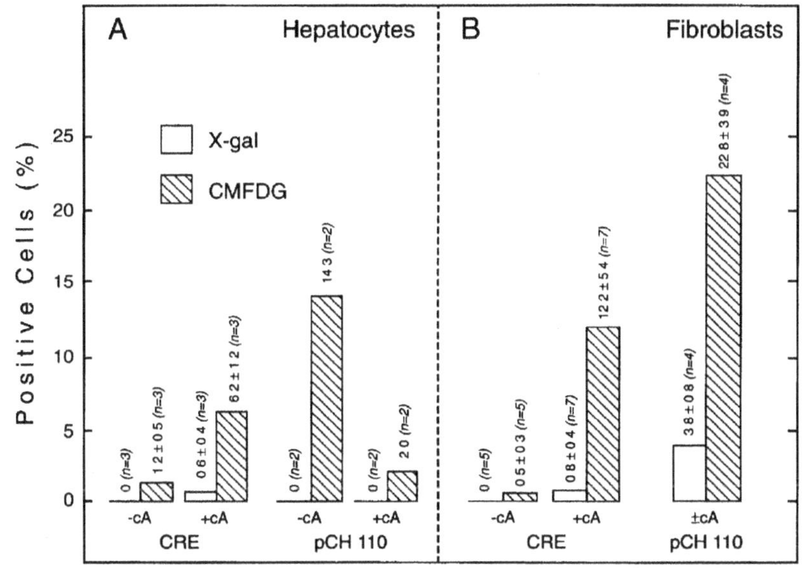

Figure 3. Comparison of microinjection of fluorogenic CMFDG and X-gal cytochemistry for detection of *lacZ*-expression in hepatocytes (A) and fibroblasts (B). The cells were first microinjected with 5xCRE *lacZ* or pCH 110. Some cultures (+cA) received cAMP challenge (10 μM forskolin and 0.5 mM IBMX) immediately after injection of DNA. 24 h after transfection the cells were assayed for expression of ß-galactosidase either by X-gal cytochemistry or by scoring the number of fluorescent cells after microinjection of the fluorogenic galactosidase substrate CMFDG. The data are given as mean ± standard error of the mean of three to seven separate experiments, except for hepatocytes microinjected with pCH 110 where the average of two experiments is given. © Academic Press.

ACKNOWLEDGMENTS

The technical assistance of Nina Lied Larsen and Erna Finsås is highly appreciated. This work was funded by the Norwegian Cancer Society (DNK), the Norwegian Research Council (NFR), and the NOVO Nordic foundation.

REFERENCES

1. MacGregor, G. R., Mogg, A. E., Burke, J. F., and Caskey, C. T. (1987) *Somat. Cell Mol. Gen.* **13**, 253-265.
2. Brustugun, O. T., Mellgren, G., Gjertsen, B.T., Bjerkvig, R., and Døskeland, S.O. (1995) *Exp. Cell Res.* **219**, 372-378.
3. MacGregor, G. R., Nolan, G. P., Fiering, S., Roederer, M., and Herzenberg, L. A. (1991) *in* Gene Transfer and Expression Protocols (Murray, E.J., Ed.), Vol. 7, pp. 217-235, The Humana Press Inc., Clifton.
4. Zhang, Y.-Z., Naleway, J. J., Larison, K. D., Huang, Z., and Haugland, R. P. (1991) *FASEB J.* **5**, 3108-3113.
5. Meinkoth, J. L., Montminy, M. R., Fink, J. S., and Feramisco, J. R. (1991) *Mol. Cell. Biol.* **11**, 1759-1764.
6. Vintermyr, O.K., Gjertsen, B.T., Lanotte, M., Døskeland, S.O. (1993) *Exp.Cell Res.* **206**, 157-161
7. Mellgren, G., Vintermyr, O. K., Døskeland, S. O. (1995) *J. Cell. Physiol.* **163**, 232-240.
8. Bigner, D. D., Bigner, S. H., Ponten, J., Westermark, B., Mahaley, M. S., Ruoslahti, E., Herschman, H., Eng, L. F., and Wikstrand, C. J. (1981) *J. Neuropath. Exp. Neurol.* **15**, 201-227.
9. Vintermyr, O. K., Mellgren, G., Bøe, R., and Døskeland, S. O. (1989) *J. Cell. Physiol.* **141**, 371-382.
10. Chalfie, M., Tu, Y., Euskirchen, G., Ward, W. W., and Prasher, D. C. (1994) *Science* **263**, 802-805.

FLUOROGENIC SUBSTRATES REVEAL GENETIC DIFFERENCES IN ALDEHYDE-OXIDATING ENZYME PATTERNS IN RAT TISSUES

Jacek Wierzchowski, Piotr Wroczynski, Elzbieta Interewicz, Izabela Orlanska and Jacek Przybylski

Department of Physical Chemistry, Faculty of Pharmacy, and
Department of Physiology,
Medical Academy,
1 Banach St.,
PL-02097 Warsaw, Poland.

INTRODUCTION

Spontaneously hypertensive rat (SHR) is an inbred rat line developing arterial hypertension early in life, accepted as a model for human essential hypertension[1]. It was found recently, that additional characteristics of SHR rats is avoidance of ethanol consumption in free-choice conditions[2]. One reason for such avoidance may be different activity of aldehyde-metabolizing enzymes, as observed in ALDH-2 deficient humans[3] and animal models[4,5]. In the present work, we examine the activity of ethanol-metabolizing enzymes, alcohol dehydrogenase (ADH, E.C.1.1.1.1.) and aldehyde dehydrogenase (ALDH, E.C.1.2.1.3.), as well as aldehyde oxidase (AldOx) in some organs of SHR rats, and compare them to those obtained for the outbred Wistar rats.

The traditional methods of ADH and ALDH detection, based on the NADH production, are not specific enough to quantify the corresponding activities in crude tissue homogenates by continuous method[6]. Several alternative assays have been proposed, usually combining spectroscopic methods with chromatography[7,8], but these procedures are rather costly and time-consuming.

We have recently developed fluorimetric assays for aldehyde-metabolizing enzymes, based on redox transformations of the fluorogenic naphthaldehydes[9,10]. In particular, 4-methoxy-1-naphthaldehyde (MONAL-41) is an excellent fluorogenic substrate for human class I ADH, allowing direct, continuous assay of this enzyme in human serum. The isomeric 6-methoxy-2-naphthaldehyde (MONAL-62) was found to be a good substrate for both ALDH and class II human ADH[9], while 7-methoxy-1-naphthaldehyde (MONAL-71) allows continuous assay of ALDH activity in human erythrocytes[10].

In the present paper, we describe application of the above naphthaldehydes to direct, continuous measurement of the ALDH activity in several rat organ homogenates. During our preliminary studies we found that some homogenates oxidized the naphthaldehyde substrates even in the absence of NAD^+, presumably via the aldehyde oxidase system. We now present data obtained for both ALDH and aldehyde oxidase activities found in the rat liver, stomach and kidney, as well as the results of the ADH determination in liver. These data indicate significant differences in enzyme activities between SHR and normotensive rat lines.

Fluorescence Microscopy and Fluorescent Probes
Edited by J. Slavík, Plenum Press, New York, 1996

MATERIALS AND METHODS

The commercial MONAL-41 (Aldrich) was used after single crystallization from hexane, and syntheses of MONAL-62 and MONAL-71 are described elsewhere[9,10]. DTT, NAD^+ and the protein assay kit were obtained from Sigma (St. Louis, MO., U.S.A.) and all other chemicals were of analytical grade. The Millipore-filtered water was used throughout.

Rat organs were collected within 2 hrs after animal decapitation and stored at -70 °C until homogenation. The homogenates (S-9 fraction, combining cytosol and microsomes, and mitochondrial fraction) were prepared according to standard protocol in isotonic buffer[11], with 0.5 mM DTT present in the medium, and examined within 5 hrs. We have observed that single freezing and thawing kills up to 70% of the initial ALDH activity.

Assay Procedure

20 µl of tissue homogenate (containing 1-30 mg protein/ml) was added to 2.9 mL of 50 mM pyrophosphate buffer, pH=8.1, containing 0.5 mM EDTA and 0.5 mM DTT. After 5 min preincubation, a fluorogenic substrate (MONAL-62 or MONAL-71) was added to the final concentration of 3-7 µM, and the fluorescence signal increase (due to aldehyde oxidase activity) recorded for 8-10 min. After this period, NAD^+ was added (final concentration 100 µM) and the signal recorded for next 2-10 min. Finally, an internal standard (purified product) was added (final concentration ~3 µM), and standard fluorescence measured.

Instrumental settings were: for MONAL-62, excitation, 310 nm, observation at 360 nm; for MONAL-71, excitation, 330 nm, observation at 390 nm, spectral bandwidth for excitation, 2.5 nm, for emission, 8 nm. The instrument was a SPEX' Fluoro-Max.

Reaction rates were calculated as previously described[10], and ALDH and AldOx reactions were assumed to be additive.

The ADH activity was measured according to the methodology developed for human serum[8], using 15 µM MONAL-41 as a fluorogenic substrate, and measuring increase of the sample fluorescence at 370 nm, with excitation at 316 nm.

The obtained rates (typically in the range 5-1000 nM/min) were then divided by protein concentration in the homogenate, measured by the Bradford method.

RESULTS AND DISCUSSION

Aldehyde dehydrogenase (ALDH) is an ubiquitous, polymorphic enzyme of relatively low specificity, previously shown to be present in nearly all rat tissues[12]. It has been recently confirmed that the activity of this enzyme is correlated with alcohol preference-avoidance shown by at least two distinct rat lines in free-choice conditions[4,5]. The somewhat unexpected discovery of alcohol avoidance by SHR line prompted us to check if there is any relation of this phenomenon to ADH/ALDH activity.

The purpose of our work is therefore twofold: (i) to check if the ALDH/ADH activity can be measured directly by continuous method in crude tissue homogenate using new fluorogenic substrates and (ii) to compare the respective values in selected organs taken from SHR and normotensive Wistar rats.

Optimisation of the Assay

MONAL-62 and MONAL-71 are fluorogenic substrates for human ALDH, and their spectral and kinetic characteristics are given elsewhere[10]. Both compounds, when exposed to a diluted rat tissue extract, undergo fast oxidation to highly fluorescent naphthoates. Typical curves showing fluorescence increase in the 6 µM MONAL-62 solution in the presence of rat kidney homogenate (protein concentration ~0.1 mg/mL) are presented on Fig.1 (left panel). It is evident that some signal drift can be observed prior to addition of NAD^+, and this we interpret as an aldehyde oxidase (AldOx, E.C.1.2.3.2.)[13] activity. After addition of 100 µM NAD^+ (saturating human ALDH), the oxidation rate increases about 10-fold for the combined S-9 fraction, and ~2-3 fold for the mitochondrial fraction, what we ascribe to ALDH reaction. The measured rates were up to 1 µM/min, well above the sensitivity limit, which we estimate as <1 nM/min[10]. Similar results were obtained for liver, stomach, lung, heart, brain and intestine extracts with both MONAL-62 and MONAL-71.

The reaction products, obtained both in presence and absence of NAD^+, were found to be spectrally identical with the corresponding naphthoates, i.e., 6-methoxy-2-naphthoate (emission maximum 360 nm) and 7-methoxy-1-naphthoate (390 nm). As an additional test, the reaction medium was acidified with sulfuric acid to pH 2, which converted both naphthoates to naphthoic acids (emission maxima at 405 and 420 nm, respectively), showing essentially no traces of the corresponding alcohols (both having maxima at 355 nm). This confirms the lack of interference by ADH. The ALDH activity was therefore calculated by subtraction of the aldehyde oxidase reaction rate, measured prior to NAD^+ addition.

To optimize assay conditions, saturation curves were measured for oxidation of both naphthaldehyde substrates by rat ALDH. The results obtained for the stomach and lung S-9 fractions are shown on Fig. 1 (right panel). It is evident that the ALDH activity, as measured with MONAL-62, is saturated at ~6 μM substrate concentration, with the apparent K_m value of 0.5 μM. By contrast, the activity measured with MONAL-71 displayed non-Michaelian kinetics, with no full saturation up to 10 μM. We were also unable to achieve saturation of the (putative) AldOx activity, even at the highest substrate concentrations used (60 μM).

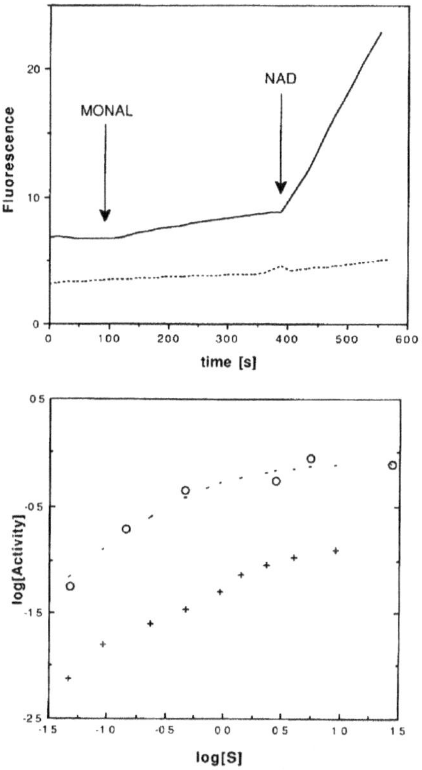

Figure 1. (Top) Typical time courses of the fluorescence increase during naphthaldehyde oxidation by rat organ homogenates. (----) mitochondrial (—) S-9 fraction. (Bottom) Saturation curves for ALDH activity, as measured with MONAL-62 (o) and MONAL-71 (+) in stomach and lung (S-9 fractions).

It is known that at least 6 ALDH isozymes were so far identified in the rat liver[14]. While we are presently unable to ascribe the measured activity to individual isozymes, it is evident that at least one cytosolic (or microsomal) and one mitochondrial form is reactive towards MONAL-62. The very low K_m value is similar to that displayed by human "low K_m" ALDH forms[10]. It is also apparent that MONAL-71, being inferior in terms of kinetic parameters, gives nevertheless a complementary information about the rat ALDH system.

ALDH and Aldehyde Oxidase Activities in Rat Organs

Preliminary experiments indicated that the above methodology is capable to detect and quantify ALDH activity in nearly all rat tissues, except for blood. Also the AldOx activity is evidently present in all tissues studied, being more pronounced in mitochondrial fractions. We present results obtained for S-9 fractions of rat stomach, liver and kidney, which were chosen for preliminary study for their relatively high ALDH activity. The results, comparing SHR and normotensive Wistar rats (n=3 for each), are summarized in Table I.

The ALDH activities found in rat liver, kidney and stomach were the highest when MONAL-62 (6 μM) was used as a substrate. Typical C.V. for homogeneous rat groups were in this case 15-20%, and the activities were comparable to those reported by Koivisto[4], who used 100 μM acetaldehyde as a substrate. MONAL-71 was much less effective as an ALDH indicator, showing both lower activities and higher variation coefficients.

Table I. Activities (Units/gram protein) of aldehyde-oxidizing enzymes, as measured by fluorimetric method in rat organ homogenates (cytosolic+microsomal fractions). Standard deviations are given in parentheses; asterisks denote significant differences.

Organ	ALDH		AldOx	
	Wistar	SHR	Wistar	SHR
Liver(A)[1]	3.73(0.37)	2.74(0.26)*?	0.48(0.09)	0.073(.005)*
Liver(B)	0.20(0.10)	0.42(0.20)	0.79(0.24)	0.031(.002)*
Kidney(A)	3.2(0.6)	3.3(0.4)	0.27	0.34
Stomach(A)	3.16(0.66)	6.7(1.9)*	0.23	0.24
Stomach(B)	0.12(0.07)	0.04(0.03)	0.026	0.039

[1]Activity measured with 6 μM MONAL-62 (A) or 3 μM MONAL-71 (B) and 100 μM NAD^+.

The most striking result of our investigation is a large difference (7-20 fold) between AldOx activities in the livers of outbred Wistar vs. SHR rats, evident in both MONAL-62 and MONAL-71 measurements. In the latter case, the AldOx activity found in normotensive line is higher than that of ALDH, and interferes in ALDH assay. Such differences are absent in other rat organs.

Significant differences between ALDH activity in the outbred and SHR rats were found in stomach and probably in liver, while no difference was observed in kidney. These differences seem to be quantitative, rather than qualitative, since saturation experiments gave similar apparent K_m values for both control and SHR rats (data not shown).

Mitochondrial fractions have shown much less ALDH activity (up to ~1.3 U/g), contrasting with literature data[4,14]. Furthermore, this activity was usually obscured by the oxidase reaction. We conclude that naphthaldehydes are relatively worse substrates for mitochondrial form(s) of the rat ALDH, similarly as it was found for human enzymes[10].

Liver ADH activities in wild and SHR rats

We compared also liver alcohol dehydrogenase (ADH) activities in the two rat lines, using previously characterized MONAL-41 as a fluorogenic substrate for human class I ADH and NADH as a cofactor. Aldehyde is the preferred ADH substrate since for most of the mammalian ADH forms, the reverse reaction (aldehyde reduction) is faster than alcohol oxidation. The present assay is free of interference by ALDH, since the corresponding naphthoate is only weakly fluorescent, while the alcohol has emission yield of 0.35[9], allowing its selective observation. For the rat ADH measurements we used conditions optimised for human enzyme, i.e., substrate concentration of 15 μM and 30 μM of NADH, obtaining average activities of 15.1±2.1 U/g for normotensive rats and 4.4±1.4 U/g for the SHR line (n=3 each). These results indicate significant difference in liver ADH activity between two rat lines, which may be related to alcohol preference/avoidance.

Concluding remarks

This work confirms, that ALDH activity can be measured in tissue homogenates by continuous method using MONAL-62 and/or MONAL-71 as fluorogenic substrates. The mammalian ALDH is of interest in connection with alcoholism studies[3], but also carcinogenesis[15] and drug activation[16], so the new fluorogenic substrates are of obvious importance. It remains to be established which isozymes of rat ADH and ALDH react with the fluorogenic naphthaldehydes, and what is degree of isozyme selectivity of the substrates.

The described method can also quantify aldehyde oxidase activity in rat organs. It is not possible at this point to precisely identify which oxidases are active towards MONAL substrates, however, the most likely candidate is the molybdenum-containing oxidase[13]. While the relation of this enzyme class to alcoholism or hypertension is not clear, it would be certainly of interest to measure their activity in other alcohol-avoiding rat lines.

The results presented here are preliminary, but in our view they unambigously indicate significant differences in alcohol and aldehyde-metabolizing enzyme activities between the control and SHR rats. The differences are, in fact, even more pronounced than those found between AA and ANA lines[4], which were selected on the basis of alcohol preference. The multi-fold difference in AldOx activity may indicate qualitative genetic difference between two lines (mutation), while differences in ALDH activities are probably quantitative.

This work was supported by Medical Academy, Warsaw, project K/23.

REFERENCES

1. Y. Yamori, Pathogenesis of spontaneous hypertension as a model for essential hypertension, Jap. Circul. J. 41:259 (1977).
2. K. Papierski et al. (Medical Academy, Warsaw), to be published elsewhere.
3. D. P. Agarwal and H. W. Goedde, The role of alcohol metabolizing enzymes in alcohol sensitivity, drinking habits and incidence of alcoholism in Orientals, in Molecular Pathology of Alcoholism, N. Palmer, ed., Oxford Univ. Press, London (1991), p.211.
4. T. Koivisto and C. J. P. Eriksson, Hepatic aldehyde and alcohol dehydrogenases in alcohol-preferring and alcohol-avoiding rat lines, Biochem. Pharmacol. 48:1551 (1994)
5. L. Carr, B. Mellenkamp, D. Crabb, L. Lumeng and T.-K. Li, A polymorphism in the rat liver mitochondrial ALDH2 gene is associated with alcohol drinking, In Enzymology and Molecular Biology of Carbonyl Metabolism, vol. 3, H. Wiener, ed., p. 61, Plenum Press, N. York (1990).
6. S. C. Guru and T. Shetty, Methodological aspects of aldehyde dehydrogenase assay by spectrophotometric technique, Alcohol 7:397 (1990).
7. A. Helander, Assay of human blood aldehyde dehydrogenase activity by high-performance liquid chromatography, Alcohol 4:121 (1987).
8. T. Yamamoto, Y. Moriwaki, S. Takahashi, Y. Nasako, K. Hiroishi and K. Higashino, Determination of erythrocyte aldehyde dehydrogenase activity by HPLC, Anal. Biochem. 215:129 (1993).
9. J. Wierzchowski, B. Holmquist and B. L. Vallee, Detection of alcohol dehydrogenase in human serum with isozyme-specific substrates, Anal. Chem. 64:181 (1992).
10. J. Wierzchowski, E. Interewicz, P. Wroczynski and I. Orlanska, Continuous fluorimetric assay for human aldehyde dehydrogenase and its application to blood analysis, Anal. Chim. Acta, in press.
11. A. Y. H. Lu and W. Levin, Patial purification of cytochromes P-450 and P-448 from liver microsomes, Biochem. Biophys. Res. Commun. 46:1334 (1972)
12. R. A. Dietrich, Tissue and subcellular distribution of mammalian aldehyde-oxidizing capacity, Biochem. Pharmacol. 16:1911 (1966).
13. C. Beedham, Molybdenum hydroxylases: biological distribution and substrate-inhibitor specificity, in Progress in Medicinal Chemistry, Vol. 24, G. P. Ellis and G. B. West, eds., Elsevier, Amsterdam (1987), p.85.
14. R. Lindahl, S. Evces, Comparative subcellullar distribution of aldehyde dehydrogenase in rat, mouse and rabbit liver, Biochem. Pharmacol. 33:3383 (1984).
15. R. Lindahl, Aldehyde dehydrogenases and their role in carcinogenesis, Crit. Revs Biochem. Mol. Biol. 27:283 (1992).
16. N. E. Sladek, Xenobiotic metabolism catalyzed by aldehyde dehydrogenase, Drug Metab. Rev. 20:697 (1989).

BINDING OF PROTHROMBIN FRAGMENT 1 TO PHOSPHATIDYLSERINE CONTAINING VESICLES: A SOLVENT RELAXATION STUDY

Rudi Hutterer[1] and Martin Hof[2]

[1]Department of Physical Chemistry, University of Wuerzburg, Marcusstr. 9/11, D-97070 Wuerzburg, Germany.
[2]Department of Physical Chemistry, Charles University, Albertov 2030, Cz-12840 Prague 2, Czech Republic.

INTRODUCTION

Prothrombin is the substrate of the prothrombinase which consists of factor Xa, factor Va, calcium ions and anionic phospholipid surfaces. A key question for understanding the prothrombinase complex is the molecular mechanism of the interaction of the vitamin K dependent proteins (factor Xa and prothrombin) with the extraordinary procoagulant phospholipid phosphatidylserine (PS). For this interaction a chelate model has been suggested: The calcium ions are thought to form a coordinate complex with the γ-carboxyglutamic acid (gla) residues of these vitamin K dependent proteins and the negatively charged headgroups of PS [1,2].

The fluorophores 6-propionyl-2-(dimethylamino)naphthalene (Prodan) and 6-palmitoyl-2-[[2-(trimethylammoniumethyl)]methylamino]naphthalene chloride (Patman) are supposed to reside in the hydrophobic-hydrophilic interface of phospholipid membranes. In polar viscous media they show a large time-dependent Stokes shift. This spectral relaxation can be attributed to the reorientation dynamics of the molecular domains surrounding the chromophore. Thus, by monitoring the time-dependence of the Stokes shift the mobility of the hydrated phospholipid headgroups can be characterised quantitatively.

In the present work the analytic approach developed for picosecond relaxation dynamics in polar solvents[3] is applied to the characterisation of the time-dependent Stokes shifts of Patman and Prodan adsorbed to small unilamellar vesicles (SUV). Using this method, the effect of calcium ions and varying lipid composition on the headgroup mobility of SUV composed of PS and phosphatidylcholine (PC) is investigated. The interaction of vitamin K dependent proteins with the negatively charged headgroups of PS is studied using a model compound, bovine prothrombin fragment 1 (BF1). BF1 is the 1-156 amino acid N-terminus and contains 9 gla residues. It has been shown to exhibit membrane binding characteristics similar to prothrombin[4].

MATERIALS AND METHODS

The phospholipids were obtained from Fluka and used as received. Prodan and Patman were purchased from Molecular Probes. Isolation and purification of prothrombin fragment 1 were carried out as described[5]. Small unilamellar vesicles of different lipid composition were prepared by sonication[6]. The fluorophores were added to the phospholipid mixture prior to sonication (molar content of fluorophore: 1%). For the BF1 binding studies SUV composed of a 60:40 PC/PS lipid mixture were used. The ratio of protein to lipid concentration was chosen large enough to achieve 95% saturation of binding sites[7-9]. All measurements were performed at 25°C, which is well above the phase transition temperature of both lipids.

Fluorescence decays were recorded with commercial single photon counting equipment (Edinburgh Instruments 199S) and analysed using an iterative reconvolution technique, as described[10]. Time-evolved emission spectra (TEES) were calculated from the fit parameters of the multiexponential decays detected from 390 to 530 nm and the corresponding steady-state intensities[3]. The TEES were fitted by log-normal functions[11]. Autocorrelation functions C(t) are calculated from the emission maxima ν(t) of the TEES at defined time t after excitation[3]:

$$C(t) = \frac{\nu(t) - \nu(\infty)}{\nu(o) - \nu(\infty)} \quad (1)$$

In all cases the solvent response cannot be satisfactorily described by a monoexponential relaxation model. In order to characterise the overall time scale of the solvation response, we use an (integral) average relaxation time:

$$\langle\tau\rangle \equiv \int_0^\infty C(t)\, dt \quad (2)$$

RESULTS AND DISCUSSION

Solvent relaxation of Prodan and Patman in PC/PS-SUV in the absence and presence of Ca^{2+}

When incorporated in fluid phase membranes both dyes show large time-dependent Stokes shifts (≈ 50 nm). The TEES of Patman are slightly blue shifted compared to those of Prodan (≈ 15 nm). For all investigated systems, the solvent response of Patman is about two-fold slower than for Prodan. Since the chromophores of both dyes are nearly identical, we conclude that Patman is incalated deeper between the lipid acyl chains sensing a less polar and / or more restricted environment. Due to the lack of a long acyl chain Prodan is localised very close to the membrane water interface. These conclusions are supported by most recent NMR-studies[12].

Addition of 3 mM $CaCl_2$ does not affect the solvent relaxation behaviour of pure PC-vesicles, as monitored by both dyes. Thus, the zwitterionic PC-headgroup does not interact with the Ca^{2+} ions at the given experimental conditions. In the absence of Ca^{2+} a small increase in the relaxation times of Patman with increasing PS-content is observed, while Prodan shows no effect. The presence of Ca^{2+} yields much stronger dependence of the solvent relaxation on the PS-content. For Prodan the average relaxation times increase from 1.1 ns for 100% PC-vesicles to 1.5 ns for vesicles composed of an equimolar PC/PS-mixture. For Patman a continuous increase from 2.1 ns to 3.4 ns is observed. These results are summarized in figure 1.

The slight deceleration of the headgroup mobility with increasing PS-content in absence of Ca^{2+} suggests a lower flexibility of the PS-headgroup. This conclusion is in agreement with NMR-studies, having shown that the rotational mobility of the PS-headgroup is lower than for phosphatidylglycerol, which is on the other hand less mobile than the PC-headgroup. In contrast to pure PC-vesicles the addition of 3 mM Ca^{2+} leads to a considerable deceleration of the solvent relaxation in PS-containing vesicles as observed by both dyes. This effect increases with increasing PS-content. Thus, the deceleration of the headgroup mobility with increasing PS-content is much more pronounced when Ca^{2+} is present. These results indicate a tighter phospholipid headgroup packing with increasing PS-content and suggest a bridging of PS-molecules by Ca^{2+} within the plane of the membrane leading to a decrease in lipid mobility. A Ca^{2+}-induced increase of the lateral packing density in bilayers has already been suggested by studies[13] investigating the phase transition of 1,2-dipentadecanoyl-3-sn-phosphatidylglycerol / 1,2-dimyristoyl-3-sn-phosphatidylcholine-SUV in absence and presence of Ca^{2+} .

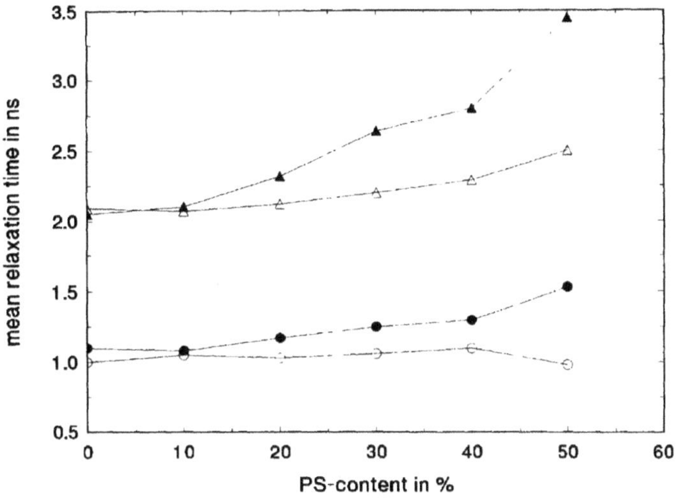

Figure 1. Average relaxation times for Prodan (circles) and Patman (triangles) in PC/PS-mixtures as function of PS-content with (filled symbols) and without (open symbols) 3 mM Ca^{2+}

Information about the relative number of excited states contributing to the emission at given times can be obtained from the halfwidths of the reconstructed TEES (data not shown). In all cases the halfwidths increase with time after excitation and reach a maximum at a time near to the average relaxation time. At longer times after the excitation we observe a decrease in the halfwidths corresponding to a reduced number of excited states. Increasing PS-content broadens the TEES at all times after excitation showing a larger heterogeneity of the dye environment. Moreover, a comparison of the halfwidths of Patman with those of Prodan indicates a less heterogeneous localisation of Patman, probably due to its long acyl chain.

Influence of Prothrombin Fragment 1 on the Relaxation Behaviour

In order to apply the solvent relaxation method described above to the BF1-Ca^{2+}-PS interaction, it had to be verified that the used fluorophores do not bind significantly to BF1 in presence of lipid. This was done comparing steady-state spectra of Prodan and Patman in buffer, in presence of 45 µM BF1, and after addition of different amounts of vesicle suspension. While the addition of BF1 did not change the emission spectra, addition of vesicles in presence of the protein led to a new blue shifted emission band caused by the binding of Prodan and Patman, respectively, to the membrane. Thus, we can be sure that no significant binding of the dyes to the protein does occur.

Binding of BF1 to vesicles composed of a 60:40 PC/PS mixture in presence of 5 mM Ca^{2+} causes a significant deceleration of the solvent relaxation of both dyes (figures 2 and 3). The mean relaxation times increase from 2.2 to 2.7 ns for Patman and from 1.1 to 2.2 ns for Prodan. These results show that the analysis of the solvent relaxation behaviour of dyes in the upper membrane region can be a new, very sensitive method for monitoring the membrane binding of these vitamin K dependent and other membrane proteins.

In contradiction to these experiments fluorescence anisotropy studies of DPH associated with the fluid phase of PC/PS vesicles did not detect any effect on the membrane order induced by BF1 binding[9]. We conclude that the formation of the coordination complex by PS, Ca^{2+} ions and the gla residues of BF1 [1,2] leads to an increased rigidity of the phospholipid headgroup region, but not to a

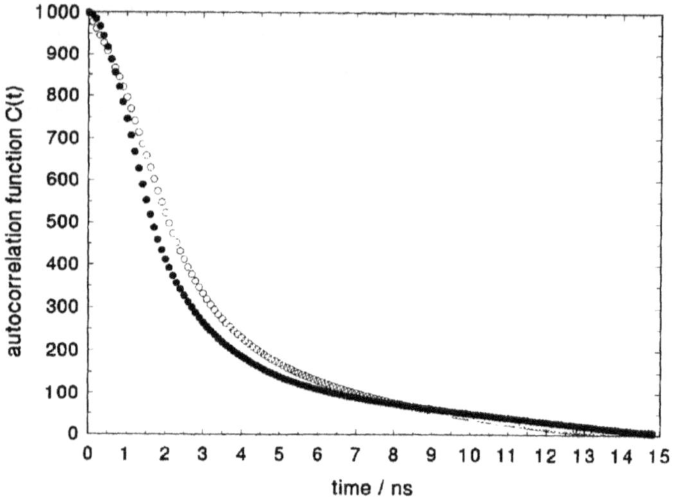

Figure 2. Autocorrelation functions for Patman in PC/PS 60.40-SUV in presence (open circles) and absence (filled circles) of 45 μM bovine fragment 1.

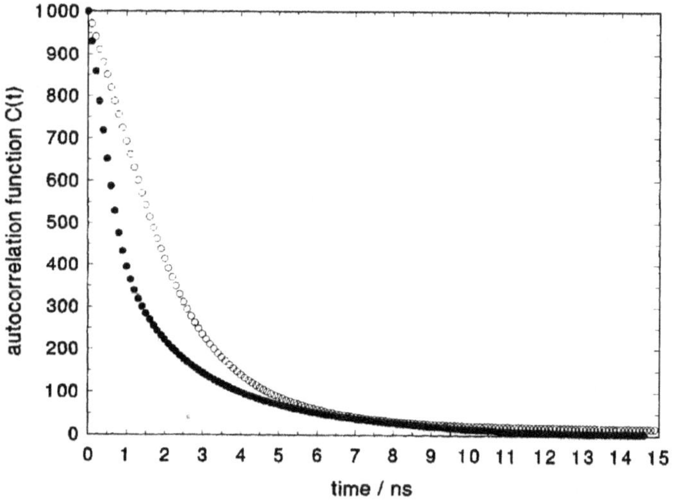

Figure 3. Autocorrelation functions for Prodan in PC/PS 60·40-SUV in presence (open circles) and absence (filled circles) of 45 μM bovine fragment 1.

higher packing density in the hydrocarbon region of the bilayer. The deceleration of the solvent relaxation process by BF1 binding is much larger for Prodan, which is bound at the surface of the membrane, than for Patman, which is located closer to the hydrocarbon region. Thus, the phospholipid moieties binding to the central ion (Ca^{2+}) and forming the chelate complex seem to be located close to the membrane water interface. Considering the chemical structure of PS, the amino and carboxyl group of the serine moiety are supposed to be directly located at the interface and, therefore, they are good candidates for the described binding. The results presented herein favour a BF1-Ca^{2+}-PS-chelate interaction[1] and seem to exclude partial penetration of the membrane by BF1.

CONCLUSIONS

In this work the solvent relaxation method developed for picosecond dynamics in polar solvents is used to characterise quantitatively the headgroup mobility in SUV. By this approach structural changes in the membrane water interface region can be monitored with higher sensitivity than it can be done by conventional fluorescence anisotropy techniques. Thus, this method is predestinated for investigating the interaction of membranes with peripheral membrane proteins. Using fluorophores which are localised slightly different in the phospholipid headgroup region, we have shown that BF1 interacts predominantly with phospholipid moieties located at the membrane water interface. Studies investigating the membrane binding of the entire protein, bovine prothrombin, should verify the existence of hydrophobic, calcium-independent binding sites in prothrombin, not present in BF1[9,14].

ACKNOWLEDGEMENT

We thank Prof. R.G. Hiskey for providing BF1, Prof. F.W. Schneider for helpful discussions and generous support and Prof. V. Fidler as well as Dr. S. Vajda for providing analysis programs. M.H. is a recipient of the "Liebig Stipendium" of the "Fonds der Chemischen Industrie".

REFERENCES

1. P. Comfurius, E.F. Smeets, G.M. Willems, E.M. Bevers, and R.F.A. Zwaal, Assembly of the prothrombinase complex on lipid vesicles depend on the stereochemical configuration of the polar headgroups of phosphatidylserine, *Biochemistry* **33** 10319-10324 (1994).
2. J. Rosing, H. Speijer, and R.F.A. Zwaal, Prothrombin activation on phospholipid membranes with positive electrostatic potential, *Biochemistry* **27**, 8-11 (1988).
3. M. Maroncelli and G.R. Fleming, Picosecond solvation dynamics of coumarin 153. the importance of molecular aspects of solvation, *J Chem Phys* **86**, 6221 (1987).
4. K.H. Pearce, M. Hof, B.R. Lentz, and N.L. Thompson, Comparison of the membrane binding kinetics of bovine prothrombin and its fragment 1, *J Biol Chem* **268**, 22984-22991 (1993).
5. M. Hof, G.R. Fleming and V. Fidler, Time-resolved fluorescence study of calcium-induced conformational change in prothrombin fragment 1,submitted to *Proteins Structure, Functions and Genetics*
6. C.L. Bashford, C.G. Morgan and G.K. Radda, Measurement and interpretation of fluorescence polarisations in phospholipid dispersions, *Biochim Biophys Acta* **426**, 157-172 (1976).
7. G.L. Nelsestuen and M. Broderius, Interaction of prothrombin and blood clotting factor X with membranes of varying composition, *Biochemistry* **16**, 4172-4177 (1977).
8. G.A. Cutsforth, R.N. Whittaker, and B.R. Lentz, A new model to describe extrinsic protein binding to phospholipid membranes of varying composition: application to human coagulation proteins, *Biochemistry* **28**, 7453-7461 (1989).
9. S.W. Tendian and B.R. Lentz, Evaluation of membrane phase behaviour as a tool to detect extrinsic protein-induced domain formation: binding of prothrombin to phosphatidylserine/phosphatidylcholine vesicles, *Biochemistry* **29**, 6720-6729 (1990).
10. M. Hof, J. Schleicher, and F.W. Schneider, Time resolved fluorescence in doped aerogels and organosilicate glasses, *Ber. Bunsenges. Phys. Chemie* **93**, 1377 (1989).
11. D.B. Siano and D.E. Metzler, Band shapes of the electronic spectra of complex molecules, *J. Chem Phys* **51**, 1856 (1969).
12. R. Hutterer, F.W. Schneider, H. Sprinz, M. Hof, Differences in localisation of Patman and Prodan in vesicles detected by time-resolved NMR- and fluorescence spectroscopy,manuscript in preparation.
13. B.R. Lentz, D.A. Alford, M.E. Jones, and F.A. Dombrose, Calcium-dependent and calcium-independent interactions of prothrombin fragment 1 with phosphatidylglycerol/phosphatidylcholine unilamellar vesicles, *Biochemistry* **24**, 6997-7005 (1985).
14. S.W. Tendian, B.R. Lentz, and N.L. Thompson, Evidence from total internal reflection fluorescence microscopy for calcium-independent binding of prothrombin to negatively charged planar phospholipid membranes, *Biochemistry* **30**, 10991-10999 (1991).

NEW THIOL ACTIVE FLUOROPHORES FOR INTRACELLULAR THIOLS AND GLUTATHIONE MEASUREMENT

Margaret E. Langmuir†, Jun-Rui Yang†, Karen A. LeCompte† and Ralph E. Durand‡

†Covalent Associates, Inc.
10 State Street
Woburn, MA 01801

‡B.C. Cancer Research Centre
600 West 10th Avenue
Vancouver, B.C.

INTRODUCTION

Cellular thiols, especially glutathione (GSH), act as scavenger nucleophiles and can protect cells against: hyperoxia, toxicity, many carcinogens and mutagenicity or transformation by radiation. A technique which combines the use of a cell membrane permeable thiol reactive fluorophore with that for cell sorting by flow cytometry (FACS) has been used as a means of quantitating glutathione in individual cells or in cell subpopulations[1,2]. These studies are aiding the understanding of the role of GSH and other protein thiols in viable cells. Although many fluorescent thiol probes exist, few are useable in viable cells. We report here the results of tests of six newly synthesized naphthopyranone fluorophore based thiol probes[3,4] in V79 Chinese hamster lung fibroblasts. Time and concentration dependence of dye uptake was evaluated, as were subcellular localization, dye cytotoxicity, and specificity of the dyes for GSH relative to other cellular thiols.

EXPERIMENTAL

Spectral data listed in Table 1 were obtained from aqueous phosphate buffered solutions of the probes containing 10% ethanol. Absorption spectra were taken on a Hitachi U-2000 spectrophotometer and fluorescence spectra were taken on a Hitachi 3010 spectrofluorometer. Fluorescence spectra were corrected using a Rhodamine B quantum counter. Quantum yields were determined in air saturated solutions by the method of Parker and Rees[5] using quinine sulfate in 1 N sulfuric acid as the quantum standard ($\phi = 0.55$).

All flow cytometric studies were performed on a Becton Dickinson FACS-IV flow cytometer with argon laser excitation at 350-360 nm. "Fluorescence relative to background" reported in the FACS studies refers to fluorescence of cells treated with the thiol probes relative to untreated cells. The probe solutions were prepared in dimethylsulfoxide (DMSO) for loading into cells.

Table 1. Optical Characteristics of Maleimides (M) and Their GSH Adducts

Compound		Absorption of maleimide		Corrected emission			ϕ_{GSH}/ϕ_M	Stokes shift, nm
		λ, nm	log ε	λ, nm	ϕ_M	ϕ_{GSH}		
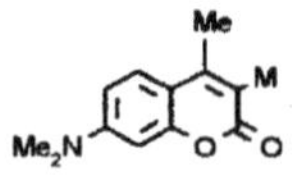	DACM	396	4.29	482	--	0.12	--	86
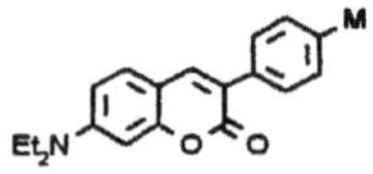	CPM	387	4.60	475	0.021	0.13	6	88
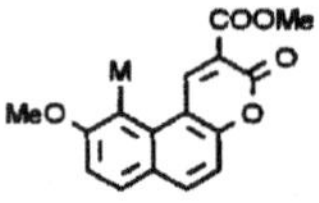	1	384	4.18	513	0.016	0.65	40.6	129
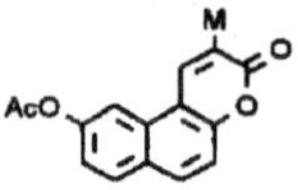	2	358	4.46	430	0.023	0.18	7.8	72
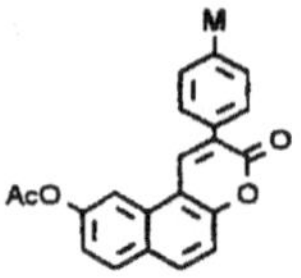	3	363	4.34	446	0.003	0.48	160	83
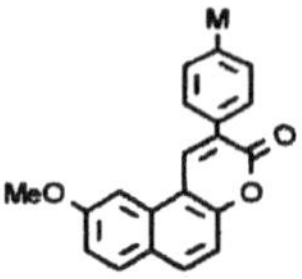	4	376	4.28	501	0.061	0.62	10	125
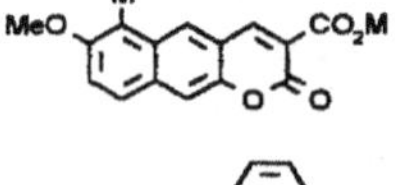	5	365	4.42	536	0.020	0.12	6	171
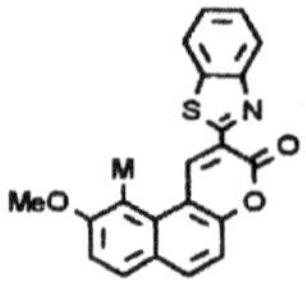	6	417	4.43	516	0.012	0.66	55	99

RESULTS

The probes were synthesized in our laboratory[3,4] and their absorption and emission spectra, extinction coefficients and quantum yields before and after reaction with GSH were determined in 10 volume percent ethanol/aqueous phosphate buffer at pH 7.0. These data are summarized in Table 1 along with those of DACM and CPM, two widely used fluorescent thiol reagents. From a standpoint of quantum yield and sensitivity the new probes **1, 3, 4** and **6** are superior to CPM and DACM. The Stokes shifts of **1, 4, 5,** and **6** are more favorable than CPM and DACM. However from the standpoint of solubility in aqeous media, compounds **4** and **6** are problematic and it is necessary to work in aqueous solutions with at least 10% ethanol, acetonitrile or dimethylsulfoxide present as a cosolvent.

Working concentrations of the probes for cellular studies were determined as shown in Figure 1, where intracellular fluorescence was evaluated after 5-minute exposures to increasing probe concentrations in DMSO. Concentrations of several μg/mL were found to be optimal for all six probes. It is of interest that increasing probe concentrations did not lead to progressive increases in the intracellular dye concentration, but rather suggested that a threshold concentration of about 1 μg/mL had to be exceeded before measurable binding was observed. A concentration of 5 μg/mL was used for all subsequent studies and was compared to the threshold value of 1μg/mL.

The subcellular localization of each of the dyes was determined with fluorescence microscopy/image analysis techniques, as indicated in Figure 2. UV excitation at 365-380 nm was employed, with emission observed above 450 nm. Diffuse, amorphous staining was seen for all compounds. To provide an indication of the relative thiol binding of each agent, the imaging camera gain was kept constant and in each case a concentration of 5μg/mL of the probe was used The combination of dye loading and quantum yield of the probe/thiol adducts in the cell indicated that dyes **1** and **4** are the most sensitive followed by **3** and **5** while dyes **2** and **6** either show little uptake or their thiol adducts have very low quantum yields in the intracellular medium.

Exposure time studies from 0.2 to 100 minutes were conducted to determine rate of intracellular uptake of the probes. With the exception of probe **2**, cell fluorescence reached a maximum within a minute and remained constant for up to 10 minutes and then declined gradually

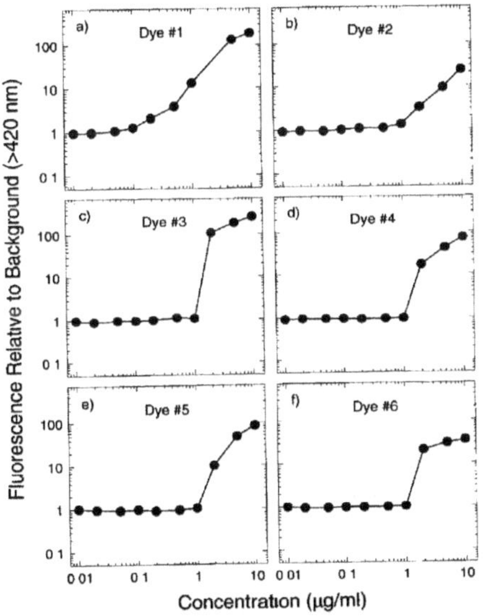

Figure 1. Relative fluorescence observed by FACS for increasing concentrations of the indicated dyes.

Figure 2. Images of dyes **1-6** in Chinese hamster cells; all images were obtained at the same camera gain. Note lack of specific intracellular localization.

over the next 90 minutes. The rapid uptake and reaction with intracellular thiols to saturation within a minute is remarkable. The decrease of fluorescence response with time beyond 10 minutes may be indicative of toxicity (non-renewal of glutathione, etc.) or gradual loss of the fluorescent adduct from the cell. This aspect of cell loading will need to be studied further.

Toxicity of the probes, shown as inhibition of proliferative potential (clonogenicity), is shown in Figures 3a and 3b; note that dose is expressed on a log scale in 3a to allow easier discrimination of survivals at low probe concentrations. All probes except **2** demonstrably killed some of the cells which were exposed, but a considerable range of toxicity was observed in addition to some variability between the two experiments. As shown in 3b, survival rates of better than 50% for 30 minutes exposures to 10μg/mL were obtained except for dye **1**. It should be remembered that only one minute exposure to 5μg/mL of these probes is necessary to react with all available intracellular thiols. The fact that probe **2** showed no toxicity suggest that it did not react with intracellular thiols. In subsequent hydrolysis experiments with this dye, we found that it is five times more rapidly hydrolyzed to the thiol inactive maleamic acid than the other five probes[6]. This is probably also responsible for the notably low response of probe **2** in Figure 2.

V79 cells were grown in the presence of DL-buthionine-S,R-sulfoximime (BSO) for 24 hours to reduce the intracellular GSH content to less than 5% that of control cells. Relative probe binding in both the untreated and treated populations is shown in Figure 4, with three dye exposure times evaluated. We also compared the probes to others which are often used to measure GSH levels in cells. These are: orthophthaldialdehyde (OPT), monobromobimane, and CPM. Probes **1, 3, 4** and **5** show the expected 40% decrease due to decreased GSH in the cells. When compared to CPM and OPT, probes **1, 3,** and **4** are more sensitive to changes in GSH levels due to higher quantum yields of their GSH adducts. They also have advantages over monobromobimane in that they are not light sensitive and react faster with GSH.

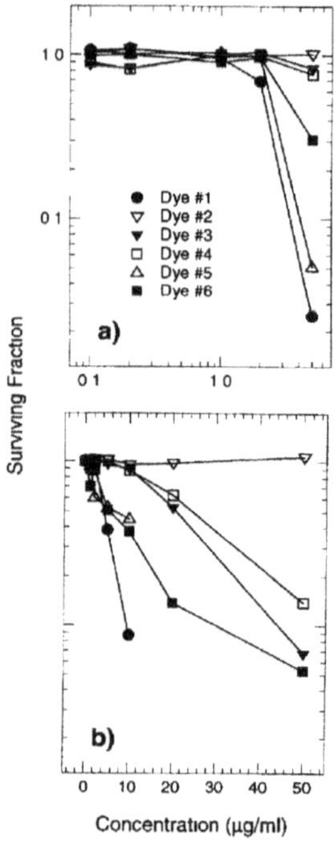

Figure 3. Cytotoxicity of the dyes in V79 cells as a function of dye concentration.

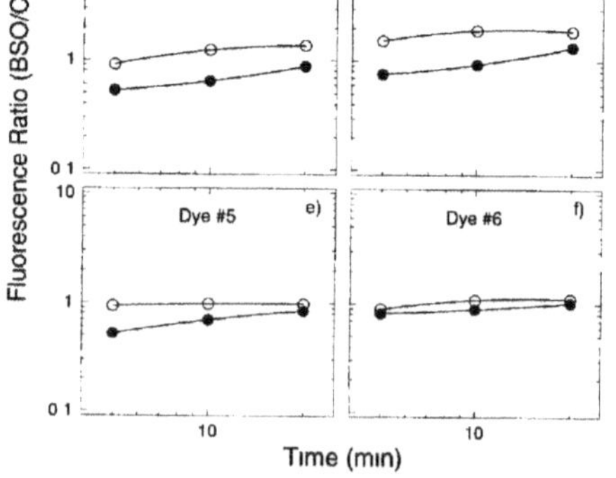

Figure 4. Fluorescence in GSH depleted cells relative to undepleted cells , both exposed to probes **1-6**.

CONCLUSIONS

One of the most important features of these probes is their extremely small fluorescence quantum yields **prior** to reaction with thiols and their large increase in quantum yield **after** reaction with thiols. This gives them high sensitivity as expressed by the quantity ϕ_{GSH}/ϕ_M in Table 1.

As with other maleimide thiol probes, hydrolysis of the maleimide ring occurs in aqueous phosphate buffer which makes them insensitive to thiols, but this reaction is a factor of two slower for probe **1** compared to DACM or CPM at pH 7. Its half life is of the order of 8 hours, much longer than the time required to accumulate the dye into the cell and react with thiols. Furthermore, with the exception of probe **2**, the corresponding maleamic acid hydrolysis products of probes **1-6** have even lower quantum yields than the maleimide probes so that they never contribute to background fluorescence. This is particularly important in cells where one cannot separate the probe reagent from the fluorescent reaction product. In contrast, many thiol reagents currently used with cells (i.e. fluorecein maleimide, tetramethylrhodamine maleimide and others) have significant fluorescence themselves and their fluorescence increases on hydrolysis. Thus it is difficult to distinguish whether the measured fluorescence is that of a thiol adduct or of the hydrolysis product of the reagent.

About one minute exposure time to the probe solution is required to accumulate the dye within a population of cells. Once in the cell, it reacts immediately with GSH and other cellular thiols. At pH 7 this reaction which results in the fluorescent adducts is much faster than that of thiols with chloro or bromobimanes. Because of the speed of the reaction, it should be possible to follow the rate of production of new intracellular GSH in BSO treated (GSH depleted) cells. Even in non BSO treated cells, an increase in fluorescence with time could be interpreted as signaling an increase in GSH with time as long as there is excess probe in the cell to react with GSH as it is formed. It should also be possible to follow the depletion of GSH in cells caused by transporter proteins provided that it is possible to first determine the rate of formation of intracellular GSH.

Of the six new probes tested here, probes **1, 3** and **4** are probably the most useful for determining where thiols exist in cells *i.e.*, for imaging. And **1** and **3** are best for discriminating GSH when used in conjunction with BSO. All three of these probes are now commercially available from Covalent Associates, Inc. under the trade names **ThioGlo™ 1, 3** and **4.** All three offer greater sensitivity for thiol measurement in cells and in other applications than existing thiol probes because of the high quantum yields of their thiol adducts.

ACKNOWLEDGMENT

This work was supported by Grant No. 5 R44-CA57067 from the National Cancer Institute of the National Institutes of Health, Bethesda, MD USA.

REFERENCES

1. P. L. Olive, J. E. Biaglow, M. E. Varnes and R. E. Durand, *Cytometry*, **3**: 349 (1982).
2. R. E. Durand and P. L. Olive, *Radiation Research*, **95**: 456 (1983).
3. J. R. Yang and M. E. Langmuir, *J. Heterocyclic Chem.*, **28**: 1177 (1991).
4. M. E. Langmuir, J. R. Yang, A. M. Moussa, R. Laura and K. A. LeCompte, *Tetrahedron Lett.*, **36**: 3989 (1995).
5. C. A. Parker and W. T. Rees, *Analyst*, **85**: 587 (1966).
6. Y. M. Wang and M. E. Langmuir, manuscript in preparation.

QUANTIFICATION OF MACROPHAGES IN THE CARDIOVASCULAR SYSTEM OF HYPERCHOLESTEROLEMIC RABBITS BY USE OF DIGITAL IMAGE PROCESSING

Jürgen Metz, Peter Pavlov, Ralf Kinscherf, Christoph Köhler, Claus Usinger

Department of Anatomy and Cell Biology III, University of Heidelberg,
Im Neuenheimer Feld 307, 69120 Heidelberg, Germany

ABSTRACT

Macrophages are suggested to play a key role in the pathogenesis of atherosclerosis in hypercholesterolemic rabbits. We characterized macrophages in the cardiovascular system of normo- and hypercholesterolemic rabbits by use of quantitative fluorescence immunohistochemistry. Colocalization of RAM11-, MHC Class I- and II-, MAC-1-, and manganese superoxide dismutase (MnSOD)-immunoreactivities could be demonstrated in a subpopulation of macrophages using differently fluorescent labelled secondary antibodies. Area density and numbers of macrophages were evaluated with an image analysis board and software developed in our laboratory. This immunohistochemically characterized subpopulation of macrophages was rarely found in the peri- and paravascular connective tissue of several organs of normocholesterolemic rabbits, but significantly increased in the heart and intimal hyperplasias, and to a lower extent in other organs of hypercholesterolemic rabbits. In conclusion, quantitative fluorescence immunohistochemistry was used to characterize a subpopulation of macrophages, which seem to respond specifically to hypercholesterolemia.

INTRODUCTION

Macrophages are thought to play an important role in arteriosclerosis. In hyperlipidemic rabbits they internalize and metabolize modified low-density lipoproteins (LDL) and assemble in the intima of the arterial wall as was especially indicated by immunostaining of atherosclerotic plaques with RAM-11 antibodies (Stadius et al., 1992). Among the mechanisms, which are presently discussed to cause LDL modification in vivo (Henning and Chow, 1988; Rankin et al., 1991) superoxide anion-induced lipid peroxidation seems to be of major importance (Cathcart et al., 1989). Since macrophages are able to bind and sequester this form of modified LDL, which was shown to be extremely cytotoxic to cells in vitro (Rosenfeld, 1991) they are likely subjected to a severe oxidative stress.

We recently were able to demonstrate that a subpopulation of macrophages within the peri- and paravascular space in the myocard, which was identified by RAM-11-immuno-

Fluorescence Microscopy and Fluorescent Probes
Edited by J. Slavík, Plenum Press, New York, 1996

reactivity, exhibited colocalization with manganese superoxide dismutase (MnSOD)-immunoreactivity (Kinscherf et al., 1995). This antioxidant enzyme builds a primary defense mechanism against oxidative damage by catalyzing the dismutation of superoxide anion radicals to oxygen and hydrogen peroxide (Dhaunsi et al., 1993; del Rio et al., 1990; Singal et al., 1993). We were interested to further characterize these macrophages using antibodies against major histocompatibility complex (MHC) encoded proteins (e. g. Daar et al., 1983; Labarrere et al., 1991).

MATERIALS AND METHODS

Animals. Male white new zealand rabbits (WNZ) weighing about 3 kg were housed individually under the same conditions, with dark-light cycles of 12h. The control group (n = 3) was given a standard rabbit diet and the experimental group (n = 4) received an atherogenic 0.5% cholesterol enriched diet for 42 days. For preparation of tissue the animals were killed by an intravenous overdoze of sodium pentobarbital at the end of the experiments. Heart and aorta were quickly removed, frozen as composite blocks in isopentane cooled by liquid nitrogen, and stored at -80° C until used. The animal studies were approved by the Referat Veterinärwesen at the Regierungspräsidium Karlsruhe.

Immunohistology: 6 µm sections were cut in a cryostat at -20° C and placed on microscope slides that had been precoated with silan adhesive subbing solution and then air dried. The sections were fixed in acetone for 10 min at -20° C and air dried. Non-specific binding was blocked with 1 % normal swine serum in 10 mM phosphate buffered saline (PBS), pH 7.2 (10 min; room temperature [RT]). Immunostaining was performed with monoclonal mouse anti-human MnSOD (Bender, Vienna, Austria), anti-rabbit macrophage [RAM-11] (Dianova, Hamburg, Germany), anti-human MAC-1 (CD 11b/CD 18), W 6/32 (MHC I), and HLA-DR (MHC II) antibodies (Boehringer Mannheim, Germany). 50 µl of antibody solutions, containing between 400 ng/ml and 160 µg/ml antibody concentrations, were added onto the tissue section and incubated (24 h; RT). Then 50 µl of biotinylated anti-mouse or anti-rat Ig (Amersham, Braunschweig, Germany) diluted 1:100 in PBS and finally 50 µl of streptavidin-fluorescein [FITC] (Amersham, Braunschweig, Germany) diluted 1:100 in PBS were applied to the section and incubated (1 h; RT) in the dark. Double staining procedures were performed using streptavidin-FITC and streptavidin-Texas red. Washing steps in PBS for 10 min at RT were included. Omission of the primary antibody abolished the immunohistochemical reaction completely.

A standard Olympus BH2 microscope fitted with a Hg light source and epifluorescence assembly was used with corresponding filter sets. Photographs were taken with an Olympus OM camera on Kodak Tri-X film. For morphometry microscope images were recorded by an Olympus HCC-3600 high gain video camera. The area of immunostained macrophages was marked as negative and contrast-enhanced images by grey level processing at a magnification of 900-fold using a computer-assisted image analysis system [VFG1 frame grabber; VIBAM-software] developed in our group. The area density (AD) was evaluated by the ratio of the area of macrophages (AM) and the myocardial reference area (RA) (AD [%] = AM/RA x 100). This measurement was repeated within five different fields of one left ventricular cross section. The mean ± SEM of the medians of both animal groups was determined and statistically analysed.

RESULTS

In the myocardium of controls and cholesterol-fed rabbits only a very few RAM-11 immunoreactive cells were found in the peri- and paravascular space. MnSOD as well as MHC I and II, and MAC-1 immunoreactive cells, which all were more numerous than the RAM-

11 immunoreactive cells in controls, showed a similar distribution. In cholesterol-fed rabbits, identical cells containing MnSOD and RAM-11 immunoreactivities increased in numbers considerably (Figs. 1 and 2). The area density of MnSOD immunoreactive macrophages in the myocardium of controls was 0.32 % + 0.045 %, while of cholesterol-fed rabbits it was 0.88 % + 0.16 %. The difference between both groups was statistically significant ($P<0.01$). Furthermore the number of MnSOD immunoreactive macrophages per mm^2 myocardium area increased from 39.5 + 4.56/mm^2 (controls) to 87.5 + 16.7/mm^2 (cholesterol-fed rabbits). MHC II immunoreactive cells increased similarly, while MHC I immunoreactive cells were not found as frequently.

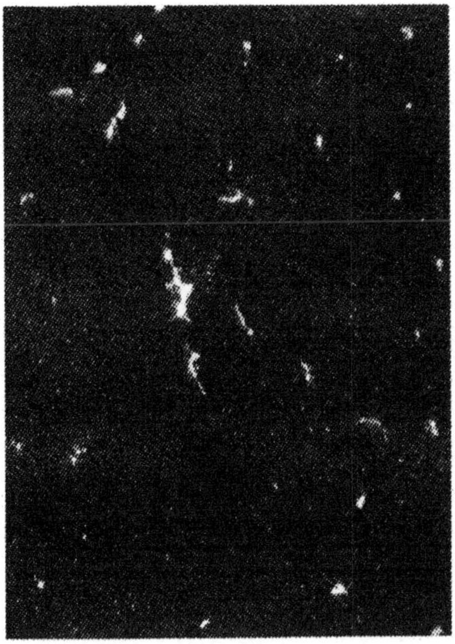

Figure 1. Myocardium of a cholesterol-fed rabbit: Numerous macrophages in the interstitium are immunostained with the RAM-11 antibody, detected with Texas Red labelled streptavidin. (x 330)

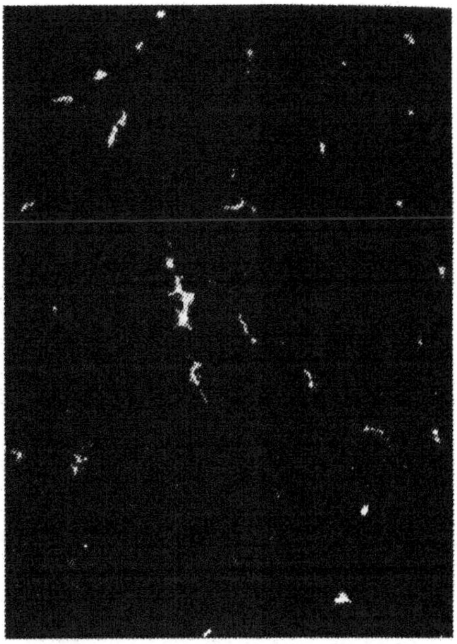

Figure 2. Myocardium of a cholesterol-fed rabbit: After double staining experiments identical cells (in comparison to Fig. 1) are immunoreactive to anti- MnSOD antibodies detected with FITC labelled streptavidin. (x 330)

DISCUSSION

We have demonstrated MHC I- and II-, and MnSOD immunoreactive macrophages in the myocard of control rabbits. Cholesterol-feeding led to a significant increase of RAM-11, MnSOD- as well as MHC I- and II-immunoreactive macrophages in the myocard, which might be due to the internalization, degradation and storage of elevated modified LDL.

In our study, staining of control rabbits' myocard with anti-MnSOD, MHC I and II antibodies showed a colocalization with RAM-11 immunoreactive cells, which have been reported to be macrophages (Rosenfeld et al., 1991; Stadius et al., 1992; Tsukada et al., 1986). However, there were more MnSOD, MHC I and II immunoreactive cells in the myocardial interstitium than RAM-11. MnSOD immunoreactivity was also found in macrophages of other species, such as in macrophages of the rat pulmonary system (Stevens and Autor, 1977). Our results suggest that MnSOD immunoreactivity is localized in a subpopulation of peripheral macrophages, which appear either to be specifically activated or are involved in specific functions.

One major hypothesis for the pathogenesis of atherosclerosis is that hypercholesterolemia involves macrophages into the internalization of modified LDL (Ball et al., 1987), which finally results in cellular cholesterol accumulation (Hoffman et al., 1992) and foam cell formation (Kaesberg et al., 1993). The cholesterol-fed rabbits in our experiments, which exhibited a ca. 50-fold increase in serum cholesterol levels, showed a significant increase ($P<0.01$) of MnSOD immunoreactive macrophages within the myocard. It could be hypothesized that (1) macrophages internalize LDL in the blood and migrate into the peri- and paravascular space of the myocard or (2) they are attracted by modified LDL in the interstitial space of the myocard (Couffinhal et al., 1993; Witztum, 1990). The uptake and intracellular processing of modified LDL in these macrophages might lead to an upregulation of the MnSOD content.

The removal of the extracellular modified LDL by macrophages provides a detoxification pathway by protecting other cells against its cytotoxic effect (Darley-Usmar et al., 1991). The upregulation of MnSOD within these macrophages appears to be a selfprotective mechanism against the effects of superoxide anions, which might be generated during the degradation process of internalized LDL, e. g. MnSOD might be necessary to maintain normal functioning of the mitochondria after oxidative stress (Janssen et al., 1993).
In conclusion, macrophages increase in number in the myocardium of cholesterol-fed rabbits, which might belong to a subpopulation of macrophages, which are involved in the metabolization of modified LDL.

ACKNOWLEDGEMENTS

The authors gratefully acknowledge the skillful technical assistance of U. Traut, C. Willich and S. Wolff.

REFERENCES

Ball, R.Y., Carpenter, K.L., and Mitchinson, M.J., 1987, What is the significance of ceroid in human atherosclerosis? Arch Pathol Lab Med 111: 1134-1140

Cathcart, M.K., McNally, A.K., Morel, D.W., and Chisolm, G.M., 1989, Superoxide anion participation in human monocyte-mediated oxidation of low-denity lipoprotein and conversion of low-density lipoprotein to a cytotoxin. J Immunol 142: 1963-1969

Couffinhal, T., Duplaa, C., Labat, L., Moreau, C., Bietz, I., and Bonnet, J., 1993, Effect of low density lipoprotein on monocyte adhesiveness to endothelial cells in vitro. Atherosclerosis 99: 35-45

Daar, A. S., Fuggle, S. V., Hart, D. N. J., Dalchau, R., Abdulaziz, Z., Fabre, J. W., Ting, A., and Morris, P. J. , 1983, Demonstration and phenotypic characterization of HLA-DR positive interstitial dendritic cells widely distributed in human connective tissues. Transplant Proc 15: 311-315

Darley-Usmar, M., Severn, A., O'Leary, V.J., and Rogers, M., 1991, Treatment of macrophages with oxidized low-density lipoprotein increases their intracellular glutathione content. Biochem J 278: 429-434

Del Rio, L.A., Sandalio, L.M., and Palma, J.M., 1990, A new cellular function for peroxisomes related to oxygen free radicals? Experentia 46: 989-992

Dhaunsi, G. S., Singh, I., and Hanevold, C.D., 1993, Peroxisomal participation in the cellular response to the oxidative stress of endotoxin. Mol Cell Biochem 126: 25-35

Henning, B., and Chow, C.K., 1988, Lipid peroxidation and endothelial cell injury: Implications in atherosclerosis. Free Radical Biol Med 4: 99-106

Hoffman, R., Brook, G.J., and Aviram, M., 1992, Hypolipidemic drugs reduce lipoprotein susceptibility to undergo lipid peroxidation: in vitro and ex vivo studies. Atherosclerosis 93: 105-113

Janssen ,Y.M., Van Houten, B., Borm, P.J., and Mossman, B.T., 1993, Cell and tissue responses to oxidative damage. Lab Invest 69: 261-274

Kaesberg, B., Harrach, B., Dieplinger, H., and Robenek, H., 1993, In situ immunolocalization of lipoproteins in human arteriosclerotic tissue. Arterioscler Thromb 13: 133-146

Kinscherf, R., Köhler, C., Kreuter, C., Pill, J., and Metz, J., 1995, Hypercholesterolemia increases manganese superoxide dismutase immunoreactive macrophages in myocardium. Histochem Cell Biology

Labarrere, C. A., McIntyre, J. A., Halbrook, H., and Faulk, W. P., 1991, Major histocombatibility antigens in transplanted human hearts before perfusion. J Heart Lung Transplant 10: 409-415

Rankin, S.M., Parthasarathy, S., and Steinberg, D., 1991, Evidence for a dominant role of lipoxygenase(s) in the oxydation of LDL by peritoneal macrophages. J Lip Res 32: 449-456

Rosenfeld, M.E., 1991, Oxidized LDL affects multiple atherogenic cellular responses. Circulation 83: 2137-2140

Rosenfeld, M.E,, Khoo, J.C., Miller, E., Parthasarathy, S., Palinski, W., and Witztum, J.L., 1991, Macrophage derived foam cells freshly isolated from rabbit atherosclerotic lesions degrade lipoproteins, promote oxidation of low-density lipoproteins, and contain oxidation-specific lipid-protein adducts. J Clin Invest 87: 90-99.

Singal, P.K., Dhalla, A.K., Hill, M., and Thomas, T.P., 1993, Endogenous antioxidant changes in the myocardium in response to acute and chronic stress conditions. Mol Cell Biochem 129: 179-186

Stadius, M.L., Rowan, R., Fleischhauer, J.F., Kernhoff, R., Billingham, M., and Gown, A.M., 1992, Time course and cellular characteristics of the iliac artery response to acute balloon injury. Arterioscler Thromb 12: 1267-1273

Stevens, J.B., and Autor, A.P., 1977, Oxygen-induced synthesis of superoxide dismutase and catalase in pulmonary macrophages of neonatal rats. Lab Invest 37: 470-478

Tsukada, T., Rosenfeld, M., Ross, R., and Gown, A.M., 1986, Immunocytochemical analysis of cellular components in atherosclerotuc lesions: Use of monoclonal antibodies with the Watanabe and fat-fed rabbit. Arteriosclerosis 6: 601-613

Witztum, J.L., 1990, The role of monocytes and oxidized LDL in atherosclerosis. Atherosclerosis News 21: 59-69

FLUORESCENCE ASSAY FOR STUDYING P-GLYCOPROTEIN FUNCTION AT SINGLE CELL LEVEL

László Homolya, Marianna Müller, Zsolt Holló, and Balázs Sarkadi

National Institute of Haematology, Blood Transfusion and Immunology, H-1113 Budapest, Hungary

INTRODUCTION

Ineffectivity of tumor chemotherapy is often caused by the resistance of malignant cells to a wide range of hydrophobic cytostatic agents. The main characteristic of these multidrug-resistant (MDR) cells is an energy dependent outward transport of drugs produced by a membrane protein, P-glycoprotein (MDR1, multidrug transporter). The functional assessment of this protein is essential for planning the proper chemotherapy. It has been previously demonstrated that MDR1-expressing cells show a decreased uptake of certain fluorescent anthracyclines, as well as fluorescent dyes as rhodamine 123, or fluo-3, and these compounds have been used to discriminate between drug-resistant and sensitive cells. We have demonstrated that hydrophobic acetoxy-methylester (AM) derivatives of various fluorescent indicators are actively extruded from cells by MDR1 (1). On the basis of this finding a quantitative assay method was developed by using calcein AM (3). Calcein AM, a non-fluorescent hydrophobic molecule, rapidly penetrates through cell membranes and becomes trapped intracellularly upon conversion into the fluorescent calcein (free acid) by nonspecific cytoplasmic esterases. In the MDR1-expressing cells, calcein AM is extruded by the multidrug transporter before its intracellular conversion to the non-MDR1 substrate free calcein (1, 2). However, when calcein AM extrusion is blocked by an agent that interferes with the MDR1 pump (e.g. verapamil), fluorescent free calcein rapidly accumulates. This assay provides an efficient experimental method for the quantitative determination of the multidrug transporter activity.

In the present work the calcein assay was applied to drug-sensitive (D12-control) and highly drug-resistant (D12 10x80) murine hepatoma cell lines.

METHODS

Flow cytometry. For immunofluorescent detection of MDR1 expression, drug-sensitive and highly drug-resistant D12 murine hepatoma cells were labeled by an MDR1-specific polyclonal antibody, 4077, then an R-Phycoerythrin conjugated anti-mouse second antibody was applied. For flow cytometric calcein assay the cells were loaded with 0.25 μM calcein in the absense or presence of 100 μM verapamil. Fluorescence intensity was followed by Cytoronabsolute flow cytometer, representative data are shown as cell number versus log PE or log calcein fluorescence, respectively.

Single cell imaging. Calcein assay was performed on the MDR1-expressing D12 10x80 cells continuously reselected with 1 μM colchicine. The cells, shown on the microscopic (DIC) picture, were settled on glass coverslip previously coated with 1 mg/ml

Fluorescence Microscopy and Fluorescent Probes
Edited by J. Slavík, Plenum Press, New York, 1996

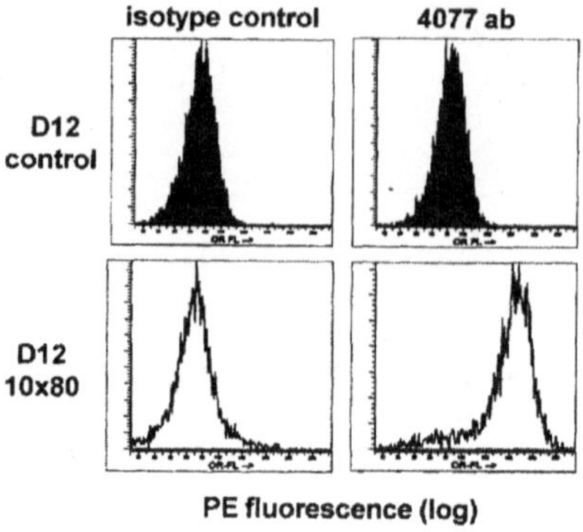

Fig. 1. Immunofluorescent detection of MDR1 expression by flow cytometry. Drug-sensitive (D12-control) and highly drug-resistant (D12 10x80) murine hepatoma cells were labeled by an MDR1-specific polyclonal antibody, 4077, then an R-Phycoerythrin conjugated anti-mouse second antibody was applied. Orange fluorescence intensity was followed by Cytoronabsolute flow cytometer. Representative data are shown as cell number versus log PE fluorescence.

poly-L-lysine. Fluorescence images were acquired in each 10 sec by PTI RF-D4021 digital imaging system based on a Zeiss Axiovert 135 microscope.

RESULTS AND DISCUSSION

Here we demonstrate an application of the calcein assay by using fluorescence digital imaging. The calcein accumulation method (2) significantly surpasses the previously used functional tests for MDR1 in many aspects: it possesses high sensitivity, low cytotoxicity, and insensitivity to the environmental changes (pH, Ca^{2+}, Mg^{2+}, etc.). The favorable optical properties of calcein make it suitable for flow cytometric, or microscopic detection of functional P-glycoprotein.

In order to detect the MDR1 expression in the D12 by flow cytometry, we used 4077 mAb for labeling the drug-resistant cells. Fig 1. shows, that the 4077 ab labeling was not observed in the case of D12 control cells, while high level of MDR1 expression was observed in drug-resistant D12 10x80 cells.

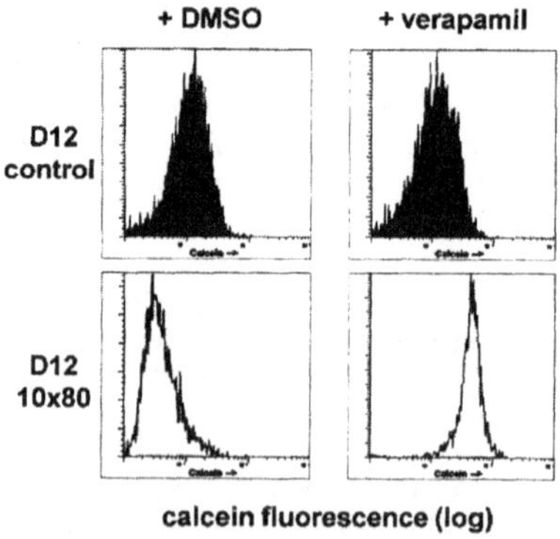

Fig.2. Calcein accumulation measurement in D12 cells by flow cytometry. The cells were loaded with 0.25 μM calcein in the absense or presence of 100 μM verapamil. Fluorescence intensity was followed by Cytoronabsolute flow cytometer. Representative data are shown as cell number versus log calcein fluorescence.

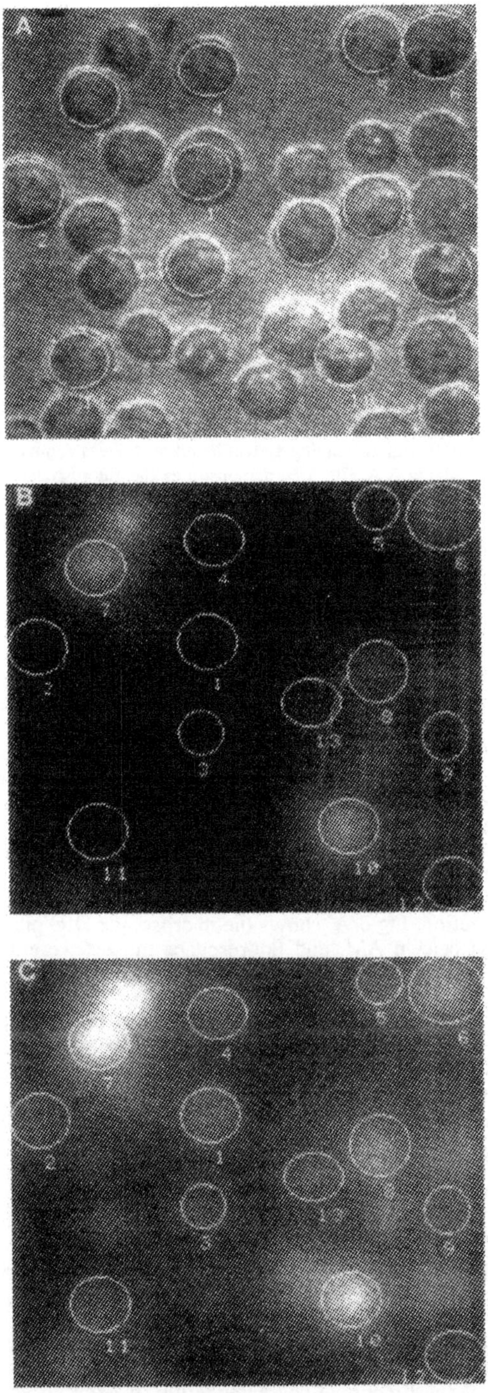
A
B
C

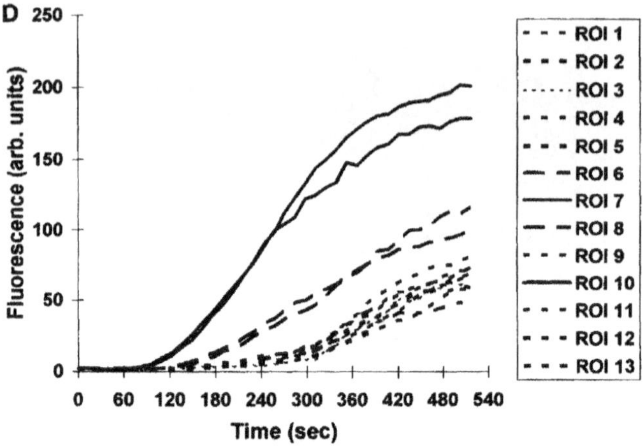

Fig. 3. Calcein assay on the MDR1-expressing D12 10x80 cells. Fluorescence images were acquired in each 10 sec by PTI RF-D4021 digital imaging system based on a Zeiss Axiovert 135 microscope. The microscopic DIC picture is in Fig 3/A. Fig 3/B. demonstrates the fluorescence image after 4 min. of incubation with calcein AM. 13 regions of interest (ROIs) corresponding to the cells were put on the image. The fluorescence image in Fig 3/C. was acquired 4 min. after verapamil addition. In Fig 3/D. the single cell photometry mean fluorescence values are plotted against time.

Calcein accumulation was measured in the same drug-sensitive (D12-control) and drug-resistant (D12 10x80) cell lines as shown in Fig.1. The cells were loaded with 0.25 μM calcein in the absense or presence of 100 μM verapamil. Fig. 2 demonstrates, that the preincubation of the cells with verapamil has no effect on calcein accumulation in control cells. In contrast, the dye accumulation is prevented in P-glycoprotein expressing D12 10x80 cells. After verapamil addition, the calcein accumulation was restored.

Due to the favorable optical properties of the dye, calcein assay is applicable also for microscopic detection of functional MDR1. Since the AM form of calcein is non-fluorescent, the dye trapping into the cells can be easily visualized. In the single cell imaging experiments calcein assay was performed on D12 10x80 cells after 2 weeks of culturing without reselection. Fig 3/A. shows the microscopic DIC picture. The cells were incubated with 0 5 μM calcein AM, and fluorescence images were acquired in each 10 sec. Fluorescence image acquired after 4 min. of incubation with calcein AM is presented in Fig 3/B. 13 regions of interest (ROIs) corresponding to 13 cells were selected and put on the image. Three different characteristic behaviors can be observed. In this field of view two cells rapidly accumulated calcein (ROIs 7 and 10). Some other cells showed slow calcein accumulation (i.e. ROIs 6 and 8), while most of the cells remained practically non-fluorescent. After 4 min. of incubation with calcein AM, 100 μM verapamil was added to the cells, and the fluorescence image in Fig. 3/C. was acquired after 4 min. of incubation with verapamil. All the cells revealed rapid calcein accumulation after the verapamil addition demonstrating that the prevented dye uptake is not due to decreased viability of the cells but the dye extrusion activity of MDR1.

After the image acquisition, single cell photometry was performed on the selected regions of interest. Mean fluorescence values were plotted against time in Fig 3/D. The photometric curves clearly demonstrate the three different types of behavior. The diversity of the calcein assay curves reflects the heterogeneity of MDR1 activity in this cell culture sample.

In these experiments presented above we show that within the same cell culture sample different levels of MDR1 expression can be detected. Thus, hopefully, the calcein assay can be applied for the functional characterization of MDR1 even in heterogeneous clinical samples.

REFERENCES

1. Homolya, L., Holló, Zs., Germann, U. A., Pastan, I., Gottesman, M. M., and Sarkadi, B. (1993) J. Biol. Chem. 268, 21493-21496

2. Holló, Zs., Homolya, L., Davis, C.W., Sarkadi, B (1994) Biochim Biophys Acta, 1191, 384-388

3. Homolya, L., Holló, Zs., Müller M., Mechetner, E.B.,and Sarkadi, B. (1995) Br. J. Cancer, submitted for publication

ALTERATIONS OF VIMENTIN-NUCLEUS INTERACTIONS AS AN EARLY PHASE IN CHOLESTEROL OXIDE - INDUCED ENDOTHELIAL CELL DAMAGE

G. Palladini[1] and G. Bellomo[2]

[1]Department of Internal Medicine, IRCCS, Policlinico S.Matteo, Pavia and [2]Department of Medical Sciences, University of Torino, Novara, Italy

INTRODUCTION

The oxidation of low density lipoprotein (LDL) is one of the key-events in the initiation and progression of the atherosclerotic disease[1,2]. Oxidized LDL exerts a series of biological effects that make it more atherogenic than its parent form. Among them is cytotoxicity to endothelial and smooth muscle cells[3,4]. The mechanisms underlying the cytotoxic effect of oxidized LDL are still under active investigation, and the biochemical identification of the factor(s) responsible for oxLDL-mediated cytotoxicity is far to be exhaustively performed. In fact, the oxidation of LDL particles generates a mixture of compounds with potential cytotoxic activity, including a variety of cholesterol oxides[5]. Among them, cholestane-3β,5α,6β-triol (CH), 5-cholesten 3β-ol-7one (KC), and 25-OH-cholesterol (COH) have received particular attention since they are generated *in vivo*[6,7,8]. Cholesterol oxide-mediated cytotoxicity has been investigated in both endothelial and smooth muscle cells[9,10,11], and CH has been reported to affect the barrier function of the endothelial layer[9].

Alterations in cytoskeleton composition, structure and organization are early events in toxic cell killing[12,13]. Toxic disruption of actin microfilaments, actin-plasma membrane interaction and microtubules may trigger a series of events that promote the detachment of adherent cells from the substrate or precipitate cell death[14,15]. The precise physiological activity of intermediate filaments is still an enigma. They are formed by various polypeptides grouped into six types based on aminoacid sequence homologies, and selectively expressed in different cell types[16]. In endothelial cells and endothelium-derived cell lines, the most abundant intermediate filament polypeptide is vimentin that appears organized in a filament meshwork connecting the nuclear periphery to the inner surface of the plasma membrane[17]. Relatively few studies have been performed in order to investigate the occurrence of selective alterations in intermediate filament structure and distribution during toxic cell injury.

The present study was carried-out in order to investigate the possible occurrence of toxic alterations in vimentin-nucleus interactions in cholesterol oxide-exposed 73/73 endothelial cells. Results are reported indicating that cholestane 3β,5α,6β-triol, 5-cholesten

3β-ol-7one and 25-OH-cholesterol promoted the rupture of vimentin intermediate filaments and the loss of their connection with the nuclear periphery.

METHODS

73/73 cells (derived from bovine aortic endothelial GM 3905A cells transformed with benzopyrene) were kindly supplied by Dr. Paolo Mignatti (Department of Genetics and Microbiology, University of Pavia). They were mantained on 75-cm^2 tissue culture flaks in D-MEM supplemented with 5% fetal calf serum, 2% antibiotics and 2 mM L-glutamine. Cells were cultured at 37 °C in a humidified atmosphere with 5% CO_2 and 95% air and the culture medium was changed every 48 hours.

Confluent 73/73 cells were harvested after incubation with a Trypsin-EDTA solution (0.05% and 0.02%) for ten minutes After washing with DMEM, cells were seeded on glass coverslips into 60 mm Petri dishes, grown in 5 % FCS for 8 hours and then washed and placed in 0.4 % FCS overnight before incubation with oxysterols. Cells were not at confluence during experiments which were carried out in the presence of 0.4 % delipidated human serum.

Freshly-prepared oxysterols (cholestane 3β,5α,6β-triol; 25-OH-cholesterol and 5-cholesten 3β-ol-7one) were dissolved in ethanol and directly added to the cultured medium at a final concentration of 10 μg/ml for six hours at 37 °C. Cells treated with 0.5 % ethanol (vehicle alone) were used as control. At the end of treatments, cells were processed according to the following immunofluorescence protocol.

To label vimentin filaments and microtubules coverslips were rinsed briefly in PBS, then fixed for 15 minutes in 3.7 % formaldehyde in PBS. Fixed cells were rinsed in PBS and then permeabilized with 0.5% Triton X-100 in PBS containing EGTA 1 mM for 15 minutes and rinsed again in 2% BSA in PBS to decrease unspecific binding. Coverslips were incubated with primary antibodies at a dilution of 1:100 in PBS (monoclonal anti-vimentin and anti-α and β-tubulin) for 45 minutes at 37 °C. After extensive washing with PBS and serum-supplemented (3 %) PBS, a FITC-conjugated secondary anti-mouse IgG antibody (at a dilution of 1:100 in PBS supplemented with 2 % BSA) was applied to coverslips for 45 minutes at 37 °C. For additional labeling of nuclei, fixed and stained cells were treated for 5 minutes with propidium iodide (50 μg/ml in PBS).

After washing with PBS, all coverslips were mounted onto glass slides using Slow Fade in glycerol/PBS and observed with a laser scanner confocal microscope (Bio-Rad MRC 600, equipped with a Nikon Diaphot inverted microscope and a Nikon PlanApo 60/1.40 oil objective). Five to seven focal frames were taken along the z axis at 1 μm interval and then merged to get a reconstructed image.

RESULTS AND DISCUSSION

All three colesterol oxides investigated in this study promoted a redistribution of vimentin filaments that took place well before cell detachement and the occurrence of any detectable sign of cell death, as revealed by the leakage of cytosolic enzimes. CH-induced alterations were characterized by the polarization of vimentin to the edges of the cell and a detectable stretching of the filaments that, in selected cells, underwent dramatic ruptures. In KC-treated cells, vimentin filaments appeared as cross-linked and formed a sort of circular network ring between the nucleus and the cell periphery COH promoted the aggregation of vimentin filaments in thick and irregular bundles that delimitated apparently empty regions.

All these changes occurred independently of gross modifications in microtubule organization, that was generally retained except for the appearance of immunoreactive tubulin spots throughout the cytoplasm.

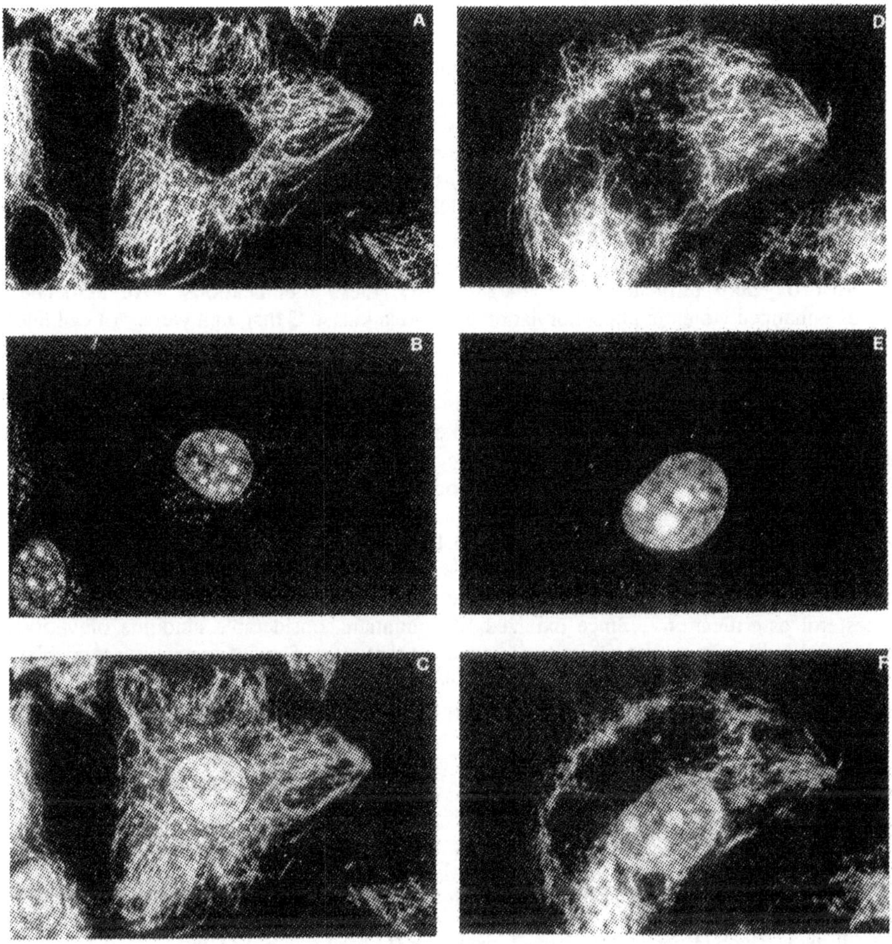

Fig. 1: Alterations in vimentin-nucleus interaction caused by cholesterol oxides.
73/73 cells were incubated for 6 hours without (panels A, B, C) and with 10 μg/ml of 5-cholesten 3β-ol-7one (panels D, E, F), fixed, immunostained for vimentin (panels A and D) and stained for DNA with propidium iodide (panels B and E) as described in Methods Panels C and F depict the superimposition of the corresponding two preceding images. Images were taken at original 600 X magnification and the pictures on the various panels were obtained by zooming (4.0 zoom) the original frames.

In order to better investigate the association between vimentin filaments and the nucleus and its disruption by cholesterol oxides, control and treated 73/73 cells were dual-labelled with both anti-vimentin antibodies and the DNA-intercalating probe propidium iodide[18] and analized by means of laser scanner confocal microscope. As reported in Fig. 1, in control, vehicle-treated cells, the fluorescent propidium iodide-positive structures were completely surrounded by immunoreactive vimentin filaments. However, the complete and close association was partially or entirely lost in KC-treated cells. Comparable effects were obtained with the other cholesterol oxides employed (not shown).

Lamin B is the major receptor for intermediate filaments in the nuclear envelope[19] and promotes an association between vimentin filaments and nuclear structures so thight that it can not be disrupted even during tissue or cell fractionation[20] or mitosis[21]. The interaction of vimentin with lamin B is maintained by phosphorylation and is weakened by dephosphorylation[21]. We have no data proving or disproving a role for an altered phopshorylation/ dephosphorylation process as a key event in cholesterol oxide-induced alterations in vimentin-nucleus interactions. However, morphological findings apparently similar to those obtained in the present study have been reported by Tang *et al.*[22] in microvascular endothelial cells exposed to the lipoxygenase metabolite of arachidonic acid 12(S)-hydroxyeicosatetraenoic acid. Paradoxically, these modifications were associated with an enhanced vimentin phosphorylation by protein kinase C that, in a variety of cell lines is structurally linked to intermediate-size filaments[23].

The physiological role of intermediate filaments is still not known in detail, and any hypothesis concerning the biological consequences of their disruption is, at least, hazardous. It can be temptatively assumed that the marked structural alterations in vimentin/nucleus interaction may be relevant for the reported cytotoxicity of cholesterol oxides[9,10,11] and in triggering a specific way of dying, known as apoptosis[24], where structural nuclear changes are a typical phenomenon. In addition, recent reports have indicate that, in selected cell types, vimentin filaments may be involved in the intracellular movement of LDL-derived cholesterol from the lysosome to the site of esterification[25]. A typical feature of oxidized LDL metabolism is their accumulation in the lysosomal compartment and the impaired cholesterol esterification[26]. Since oxidized LDL contain considerable amounts of various cholesterol oxides, it can be temptatively assumed that alterations of vimentin intermediate filaments caused by cholesterol oxides could play a role in the altered intracellular processing of oxidized LDL. However, further investigation in this direction is needed and is actually in progress in our laboratory.

REFERENCES

1. D Stenberg, S Parthasarathy, TE Carew, JC Khoo and JL Witztum - Beyond cholesterol: modifications of low density lipoprotein that increase its atherogenicity *N.Engl.J.Med.* 320: 915 (1989)
2. JL Witztum - The oxidation hypothesis of atherosclerosis *The Lancet* 344: 793 (1994)
3. DW Morel, JR Hessler and GM Chisolm - Low density lipoprotein citotoxicity induced by free radical peroxidation of lipids *J Lipid Res.* 24: 1070 (1983)
4. GM Chisolm, MA Guoping, KC Irwin, LL Martin, KG Gunderson, LF Linberg, DW Morel and PE DiCorleto - 7β-hydroperoxycholest-5-en-3b-ol, a component of human atherosclerotic lesions, is the primary cytotoxin of oxidized human low density lipoprotein *Proc.Natl.Acad.Sci.USA* 91: 11452 (1994)
5. H Esterbauer, J Gebicki, H Puhl and G Jurgens - The role of lipid peroxidation and antioxidants in oxidative modifications of LDL *Free Radic.Biol.Med.* 13: 341 (1992)
6. PB Addis, HA Emanuel, SD Bermann and JH Zavoral - Capillary GC quantitation of cholesterol oxidation products in plasma lipoproteins of fasted humans *Free Radic.Biol. Med.* 7: 179 (1989)
7. CJW Brooks, WA Harlan andG Steel - Squalene, 26-hydroxycholesterol and 7-ketocholesterol in human atheromatous plaques *Biochim.Biophys.Acta* 125: 620 (1966)

8. HN Hodis, DW Crawford and A Sevanian - Cholesterol feeding increases plasma and aortic tissue cholesterol oxide levels in parallel: further evidence for the role of cholesterol oxidation in atherosclerosis *Atherosclerosis* 89: 117 (1991)
9. GA Boissonneault, B Henning and CM Ouyang - Oxysterols, cholesterol biosynthesis and vascular endothelial cell monolayer barrier function *Proc.Soc.Exper.Biol.Med.* 196: 338 (1991)
10. Q Zhou, TL Smith and F Kummerow - Cytotoxicity of oxysterols on cultured smooth muscle cells from human umbelica arteries *Proc.Soc.Exper.Biol Med.* 202: 75 (1993)
11. H Hughes, B Mathews, ML Lenz and JR Guyton JR - Cytotoxicity of oxidized LDL to porcine aortic smooth muscle cells is associated with the oxysterol 7-ketocholesterol and 7-OH cholesterol *Arterioscler.Thromb.* 14. 1177 (1994)
12. G Bellomo, F Mirabelli, M Vairetti, W Malorni and F Iosi - Cytoskeleton as a target in menadione-induced oxidative stress. Biochemical and immunocytochemical features *J.Cell.Physiol.* 143: 118 (1990)
13. G Bellomo and F Mirabelli - Oxidative stress and cytoskeletal alterations *Ann.N.Y.Acad.Sci.* 663: 97 (1992)
14. G Bellomo, F Mirabelli, M Vairetti, F Iosi and W Malorni - Morphological and Biochemical Investigations on plasma membrane blebbing during cell injury *Methods Toxicol.* IB, 58 (1994)
15. E Bonfoco, S Ceccatelli, L Manzo and P Nicotera - Colchicine induces apoptosis in cerebellar granule cells *Exp. Cell Res.* 218: 189 (1995)
16. E Lazarides - Intermediate filaments as mechanical integrators of cellular space. *Nature* 283: 249 (1980)
17. WW Franke, E Schmid, M Osborn and K Weber - Intermediate-sized filaments of human endothelial cells *J.Cell Biol.* 81: 570 (1979)
18. DM Arndt-Jovin, M Robert-Nicoud and TM Jovin - Probing DNA structure and function with a multi-wavelength fluorescence confocal laser microscope *J.Microscopy* 157: 61 (1990)
19. SD Georgatos and G Blobel - Lamin B constitutes an intermediate filament attachment site at the nuclear envelope *J.Cell.Biol.* 105: 117 (1987)
20. M Staufenbiel and W Deppert - Intermediate filament systems are collapsed onto the nuclear surface after isolation of nuclei from tissue culture cells *Exp Cell Res.* 138: 207 (1982)
21. C Maison, H Horstmann and SD Georgatos - Regulated docking of nuclear membrane vesicles to vimentin filaments during mitosis *J.Cell Biol.* 123: 1491 (1993)
22. DG Tang, J Timar, IM Grossi, C Renaud, VA Kimler, CA Diglio, JD Taylor and KV Honn - The lipoxygenase metabolite 12(S)-HETE indues a protein kinase C-dependent cytoskeletal rearrangement and retraction of microvascular endothelial cells *Exp. Cell Res.* 207: 361 (1993)
23. KG Murti, K Kaur and RM Goorha - Protein kinase C associates with intermediate filaments and stress fibers *Exper. Cell Res.* 202. 36 (1992)
24. G Palladini, F Taddei and G Bellomo - Cell death by apoptosis: morpholigical, genetic and biochemical features and toxicological implications. *in* "Modulation of Cellular Responses in Toxicity" CL Galli, AM Goldberg and M Marinovich editors, Springer, Berlin (1995)
25 AJ Sarria, SR Panini and RM Evans - A functional role for vimentin intermediate filaments in the metabolism of lipoprotein-derived cholesterol in human SW-13 cells *J Biol.Chem.* 267: 19455 (1992)
26. I Maor and M Aviram -Oxidized low density lipoprotein leads to macrophage accumulation of unesterified cholesterol as a result of lysosomal trapping of the lipoprotein hydrolyzed cholesteryl ester *J.Lipid Res.* 35: 803 (1994)

Fluorescence Microscopy of Rye Cell Walls from Kernels to Incubated Doughs

Koskinen, M., Parkkonen, T. and Autio K.

VTT Biotechnology and Food Research, P.O. Box 1500, FIN-02044 VTT, Finland

Introduction

The cell walls are important component in rye dough, as they affect the water distribution and gas retention.

Aim

Aim of this work was to compare the microstructure of kernels of three different rye varieties: Ensi, Anna and Amando, and to study the microstructural changes from kernels to incubated doughs.

Materials and methods

Ensi and Anna are typical Finnish rye varieties, but Amando is a hybrid rye developed in Germany but cultivated in Finland. Doughs were made by mixing 200 g rye flour, 160 g water and 2 % salt. No pH adjustments were done. Doughs were incubated at 36 °C and 80 % humidity for 70 min. Structural changes in incubated doughs were studied by viscoelastic measurements and by fluorescence microscopy (fig. 1) and image analysis.

Results and discussion

These rye samples differed greatly in falling number (table 1). Ensi had the lowest falling number and highest α-amylase activity. Whereas Amando had the highest falling number and least

Table 1. Falling number and α-amylase and β-glucanase activity of rye varieties.

Sample	Falling number	α-Amylase activity (U/g d.m.)	β-Glucanase activity (U/kg d.m.)
Ensi	64	6.4	71
Anna	174	0.4	66
Amando	276	0.3	66

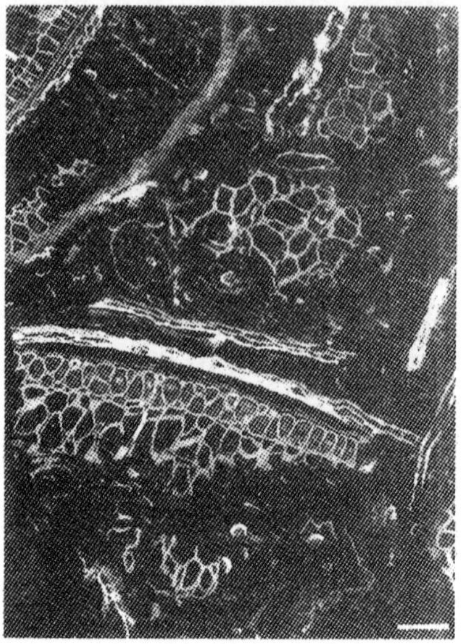

Figure 1. Photomicrograph of rye dough. Acid fuchsin and Calcofluor staining. bar 100 µm

Table 2. Chemical analysis of rye varieties

Sample	Moisture content (%)	Ash content (%)	Starch content (%)	Fat content (%)	Proteins (%)	Total fiber content (%)	Total arabino-xylans (%)	Soluble arabino -xylans (%)
Ensi	11.3	1.8	49 9	2 4	13.5	16.1	8.5	2.3
Anna	10.4	1.7	53.3	2.5	10.8	14.7	7.4	1.5
Amando	11.5	1.7	57.5	2.4	8.6	15.6	8.4	1.7

enzymatic activity. Chemical analysis showed that Amando had the highest starch and the lowest protein content (table 2).

Blue fluorescence of cell walls (in black and white figures as light grey) was disappeared from the subaleurone layer of some kernels of Ensi and Anna (fig. 2) which had lower falling number than Amando. Amando had intact cell walls.The area of cell walls after mixing was highest with Amando and lowest with Ensi (table 3). Amando dough was more rigid than Ensi dough suggesting that the structure of cell walls affects the rheological properties of the doughs. For all samples studied the area of cell walls decreased during fermentation and the doughs became softer (table 4).

Table 3. Image analysis of rye doughs.

Sample	Area of cell walls$_{mix}$ (%)	Area of cell walls$_{incub}$ (%)	Change in the area of cell walls during incubation (%)
Ensi	6.8 ± 1	4.8 ± 1	29.4
Anna	8.8 ± 1	7.9 ± 1	10.2
Amando	11.3 ± 2	8.6 ± 1	23.9

Table 4. Viscoelastic measurements of rye doughs. Three different doughs of the whole meal and three different measurements of each dough.

Sample	G'_{mix} (kPa)	G'_{incub} (kPa)	$(G'_{mix}-G'_{incub})/G'_{mix}$ (%)
Ensi	61.0 ± 4	47.8 ± 6	21.6
Anna	63.4 ± 4	54.4 ± 2	14.2
Amando	70.7 ± 4	62.0 ± 4	12.3

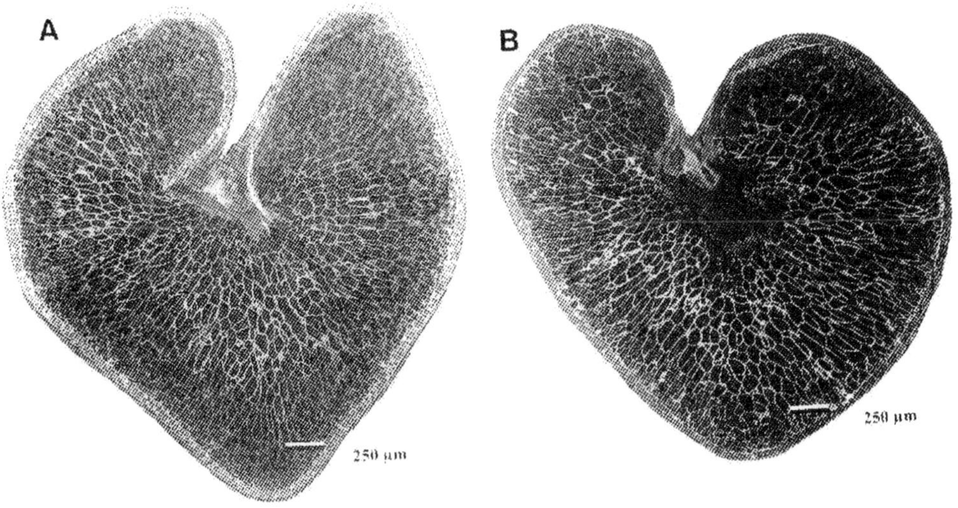

Figure 2. Photomicrographs of a) Ensi and b) Anna kernel. Acid fuchsin and Calcofluor staining.bar 250 µm

The changes that occurred in the magnitude of storage modulus (G') during incubation and relative change in the area of cell walls were highest for Ensi with the lowest falling number. The area of cell walls of Amando dough decreased substantially during fermentation. Instead the change in storage modulus was minor suggesting that other structural changes occurred which might have increasing effects on the storage modulus.

PRACTICAL APPROACH FOR IMMUNOHISTOCHEMICAL STAINING OF MUSCLE BIOPSIES

Vera Maravić-Stojković and Vladimir Baltić

Institute of Oncology, Department of Experimental Oncology
Institutski put 4, 21204 Sremska Kamenica, Yugoslavia

INTRODUCTION

As it is proposed by Stevenson et al.[1] it is best that an experienced histopathologist handle the distribution of tissue: a touch preparation can be useful guides to further investigations (cytology, intracellular markers), frozen sections (surface markers, intracellular markers), formalin fixed sections (cytology, intracellular markers) and cell suspension (surface markers, intracellular markers, functional and special assessments).

Previously, we reported that staining fibronectin by immunofiuorescence[2] is a good method for demonstrating capillaries and sarcolemma in skeletal muscle tissue. Fibronectin, as a major connective tissue protein, offers a better insight in muscle ultrastructure in general. Its distribution in basal membrane of sarcolemma[3] and matrix of endomysium, perimysium and perineuryum[4] enable a precise localization of interstitium. Fibronectin is a content of endothelium of all vascular vessels, from capillaries to arterial and veins walls. Visible are lamina interna, lamina muscularis and even a lumen of arterial vessels. There are two concomitant veins without smooth muscle ring of lamina muscularis with consequently flattened walls.

MATERIALS AND METHODS

Preparation of the Specimens. The specimens of human heart muscle biopsy (during endo-myiocardial biopsy) were deposited on microscope slide and performed two or three imprints. Air-dried touch preparations were fixed for 30 sec in buffered formol-acetone before May-Grünwald-Giemsa or immunocytochemical staining[5].

Human skeletal muscle (vastus lateralis) during the biopsy was divided into two portions. Fresh frozen unfixed sections 10μm-thick were compared with buffered paraformaldehyde fixed sections. Both sections were incubated 60 min. with serial dilutions of rabbit anti-human fibronectin antibodies (Boehring, Germany). Primary antibodies were diluted 1:10, 1:20, 1:40, 1:80, 1:100 and 1:200. After washing, samples were incubated 30 min with goat-anti-rabbit IgG conjugated by FITC (1:20). Between all steps of the procedure, slides were washed in 0.1 mol/L of phosphate-buffered saline at a pH of 7.2.

RESULTS

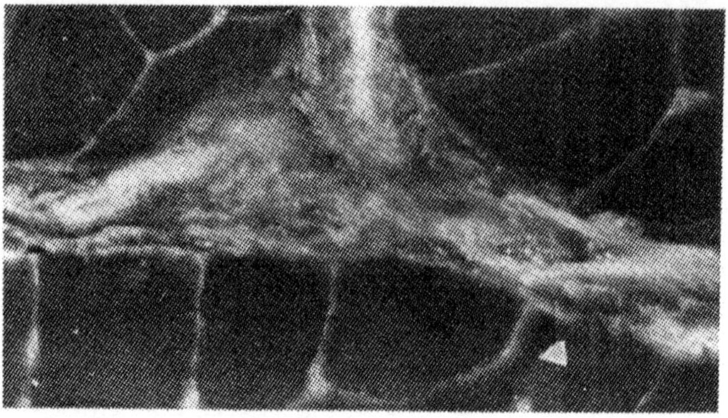

Figure 1: Frozen section of skeletal muscle stained for fibronectin by immunofluorescence. Between muscle fibers sometimes appears splitting (arrow) which may disturb stereological analyze (x 400).

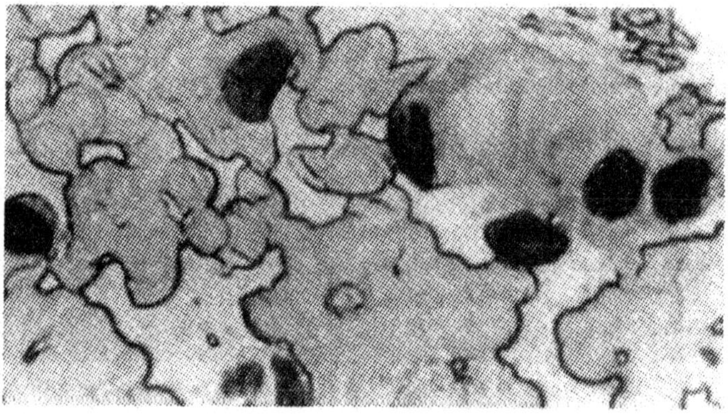

Figure 2: May-Grünwald-Giemsa on touch preparation of human heart muscle (x 400).

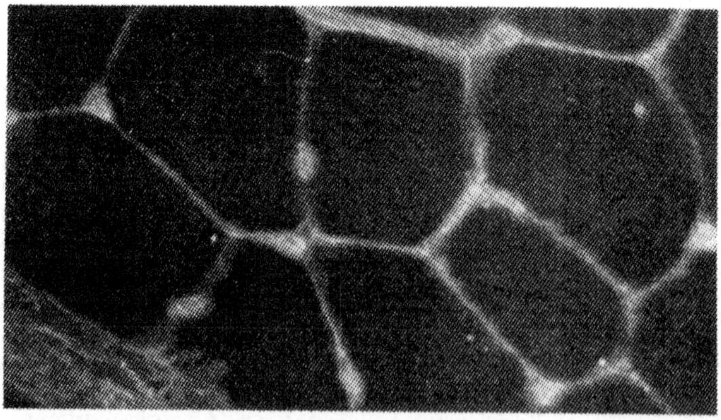

Figure 3: Immunofluorescence staining of fibronectin on formalin-fixed section of skeletal muscle was compared with frozen section by serial dilutions of primary antibodies. No differences (x 400).

DISCUSSION

Frozen sections offer access to both surface and intracellular markers, provided the appropriate technique is followed in each case. It is important that all biopsy material is received fresh, preferably within 15 min. of operation.

Cytochemical determinations of those enzymes which are relevant for us are sometimes carried out on frozen or formalin-fixed sections, but they are more satisfactorily undertaken on touch preparations[1]. Due to the fact that autolysis began within 60 seconds following tissue extraction the touch preparat has been made within 90 sec. It is possible to make more than one imprint from the same surface. Various staining techniques can be used: cytochemical, immunofluorescence or immunoenzymatic labeling[6].

Formalin-fixed sections are suitable for intracellular markers and cytology, as well as cell suspension, but processing the tissue for assessment immunological markers is always doubtful.

CONCLUSION

This approach enable precise and exact analyze of biopsy specimens from heart, skeletal or smooth muscles. Touch preparation cytology, frozen section and finding in the permanent histologic sections is a protocol aimed to demonstration the local cellular status and morphological image analysis.

ACKNOWLEDGMENT

We thank Prof. D. Stojšić, Institute of cardiovascular disease, Clinic of cardiology in Sremska Kamenica, for endomyocardial biopsy matherial and D. Vučković, M.D., Institute of TBC in Sremska Kamenica, for technical assistance.

LITERATURE

1. Stevenson GT, Smith JL, Hamblin TJ. Practical methods in clinical immunology. Immunological investigation of lymphoid neoplasms. Ed Nairn RC, Melburn (1983).

2. Erzen I, Maravić V. Simultaneous histochemical demonstration of capillaries and muscle fibre types. Histochem 99: 57-60 (1993).

3. Lehto M, Kvisto M, Vieno T, Jozsa L. Macromolecular composition of the sarcolemma
and endomysium in rat. Acta Anat 133:297-302 (1988).

4. Stennman S, Vaheri A. Distribution of a major connective tissue protein, fibronectin, in normal human tissue. J Exp Med 147:1054-63 (1978).

5. Cordell JL, Falini B, Erber WN et al. Immunoenzymatic labeling of monoclonal antibodies using immune complexes of alkaline phosphatase and monoclonal anti-alkaline phosphatase. J Histochem Cytochem 32:219-229 (1984).

6. Pinto MM, Meriano FV, Afridi S, Tauhin HL. Cytodiagnosis of Campylobacter pylori in Papanicolau-stained imprints of gastric-biopsy specimens. Acta Cytol 35:204-6 (1991).

RAPID AUTOMATIC SEGMENTATION OF FLUORESCENT AND PHASE-CONTRAST IMAGES OF BACTERIA

Michael H.F. Wilkinson

Department of Medical Microbiology,
University of Groningen,
Oostersingel 59,
9713 EZ Groningen, The Netherlands
E-Mail: m h.f.wilkinson@med.rug.nl

INTRODUCTION

Image segmentation, i.e. the division of an image into regions with different properties, is one of the most important steps in image analysis. All results derived from all further steps, such as morphometry or fluorimetry, depend critically on the segmentation algorithm's ability to determine the exact borders between, e.g. objects and background. In many, if not most cases, image segmentation is performed by manual setting of a threshold All pixels with grey levels above the threshold are considered part of an object, all others background (or vice-versa, depending on the stain). Though simple to implement and reasonably fast, it is by no means the method of choice, since it is not objective and therefore not very reproducible. Besides, if different objects have been stained with different intensities it is impossible to set a single threshold for all objects within one field of view.

There are two ways to solve these problems: (i) compute a different threshold for different areas of the image, and (ii) pre-process the image in such a way that a global, fixed or automatic threshold can be used. An example of the former approach is given by Kittler *et al.*[1] The latter approach has been used with some degree of success by Viles and Sieracki on fluorescent images of bacteria.[2] Since their method was too computationally intensive for interactive use, an alternative has been sought for segmentation of both fluorescent and phase-contrast images of bacteria, for use in the image processing system used in our laboratory.[3,4] It is shown that Robust Automatic Threshold Selection (RATS) introduced by Kittler *et al.*[1] is both fast enough and more reproducible than manual methods.

Fluorescence Microscopy and Fluorescent Probes
Edited by J. Slavík, Plenum Press, New York, 1996

MATERIALS AND METHODS

Robust automatic threshold selection (RATS)

The RATS algorithm uses two simple statistics to decide whether a region of the image contains a threshold, and what the optimum threshold is. The statistics are based on a gradient operator $e(x,y)$ which is defined as:

$$e(x,y) = \max\begin{Bmatrix} |p(x+1,y)-p(x-1,y)|, \\ |p(x,y+1)-p(x,y-1)| \end{Bmatrix} \tag{1}$$

in which $p(x,y)$ is the grey level of the pixel at (x,y). It has been shown by Kittler *et al.*[1] that the optimum threshold T for an area A is determined by the statistic:

$$T = \frac{\sum_A e(x,y)p(x,y)}{\sum_A e(x,y)} \tag{2}$$

To determine whether an edge is present within A, the statistic C may be used:

$$C = \sum_A e(x,y) \tag{3}$$

If C is significantly above the value expected for a flat, noisy image, an edge is present, and threshold T should be used (for details see Kittler *et al.*[1,5]). An obvious advantage of the use of C is that it is just the denominator of (2), so it needs to be computed anyway. Alternatively, any bias in T caused by noise can be reduced by modifying the gradient $e(x,y)$ itself. Following Kittler *et al.*,[5] this modified form $e_t(x,y)$ is defined as:

$$e_t(x,y) = \begin{cases} e(x,y) & \text{if } e(x,y) > \lambda\cdot\sigma \\ 0 & \text{otherwise} \end{cases} \tag{4}$$

in which σ is the expected noise (per pixel) and λ is a multiplication factor limiting the sensitivity of the method to both noise related and object related gradients. Following Kittler *et al.* the value of λ=3.974 was used to eliminate 99% of the noise pixels. The product $\lambda\cdot\sigma$ (which also constitutes the detection limit) was rounded to the nearest integer number (5 for fluorescence and 8 for phase-contrast images).

Local thresholding using RATS

In our approach the image is subdivided into a quadtree or image pyramid hierarchy of rectangular areas (see fig. 1). At the highest level of the hierarchy lies the entire image, which is subdivided into four "child" areas, each of which in turn are divided into four, etc. down to subdivisions with sizes in the orders of those of the objects of interest. If an area at

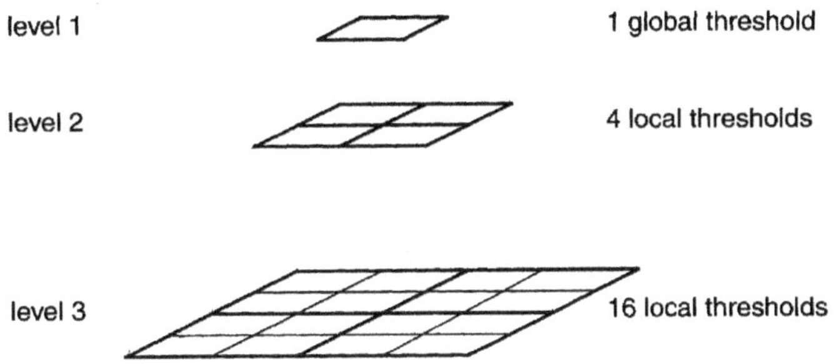

Figure 1. The image pyramid or quadtree structure used in this implementation of RATS

the lowest level cannot be assigned a threshold, its "parent" is consulted recursively if necessary to the highest level. Step by step:

1. Divide the image into a quadtree, in which the lowest level of subdivisions is about the same size as the objects of interest.
2. For each lowest level area A compute the enumerator and denominator (=edge criterion C) of (2).
3. Compute the enumerator and denominator of (2) for all higher levels in the quadtree. Since these are simple summations, each parent area gets assigned the sum of its children.
4. For each area A
   a. If C (the denominator) is larger than expected from noise, use the local threshold T.
   b. otherwise use parent threshold.

   This is done recursively until either a parent with a sufficiently high value of C is found, or the top level is reached, in which case the global threshold is assigned.
5. The thresholds T of each area are assigned to the centre pixel. To compute a local threshold for each pixel in the image, bilinear interpolation is used.

Specimens and Images

Slides of a pure culture of *Bifidobacterium longum* and of faecal bacteria, fluorescently stained using 16S rRNA targeted fluorescence *in situ* hybridization (FISH) with a fluorescein-isothiocyanate (FITC) labelled DNA-probe which stains all Eubacteria (EUB-338: for details of staining procedure see Langendijk *et al.*[6]), were recorded with an integrating video rate CCD-camera system developed in our laboratory.[7] Two images were recorded for each field of view: one phase-contrast and one fluorescent. The former was acquired at normal video rates, the latter at an integration time of 8 seconds. All images were corrected for dark current and shading. The resulting images were 512x480 pixels in size and had 8 bit per pixel resolution. A typical fluorescence image is shown in figure 2.
Due to large differences in fluorescence intensity which hampered manual segmentation, the images of faecal samples had to be subdivided into smaller images containing only a single bacterium.
In all 20 full field of view images and 60 (single bacterium) subimages were used.

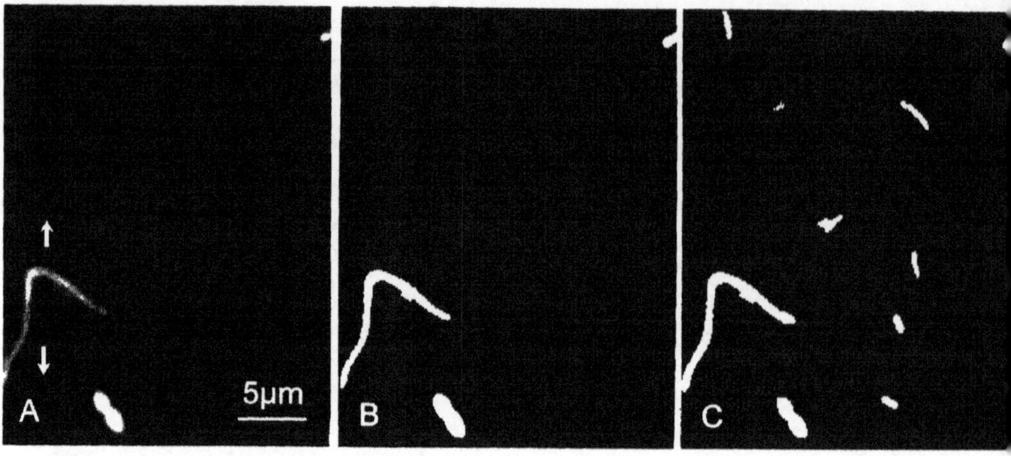

Figure 2. Part of a fluorescence image of faecal bacteria: (a) original image: note the two extremely faint objects at the arrows; (b) segmented using RATS algorithm globally: only the brightest objects are detected; (c) segmented using the same algorithm locally with a 5 level image pyramid: only the two faintest objects are lost in the noise.

Segmentation and reproducibility testing

In spite of a number of problems with manual segmentation, it is still considered a benchmark against which to test segmentation algorithms.[2] To test RATS, each of the images was segmented both manually and using RATS, using 5 levels of subdivision for full field of view images, and no subdivision for the single bacterium images. Following manual (2x) and automatic segmentation, the bias of the RATS algorithm with respect to manual thresholding is computed by comparing number of white pixels in each segmented image. This value is expressed as a percentage of the white pixels in the RATS-segmented image. Furthermore, in the case of full field of view images, the number of pixels assigned differently by the two algorithms is computed. If the bias is zero, this yields the error in the manual method caused by global as opposed to local thresholding. This too is expressed as a percentage of the number of white pixels in the RATS-segmented image.

Table 1. Errors manual vs. RATS segmentation

Source	Illumination	Surface area error manual vs. RATS (%)	
		bias (SD)	total (SD)
Bif. longum	fluorescent	1.2 (6)	18.0 (4.0)
	phase-contrast	0.3 (12)	13.7 (5.7)
faecal sample	fluorescent	-0.1 (21.7)	N.D.*
	phase-contrast	-0.3 (18.8)	N.D.*

* Not determined since the RATS algorithm was used without subdivisions

Table 2. Errors in manual segmentation

Source	Illumination	error manual1 vs. manual2 (SD)	
		surface area (%)	threshold (grey levels)
faecal sample	fluorescent	5.3 (28.0)	-0.5 (3.1)
	phase-contrast	-4.2 (18.5)	-1.1 (4.5)

RESULTS

Figure 2 shows the performance of RATS as a local as opposed to global threshold segmentation method. The bacteria vary considerably in brightness over the image (5 to 85 grey levels above background). When using local thresholding, only two very faint objects (grey levels 5 and 7) are lost in the noisy background. The faintest object detected was 10 grey levels above background. Used globally with the same noise estimate, only the very brightest objects are detected.
The results of the first bias test are shown in table 1. None of the biases differ significantly from zero. Table 2 lists the (means and SDs) of the differences between the two manual segmentations performed on each single bacterium subimage. The errors between the two manual segmentations are larger than between RATS and the manual thresholds.

DISCUSSION

The RATS algorithm is a fast, local thresholding algorithm, which shows no significant bias with respect to manual thresholding. The total segmentation errors listed in table 1 are probably the result of the poor performance of manual thresholding when the objects are not stained equally, or when the background shows (even minor) fluctuations. The increased (manual 1 vs. manual 2) errors in table 2 support this. To evaluate the algorithm's absolute accuracy, testing on objects of known dimensions would seem to be necessary.

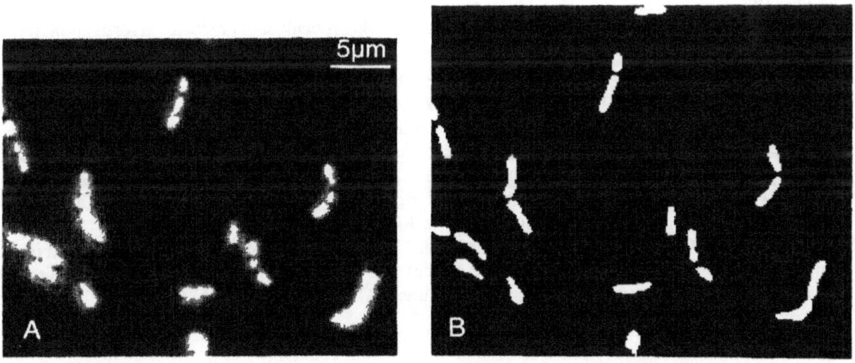

Figure 3. Overlay of phase-contrast and FISH images of the same field of view of *Bifidobacterium longum*: (a) original images; (b) segmented images. The fluorescent objects are smaller than the phase-contrast objects in both (a) and (b).

However, since manual thresholding has been shown to produce only small systematic errors when images of single objects are thresholded,[2] the lack of bias of RATS with respect to manual thresholding should be a good indication of its absolute accuracy. Segmentation of fluorescent images yielded slightly smaller surface areas than in the case of phase-contrast images of the same field of view. This can be explained by the fact that it is the ribosomes, which do not fill the entire cell, which are stained fluorescently. This can clearly be seen in figure 3.

In our laboratory, applications of the algorithms have included Gram-stained and nigrosin counter-stained bacteria,[3] and Hoechst 33342 fluorescently stained (human) cellular nuclei indicating the flexibility of the algorithm.

The speed of 2.5 s for 512x512 images on a 66MHz, 80486 DX2 based computer is sufficient for interactive image analysis. Even on a 16 MHz 80286 the time required is only 12.5 s. The memory requirements of the algorithm are modest: under 32 kB. The fact that no histogram is needed means the algorithm can be modified to accommodate 16 to 32 bit per pixel images easily. Such histograms would require 256 kB and 16 GB(!!) respectively.

The main drawback of the algorithm is that it can only segment into two categories: background and object. There are however many instances of images requiring just such segmentation. In all such cases, RATS should be considered.

ACKNOWLEDGEMENTS

This work has been made possible by financial support from the Institute for Microbiology and Biochemistry in Herborn, Germany, and the Dutch Foundation for Micromorphological Systems development. I would also like to thank F. Schut and G.C. Raangs for the use of their slides of bacteria.

REFERENCES

1. J. Kittler, J. Illingworth, J. Föglein, Threshold selection based on a simple image statistic, *Comp. Vision Graph. Image Proc.* 30:125 (1985).
2. C.L. Viles and M.E. Sieracki, Measurement of marine picoplankton cell size using a cooled, charge-coupled device camera with image-analyzed fluorescence microscopy. *Appl. Environm. Microbiol.* 58:584 (1992).
3. B.C. Meijer, G.J. Kootstra and M.H.F. Wilkinson, A theoretical and practical investigation into the characterisation of bacterial species by image analysis, *Binary Comp. Microbiol.* 2:21 (1990).
4. M.H.F. Wilkinson, G.J. Jansen and D. van der Waaij, Computer processing of microscopic images of bacteria: morphometry and fluorimetry. *Trends Microbiol.* 2:485 (1994).
5. J. Kittler, J. Illingworth, J. Föglein and K. Paler, An automatic thresholding algorithm and its performance, *in*: "Proc. 7th Int. Conf. on Pattern Recognition", Montreal 1984, (1985).
6. P.S. Langendijk, F. Schut, G.J. Jansen, G.C. Raangs, G. Kamphuis, M.H.F. Wilkinson and G.W. Welling, Quantitative fluorescence *in situ* hybridization of *Bifidobacterium spp.* with genus specific 16S rRNA targeted probe, application in faecal samples. *Appl. Environm. Microbiol.* 61:3069 (1995).
7. M.H.F. Wilkinson, G.J. Jansen and D. van der Waaij, Very low level fluorescence detection and imaging using a long exposure charge coupled device system, *in*: "Biotechnology Applications of Microinjection, Microscopic Imaging, and Fluorescence", P.H. Bach, C.H. Reynolds, J.M. Clark, P.L. Poole and J. Mottley, eds, Plenum Press, New York, (1993).

USE OF CONFOCAL MICROSCOPY FOR ABSOLUTE MEASUREMENT OF CELL VOLUME AND TOTAL CELL SURFACE AREA

C. P. Rubbi and D. Rickwood

Biology Department
University of Essex
Colchester CO4 3SQ - UK

INTRODUCTION

Confocal microscopy allows the real-time production of 3-dimensional images of live specimens. These images can in turn be used for measurement of specimen sizes in all three dimensions, provided that: 1) the system is properly calibrated; 2) the images accurately reproduce the observed object and 2) an appropriate algorithm is applied to extract the desired information from the images. In this paper, we analyse the precision with which cell size and surface area can be measured by confocal microscopy. Since there is no alternative way to check the results on a single cell basis, and thus the cytometric parameters cannot be confirmed by an alternative technique, we separately analysed the sources of errors. The confocal system employed is a BioRad MRC600 equipped with a Kr/Ar mixed gas laser and attached to an Olympus IMT-2 inverted microscope.

RESULTS

The following three sources of errors were separately analysed:

1) Apparent axial displacement due to different refractive indices and spherical aberration. The actual axial position of an imaged point will differ from that expected by absolute displacement whenever imaging is performed through layers of different refractive indices. It is considered that this difference can be accounted for by introducing a z-correction factor (Gahm and Witte, 1986). However, differences in refractive indices also introduce aberrations, especially when imaging is performed deep into specimens. For these reasons, we determined our z-correction factor from an experimental estimation based on confocal imaging in reflection mode. To do this, medium was mounted between slide and coverslip, leaving and air bubble. Axial reflection mode profiles were obtained for both glass/air/glass (bubble) and glass/medium/glass interfaces and distances between the reflection maxima were measured. When working with air objectives, the distance measured between reflection maxima in the glass/air/glass system (bubble) was taken as the real distance (no z-correction). The real

Fluorescence Microscopy and Fluorescent Probes
Edited by J. Slavík, Plenum Press, New York, 1996

distances measured with dry objectives were used for a X60 1.4NA oil immersion objective. The z-correction factor was calculated as:

real distance = z-factor x apparent distance

For theoretical estimation of the z-correction factor, the equation proposed by Gahm and Witte (1986) was used:

$$\text{z-factor} = \sqrt{[(n_i/n_{ob})^2 - \sin^2(\alpha)]} \,/\, \sqrt{[1 - \sin^2(\alpha)]}, \text{ where } \alpha = 2/3\ \alpha_{max}$$

Measured values of the z-correction factor are shown in Table 1. Figure 1 shows the reconstruction of a vertical section of a fluorescent bead (only objective-side of the bead considered) using our z-correction factor. Table 2 compares the accuracies with which both z-factors reconstruct the shape of the bead shown in Figure 1.

Table 1. Theoretical and experimental values of the z-correction factor

Objective	NA	n_{ob}	Theoretical	Experimental
X20	0.4	1	1.358	1.433
X40	0.6	1	1.395	1.435
X60	1.4	1.516	0.743	0.935 (X20)
				0.957 (X40)

Medium: PBS (n = 1.3349). n_{ob}: refractive index of the objective's immersion medium.

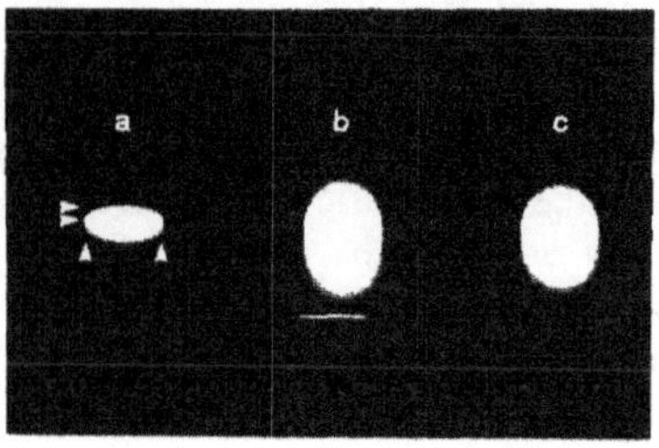

Figure 1. Vertical section of a fluorescent bead suspended in 1.5% agarose (n=1.3358). a) original XZ section; b) XZ section corrected for square pixels and c) full correction using the experimental z-correction factor. Bar: 10μm.

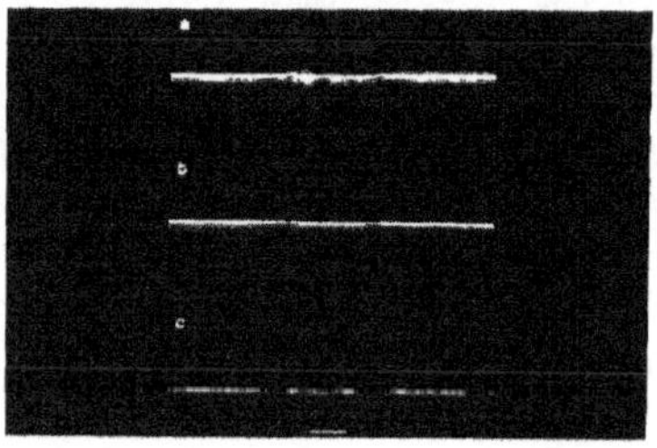

Figure 2. Axial reflection profile of a glass/medium interface with a bead settled on to it. Media are: a) PBS (n=1.3342); b) Foetal Calf Serum (n=1.3475) and c) 15% Ficoll (n= 1.3535). Bar: 10μm.

Table 2. Experimental and theoretical z-correction over the fluorescent bead of Figure 1

	Number of pixels	Length (μm)		Difference with X radius (%)
		Absolute	Corrected	
X diameter	88	11.48	11.48	--
Z-radius - experimental	16	6.4	5.76	+0.35%
Z-radius - Gahm-Witte	16	6.4	4.77	-17%

XY pixel size: 0.1304μm. Distance between Z sections: 0.4μm. Experimental z-correction factor: 0.9. Theoretical z-correction factor: 0.745

2) Distortion of regions behind a live cell due to differences in refractive index between object and medium. We adapted the method of Cheng and Summers (1990) and analysed the deformation that a cell would introduce on the reflection image of a medium/glass interface. Figure 2 shows the axial deflection of the reflection of a glass/medium layer with a cell in between. It can be seen that refractive index differences usually found between live cells and surrounding medium (PBS) do not severely affect the image accuracy; yet, an improvement can be made by increasing the refractive index of the medium.

3) Algorithms for extracting geometric data from scanned images. For measuring cell volume and surface area we segmented the images of cells whole-volume stained with fluorescein diacetate into inside-the-cell and outside-the-cell pixels. The cell volume measurement algorithm performs summation of all positive pixels and multiplies the value by the pixel volume. The surface area measurement algorithm performs sequential convolution/thresholding to extract perimeters of sections and, from these, calculates the cell surface area. Algorithms were checked against computer-generated images and the differences between the theoretical and experimental values were in the range of 0.4 - 3%.

Cell volumes were calculated from binarised images (p(x, y, z)) according to:

$$\text{Vol} = \text{pixel\_volume} \times \Sigma_{xyz}\, p(x,y,z)$$

and the calculation was performed as part of the measurement of cell surface.

The total area was calculated as: $A = S_1 + S_N + \Sigma_N A_i$, where S_1 and S_N correspond to the areas of the first and last sections, respectively. Figure 3 describes the calculation of the A_i terms. Figure 4 describes the convolution/thresholding algorithms for calculation of section perimeters. Peripheral pixels were classified as "row-column" (pixel distance = 1)or "angle" (pixel distance = $\sqrt{2}$). Also, pixels were subdivided in 1-neighbour and 2-neighbour pixels, the former contributing ½ of the pixel distance, and the latter 1.

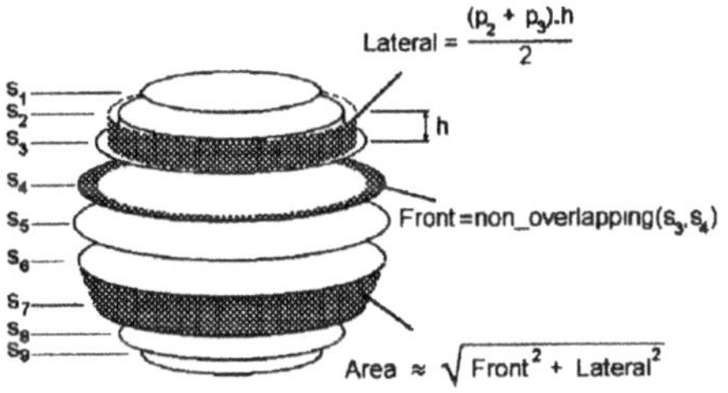

Figure 3. Scheme of the picture elements used for the calculation of the area of each pair of sections (Ai)

$$\textbf{Perimeter} = \textbf{Binary} \otimes \begin{vmatrix} 1 & 5 & 1 \\ 5 & 24 & 5 \\ 1 & 5 & 1 \end{vmatrix}$$

$24 < p(x,y) < 44 \rightarrow$ peripheral

$$\textbf{Alignment} = \textbf{Perimeter} \otimes \begin{vmatrix} A & R & A \\ R & 4 & R \\ A & R & A \end{vmatrix}$$

$4 < p(x,y) < 6 \rightarrow$ 1-neighbour

$6 < p(x,y) \rightarrow$ 2-neighbour

Figure 4. Processing for generating the images used for the calculation of section perimeters. Starting image is assumed to be binarised to 0-1. For simplicity, results are not normalised. With A=1, R=0, Alignment contains "angle" pixels; the inverse produces "row-column" pixels.

The calculations were applied to the bead shown in Figure 1, using the experimental z-correction factor. Table 3 shows the spherical parameters obtained.

Table 3. Spherical parameters and associated radii calculated for the bead in Figure 1

		Measured	Expected	% difference
Perimeter	**value**	38.7073	36.065	+7.3
	$r_{perim.}$	6.16	--	+7.3
Section area	**value**	106.089	103.508	+2.5
	$r_{sect.}$	5.81	--	+1.2
Surface area	**value**	439.676	414.032	+6.2
	$r_{surf.}$	5.92	--	+3
Volume	**value**	858.412	792.181	+8.4
	$r_{vol.}$	5.90	--	+2.8

Expected values are referred to a radius of 5.74, as measured for the equatorial section in Figure 1. Distances, areas and volumes are in μm, μm^2 and μm^3, respectively.

Next, measurements were performed over a group of 92 fluorescein diacetate stained cells shown in Figure 5. The corresponding single-cell parameters are plotted against each other in Figure 6. Sphericity and shape factor are defined as:

$$\text{Sphericity} = 36\pi \frac{\text{Volume}^2}{\text{Area}^3} \qquad \text{shape\_factor} = \frac{\text{cell\_height}}{\sqrt{\text{projected area}}}$$

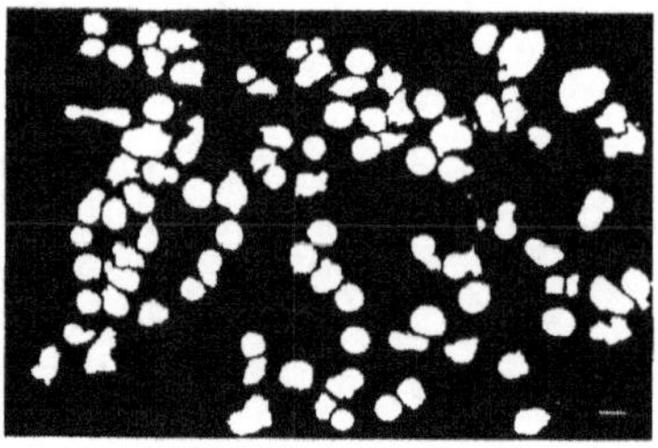

Figure 5. A group of 92 fluorescein diacetate stained HL-60 cells used for the calculations. Bar: 10μm.

CONCLUSIONS

A confocal microscope is the only equipment capable of yielding the amount of absolute cytometric information shown here. Since there is no other equipment against which to check the microscope measurements over single viable cells of random shape, we separately analysed the following sources of error: i) axial displacement due to different refractive indices and spherical aberrations ii) distortion of regions behind a live cell and iii) extraction of geometric information from the 3-D images by simple algorithms.

We consider that, all those sources of errors being analysed, the above results show that the approach that we have followed allows the calculation of absolute values for several geometric parameters of cells. The precision with which those values are calculated is within the order of magnitude of the resolution of the imaging system.

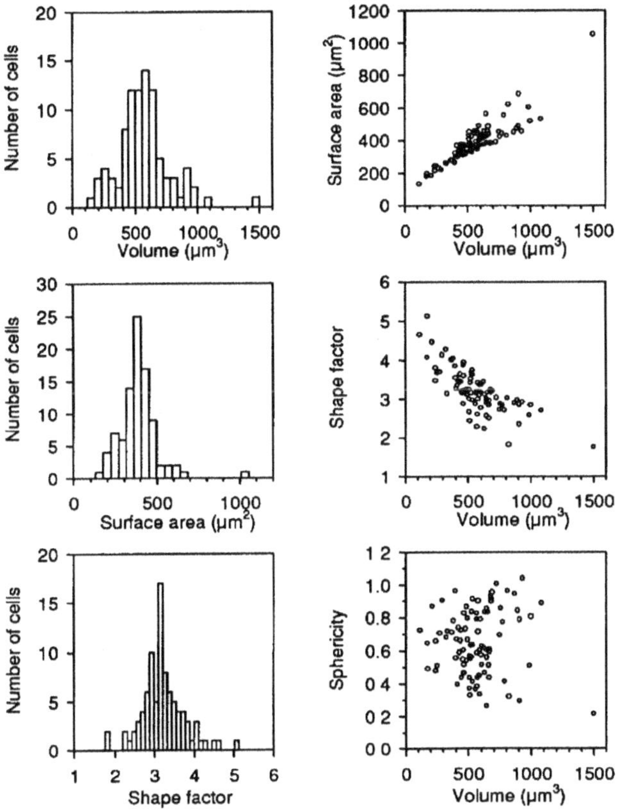

Figure 6. Cytometric parameters measured over the cells shown in Figure 5. Left column: population distributions for cell volume, surface area and shape factor. Right column: comparison of cell surface area, shape factor and sphericity against cell size (volume).

Acknowledgements

We are grateful to Mr. P. Morris for his help with photography. This work was funded entirely by the MRC.

REFERENCES

Cheng, P. C. and Summers, R. G. (1990). Image contrast in confocal light microscopy. In: Pawley, J. B., Ed. Handbook of biological confocal microscopy. Plenum Press, New York.

Gahm, T. & Witte, S. (1986). Measurement of optical thickness of transparent tissue layers. J. Microsc. **141**:101.

CELL VOLUME MEASUREMENTS USING CONFOCAL LASER SCANNING MICROSCOPY

Hanna Tinel, Frank Wehner, and Rolf K.H. Kinne

Max-Planck-Institut für Molekulare Physiologie
44139 Dortmund
Germany

INTRODUCTION

To survive, cells have to avoid excessive alterations of their volume. Upon exposure to anisotonic conditions most cells initially behave as osmometers and, thereafter readjust their volume to the initial value. Many cell types have developed a complex machinery of cell volume regulatory mechanisms comprising transport across the cell membrane and/or metabolism. Because swelling or shrinking of cells plays a crucial role in the overall regulation of cell function, determinations of cell volume play a major role in cell biology. In this contribution the application of confocal laser scanning microscopy to volume measurements is reviewed.

RESULTS

A confocal laser scanning microscope (CLSM, MRC-600, BioRad, Hemel Hempstead, UK) was used to study volume regulation of kidney cells (inner medullary collecting duct cells: IMCD cells; isolated or in primary culture) and of hepatocytes in primary culture. Cells were labelled with different fluorescent dyes (see below), which were excited by the use of the 488 nm band of an argon ion laser (Ion Laser Technology, Salt Lake City, UT) or the 442 nm band of a helium-cadmium laser (4310 N, Liconix, Santa Clara, CA) on a confocal laser scan unit This device was coupled to a standard microscope (Diaphot, Nikon, Düsseldorf, Germany). Three different techniques were used:

1. Dye Concentration Analysis

Cells were loaded for 30 min with 10 µmol/l of BCECF in form of AM ester. The dye was excited at 442 nm which is the wavelength, where the fluorescence solely depends on the concentration of the dye inside the cell Figure 1 shows the volume changes and the concomitant alteration in dye concentration and increase or decrease of the fluorescence

Fluorescence Microscopy and Fluorescent Probes
Edited by J. Slavík, Plenum Press, New York, 1996

signal detected by use of CLSM (objective 25 x, numerical aperture 0.8). The relative cell volume could be calculated taking into account dye leakage and photobleaching determinated in separate experiments.[1]

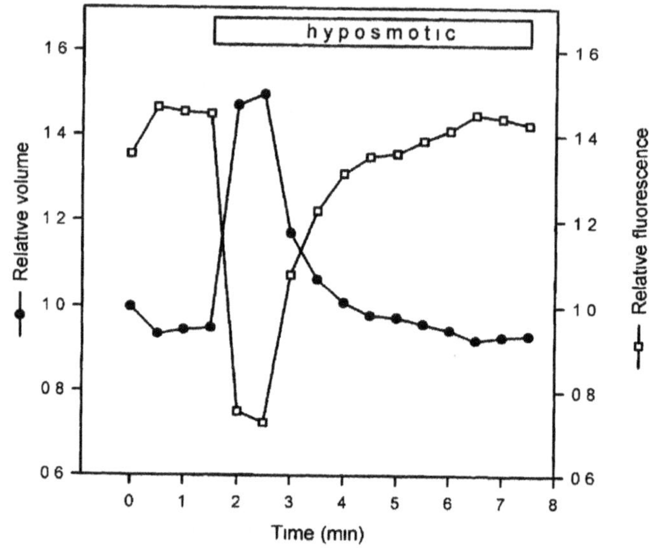

Figure 1. Volume changes determined by dye concentration measurements in isolated IMCD cells. Cell volume was changed by a decrease of the medium osmolarity from 600 to 300 mosmol/l. The mean intensity of BCECF excited at 442 nm (□) decreased and from this the relative cell volume changes (●) could be calculated (single experiment). A rapid cell swelling and regulatory volume decrease is observed. From [1] with kind permission.

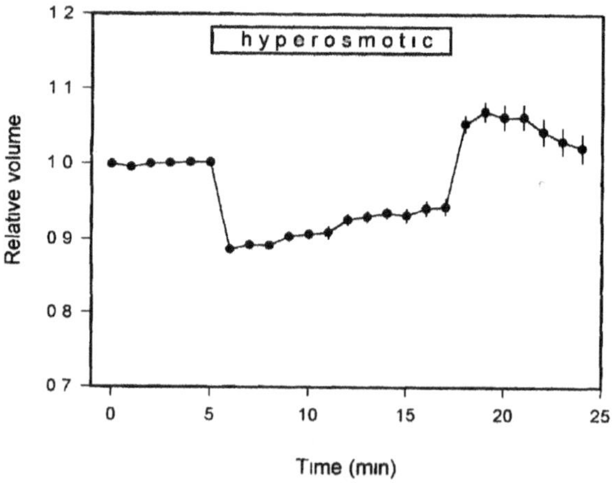

Figure 2. Effect of increase in osmolarity on the cell volume of rat hepatocytes in confluent monolayers. 100 mmol/l sucrose was added to the normosmotic superfusate of 300 mosmol/l for the time indicated (n = 20) The relative changes of cell volume were determined by quantifying calcein fluorescence in a single plane. Reproduced from [2] by copyright permission of The Rockefeller University Press.

2. Self Quenching of Fluorescence

Cells were loaded for 45 min with 10 µmol/l calcein (in its acetoxymethyl ester form) as a volume marker of aqueous compartments which (under physiological conditions) is pH- and Ca^{2+} insensitive. Cell volumes were determined by quantifying calcein fluorescence (objective 20 x, numerical aperture 0.65) in a single plane. Calcein exhibits a pronounced self quenching. Therefore fluorescence decreases upon concentration and increases upon dilution of the compound. The fluorescence was related to cell volume in a set of pilot experiments, in which cells were exposed to different osmolarities. Percent changes in calcein fluorescence were then compared to percent changes in cell volume determined with method 3. The two parameters were linearly correlated (r = 0.9987) with a slope of 0 601 ± 0.018 (95% confidence).[2] Figure 2 shows volume changes of rat hepatocytes in primary culture determined by means of this method. Increasing extracellular osmolarity from 300 to 400 mosmol/l decreased the cell volumes to 88 ± 0 5%. Thereafter hepatocyte volume increased again to 94% of control, equivalent to a regulatory volume increase (RVI) of 45%.

3. Optical Sectioning of Cells

IMCD cells in primary culture were loaded with BCECF as mentioned above. Measurements were performed at an excitation wavelength of 442 nm, where BCECF fluorescence is pH insensitive (isosbestic point). For the experiments, a 20 x objective (Nikon) with a numerical aperture of 0 65 was chosen. The theoretical value for the z-resolution of the microscope was calculated to be 0.91 µm. This value could be confirmed in calibration experiments with fluorescent microbeads of 1 µm in diameter. Single cells were optically sectioned from the basolateral side to the apical pole by 1 µm steps in the z-direction. Five to six optical sections were performed per cell. A set of optical sections through the cells was acquired within 5 s. Cell volume was determined quantitatively by measuring surface areas of every cross section and calculating the total cell volume.[3]

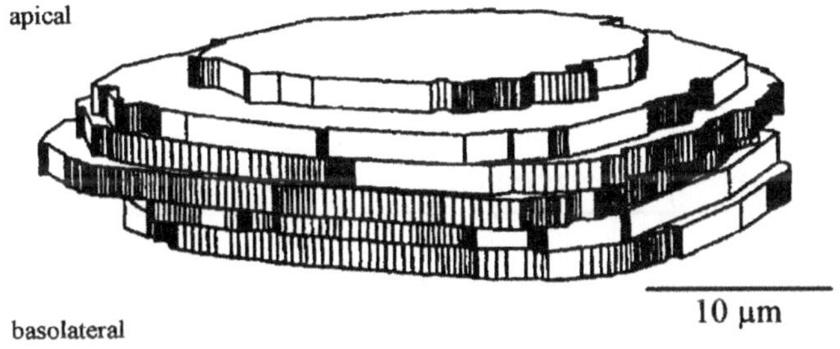

Figure 3. Reconstruction of cell volume determinated by optical sectioning A single IMCD cell in primary culture loaded with BCECF was optically sectioned from the basolateral side to the apical pole by 1 µm steps in the z-direction. Six optical sections were obtained per cell within 5 s. Individual sectional areas of a single cell were determined by computer planimetry Sum of individual section-volumes yields total cell volume[4].

Figures 4 and 5 show volume changes of rat IMCD cells in primary culture determinated using technique 3. Under hypotonic stress cells swelled within 1 min by 35 ± 3% above the control value (Fig 4). Within 5 min, cells then underwent a partial regulatory volume decrease (RVD) by 73 ± 6% This volume regulation of IMCD cells was completely abolished by 100 µmol/l TMB-8 (8-(N,N-diethylamino)octyl-3,4,5-trimethoxybenzoate hydrochloride) (Fig. 5) which is an effective inhibitor of intracellular Ca^{2+} release. The experiments indicate that intracellular Ca^{2+} release regulates RVD in IMCD cells.

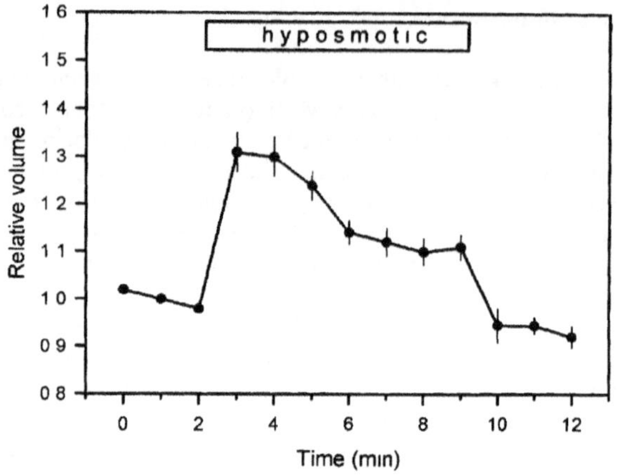

Figure 4. Changes in cell volume of IMCD cells in primary culture upon reduction of osmolarity from 600 to 300 mosmol/l Data are normalized to control values 60 s before the omission of sucrose; n = 9 for each data point. From [3] with kind permission.

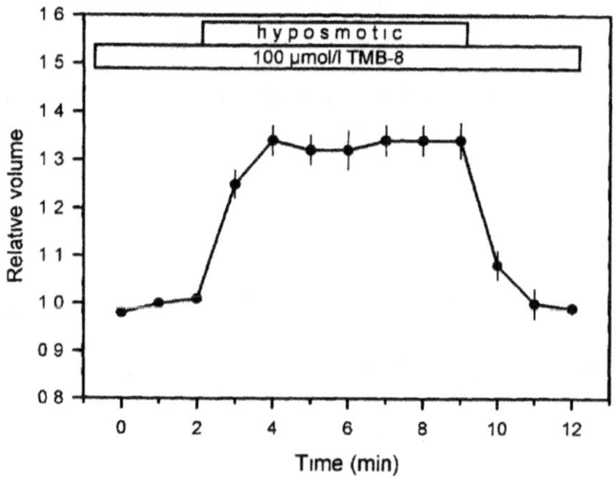

Figure 5. Effects of TMB-8 (inhibitor of Ca^{2+} release from intracellular calcium stores) on the cell volume changes of IMCD cells during hyposmotic stress. Cells were preincubated with 100 µmol/l TMB-8 for 3 min. Osmolarity was reduced from 600 to 300 mosmol/l as indicated. From [3] with kind permission.

CONCLUSIONS

We presented three different techniques for the determination of cell volume using a confocal laser scanning microscope Methods 1 and 2 allow a dynamic measurement but require additional calibration curves for the total cell volume calculation. Method 3 yields absolute values of cell volume, however, photobleaching limits the temporal resolution to 1/min

REFERENCES

1 M. Weinlich, G. Capasso, and R.K.H. Kinne, Confocal microscopy, *in*: "Methods in Membrane and Transporter Research," J.A. Schafer, G. Giebisch, P Kristensen, H.H. Ussing eds., pp 215-272, R.G. Landes Co., Austin (1994)

2. F. Wehner, H. Sauer, and R.K.H. Kinne, Hypertonic stess increases the Na^{+} conductance of rat hepatocytes in primary culture, *J. Gen. Physiol.* 105:507-535 (1995).

3 H. Tinel, F Wehner, and H. Sauer, Intracellular Ca^{2+} release and Ca^{2+} influx during regulatory volume decrease in IMCD cells, *Am. J. Physiol.* 267.F130-F138 (1994).

4. H Tinel, Kalziumfreisetzung aus intrazellulären Speichern und Kalziumeinstrom während der hypoosmotischen Volumenregulation papillärer Sammelrohrzellen der Rattenniere: Eine fluorometrische Analyse der Signaltransduktion (PhD thesis), Ruhr-Universität Bochum, Bochum, Germany (1995).

SUBCELLULAR CYTOFLUOROMETRY IN CONFOCAL MICROSCOPY

Denis Demandolx and Jean Davoust

Centre d'Immunologie CNRS-INSERM de Marseille-Luminy
Case 906
F-13288 Marseille Cedex 9
E-mail address: demandol@ciml.univ-mrs.fr

INTRODUCTION

The three-dimensional sampling capacity of confocal laser scanning microscopy can be used to analyze simultaneously the distribution of multiple proteins and nucleic acids. Confocal microscopes filter out most of the light emanating outside the plane of focus [1-8] and multiwavelength instruments are now able to discriminate several fluorescent molecules using two or more laser lines. The concurrent detection of several structures labeled with specific probes has potential applications such as the segmentation of double positive and single positive cells on tissue sections, or of organelles at the subcellular level [9,10]. We are describing here two methods used to compare closely related fluorescence micrographs. First, we are presenting a new graph, called cytofluorogram, which estimate the joint probabilities linking fluorescence intensity distributions of multiple fluorescent probes [11]. Such graphs provide a semi quantitative method to analyze the sites of coincident fluorescence labeling, the fluorescence background and the single labeled structures. Locally, the cytofluorograms are related to the notion of image correlation [12]. Next, we have applied a method based upon statistical correlation between windows defined on the raw images. Local image correlation reveals highly related fluorescent distributions in double fluorescence micrographs and eliminates superimposed fluorescent objects of unequal profiles.

MATERIALS AND METHODS

Cells, Antibodies

MHC Class II IA^k and Ii chain positive fibroblast cells were obtained after transfection using the calcium phosphate precipitation method as described [13,14]. Indirect immunofluorescence was performed after fixation and permeabilization using fibroblast cells cultured for 48 h on glass coverslips as described previously [13,14]. The labeled cells adherent to the glass coverslips were mounted on glass slides with a Mowiol embedding medium [13,14].

Confocal Microscopy and C Language Programs

Confocal laser scanning microscopy was performed using a Leica TCS 4D (Leica Lasertechnik, GER) instrument based on a Leitz DM microscope interfaced with an Argon-Krypton Laser adjusted at 488 nm and 568 nm. With the 1.4 numerical aperture 63 PL APO objective, the final resolution is close to 0.2 micron in the plane of section and 0.5 micron along the *z*-axis. Acquisition time of 0.5 second per scan and 16 scans of 512x512 images were selected as a standard procedure for all the applications. The residual axial shifts between 488 and 568 nm lines is not greater than 50 nm. Focal series of four horizontal planes of section spaced by 0.3 micron have been monitored simultaneously for FITC and Texas red, using a double dichroic mirror for the excitation beam, an FITC band pass 520-560 nm barrier filter and a long pass barrier filter above 580 nm for Texas red detection, using respectively green sensitive and red sensitive photomultipliers. Image processing tools and cytofluorogram algorithms were implemented in C programming language using Microware's C Language Compiler System (Microware, USA) then linked with the Leica image handling library. Our current programs run on a Motorola 68040 microprocessor system (Motorola, USA) with Microware's OS-9 operating system.

RESULTS

Presentation of Micrographs

We have acquired confocal images from fibroblast tissue culture cells, transfected here with $IA^k\alpha\beta$ MHC class II molecules labeled indirectly with FITC complexed antibodies and with the MHC class II associated invariant chain labeled indirectly with Texas red complexed antibodies [13,14]. Panels A and B of figure 1 show a typical plane of section recorded one micron above the solid glass support used for cell culture and simultaneously observed with FITC and Texas red acquisition channels. Field of 50x50 microns (512x512 pixels) is visualized with a linear black and white look-up table. In figure 1A, the plasma membrane and a set of internal vesicles are labeled by FITC indirect immunofluorescence, revealing respectively the surface and the intracellular expression of MHC class II molecules. Figure 1B shows the Texas red labeled antibodies directed against the transfected invariant chain construct. A large surface of the micrograph, represented in black, corresponds to background fluorescence.

Cytofluorograms

The number or frequency of each gray level pixel of an image can be represented on a pixel histogram [12] providing an estimate of their probability density [11] (data not shown). It is possible to extend this representation to a two-dimensional one by estimating the joint probability density linking two fluorescence distributions. We have constructed in figure 2A the two-dimensional histogram of registered pixels from images A and B of figure 1. Pixel values of images 1A and 1B are placed respectively along the *x* and *y*-axes. The logarithm of the number of pair of corresponding pixel values are represented along the *z*-axis with a black and white look-up table (on a white background). The sharp peak at the origin of this two-dimensional pixel histogram corresponds to the low luminance levels of the double fluorescence labeling and, by analogy with flow cytometry, we are proposing to call this representation a cytofluorogram [15,16]. In figure 2B, we have represented in white the pixels which are above a threshold in each fluorescence channel. The extraction on the images, of zones of interest defined on the cytofluorogram, can provide the basis for a segmentation of the images. Cytofluorograms cannot be used to distinguish a real correlation from the superimposition of non-correlated fluorescence profiles.

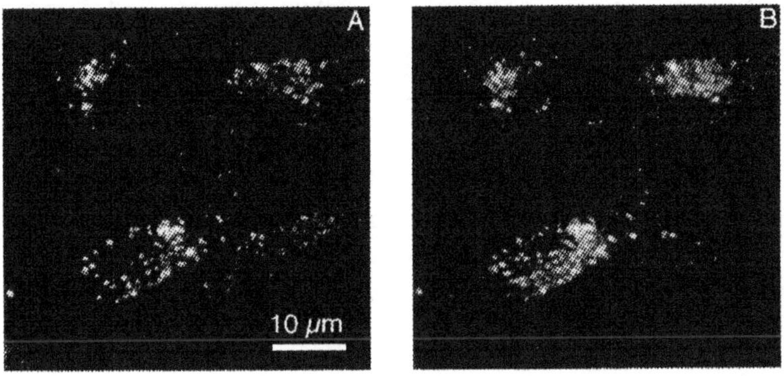

Figure 1. Double immunofluorescence labeling. of fibroblast tissue culture cells observed by confocal microscopy. In panel A, $IA^k\alpha\beta$ MHC class II molecules are labeled indirectly with FITC complexed antibodies. In panel B, the MHC class II associated invariant chain is labeled indirectly with Texas red complexed antibodies.

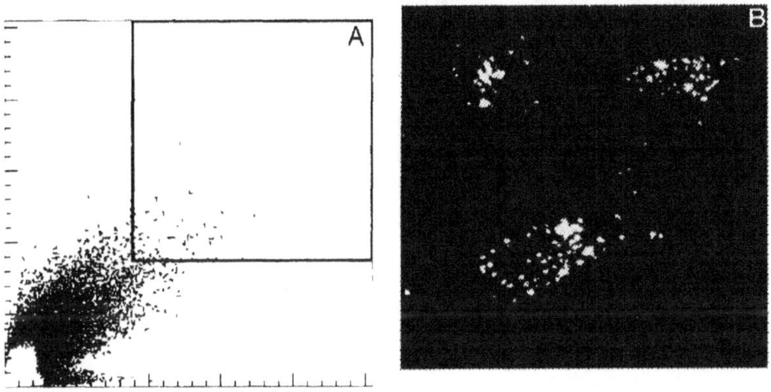

Figure 2. Two-dimensional analysis of double labeling. In panel A, we have constructed the cytofluorogram representation of the pixel distribution of FITC (x-axis) and Texas red (y-axis) from figure 1. The pixels, belonging to the square-area in the cytofluorogram of A, are represented on the binary image of panel B.

Local Image Correlation

In order to draw attention on correlated fluorescence profiles in multiple immunofluorescence micrographs, we had to extract more information from neighboring pixels and therefore to define a rolling window of interest. Gaussian windows are preferable because of their isotropic properties. It is possible to consider local cytofluorogram on such windows. If the pixel pairs within the image are highly related, the entries on cytofluorogram will be clustered along the diagonal of the graph. The local correlation, measured on the cytofluorogram, depends on the orientation and the dispersion of the cloud of dots.

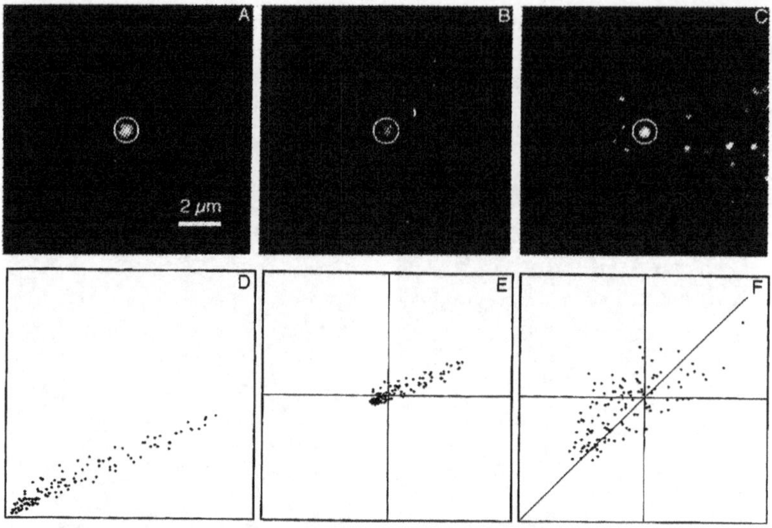

Figure 3. A small area of interest has been circled in panel A (FITC) and B (Texas red) to show the principle of the local correlation method (panel C). Panels D, E, and F show the cytofluorograms of the selected vesicle for successive steps of the local correlation.

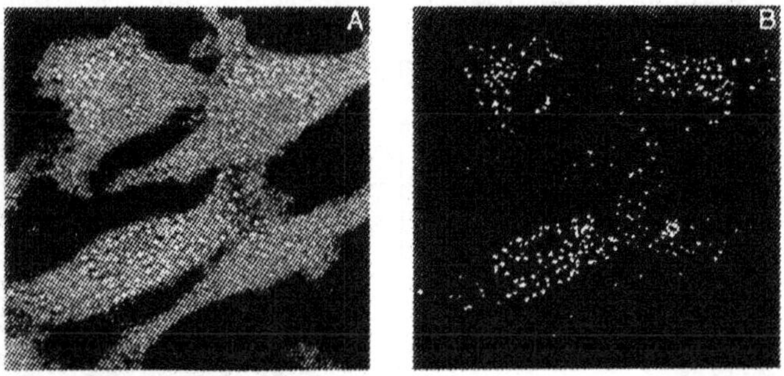

Figure 4. Correlation maps. A: The complete correlation map corresponding to the covariance of the fluorescence micrographs of figure 1. B: Thresholded version of the correlation map .

To illustrate the principle of a correlation method, we have chosen a small area of interest (about 100 pixels) from figure 1 encircled in figure 3A (FITC channel) and 3B (Texas red channel). The corresponding cytofluorogram is shown in figure 3D. The very stretched distribution of points shows a correlated area. To complete a statistical differencing [12] of each image, we have first subtracted to each pixel $F(i,j)$ of each raw image, the mean value $\overline{F}(i,j)$ of the surrounding pixels computed with a 9-pixel-half-width gaussian filter. The resulting cytofluorogram of panel 3E is therefore centered. Second, we have divided each pixel value of images by its measured statistical standard deviation $\sigma(i,j)$ computed over the same window at the coordinate (i,j). The resulting cytofluorogram has now the same standard deviation in x and y directions (panel 3F). Non correlated structures would have given rise to cloud of dots dispersed in all directions. To visualize the covariance, we have multiplied and locally averaged the two normalized images. This can also be interpreted as a logical "AND" operator. The overall operation is close to a mathematical correlation performed on defined windows. In figure 4A, we have displayed the resulting correlation image for the whole field of the initial micrographs. In figure 4B, this image has been thresholded to highlight the correlated zones of interest in the raw images.

CONCLUSION

The cytofluorograms are representing on two-dimensional graphs the frequency of registered pixels from double fluorescence digital images. They can be used to classify background, single-labeled and double-labeled areas. More elaborate cytofluorograms can be performed after offsetting the fluorescence value to remove non-specific fluorescence levels. Cytofluorograms are also of importance to evaluate the crosstalk between fluorescence channels and to perform the appropriate compensation. The local image correlation, estimating the local similarities of two fluorescence distributions, can be used to identify colocalized structures. Image correlation is therefore an important attribute to perform the segmentation of multiple fluorescence micrographs.

ACKNOWLEDGMENTS

This work was supported by CNRS, INSERM, and ARC. One of the author (D.D.) is the recipient of a predoctoral fellowship from the Conseil Régional PACA.

REFERENCES

1. J.B. Pawley. "Handbook of biological confocal microscopy," Plenum Press, New York (1995).
2. M. Petran and A. Boyde, New horizons for light microscopy, *Science* 230:1258 (1985).
3. M. Robert-Nicoud, D.J. Arndt-Jovin, T. Schormann, and T.M. Jovin, 3D imaging of cells and tissues using confocal laser scanning microscopy and digital processing, *Eur.J.Cell Biol.* 25:49 (1989).
4. C.J.R. Sheppard and C.J. Cogswell, 3-Dimensional image-formation in confocal microscopy, *J.Microsc.* 159:179 (1990).
5. D.M. Shotton, Confocal scanning optical microscopy and its applications for biological specimens, *J.Cell Sci.* 94:175 (1989).
6. J.G. White, W.B. Amos, and M. Fordham, An evaluation of confocal versus conventional imaging of biological structures by fluorescence light microscopy, *J.Cell Biol.* 105:41 (1987).
7. R.W. Wijnaendts-van-Resandt, H.J.B. Marsman, R. Kaplan, J. Davoust, E.H.K. Stelzer, and R. Stricker, Optical fluorescence microscopy in three dimensions: microtomoscopy, *J.Microsc.* 138:29 (1985).
8. T. Wilson. "Confocal microscopy," Academic Press, London (1990).
9. T.C. Brelje, et al, Multicolour laser scanning confocal immunofluorescence microscopy: practical applications and limitations, *Methods Cell Biol.* 38:97 (1993).
10. S.W. Paddock, et al, Three-color immunofluorescence imaging of Drosophila embryos by laser scanning confocal microscopy, *BioTechniques* 14:42 (1993).
11. W. Feller. "An introduction to probability," John Wiley and sons, New York (1971).
12. W.K. Pratt. "Digital image processing," John Wiley and sons, New York (1978).
13. M. Humbert, G. Raposo, P. Cosson, H. Reggio, J. Davoust, and J. Salamero, The invariant chain induces compact forms of class II molecules localized in late endosomal compartments, *Eur.J.Immunol.* 23:3158 (1993).
14. J. Salamero, M. Humbert, P. Cosson, and J. Davoust, Mouse B lymphocyte specific endocytosis and recycling of MHC class II molecules, *EMBO J.* 9:3489 (1990).
15. D. Demandolx and J. Davoust, Multicolor analysis in confocal immunofluorescence microscopy, *J.Trace Microprobe Tech.* JTMT 95022 (1995) (In Press).
16. H.M. Shapiro. "Practical flow cytometry," Alan R. Liss, West Newton, Massachusetts (1988).

APPLICATION OF CONFOCAL MICROSCOPY TO 3-D RECONSTRUCTION AND MORPHOMETRICAL ANALYSIS OF CAPILLARIES

Lucie Kubínová[1], Marie Jirkovská[2], Petr Hach[2], Daniel Palouš[2], Petr Karen[1] and Ivan Krekule[1]

[1]Institute of Physiology, Academy of Sciences of the Czech Republic Vídeňská 1083, CZ-14220 Prague, Czech Republic
[2]Institute of Histology and Embryology, 1st Medical Faculty, Charles University, Prague, Czech Republic

INTRODUCTION

The main advantage of a confocal laser scanning microscope (CLSM) consists in its possibility to capture images of thin, perfectly registered serial optical sections of the studied specimen. Such stacks of sections are useful in 3-D rendering of objects contained in the specimen as well as in implementing some of the recent efficient stereological methods. However, when using CLSM, a suitable fluorescence staining of the components of interest must be found.

The application of confocal microscopy is demonstrated on the 3-D study of terminal villi in human term placenta. Terminal villi form a predominating structure in the placenta at term. They contain majority of placental capillary bed which plays a key role in fetomaternal transport. At their level some pathological changes also take place, e.g., in diabetes. Therefore, information about the spatial organization of their capillary bed and its surface area, volume or topological characteristics is valuable in the comparison studies of normal and pathologic placenta.

MATERIAL AND METHODS

Tissue blocks were taken from the central region of the placenta of healthy mothers after spontaneous delivery. Samples were fixed in 4% formalin solution and embedded in paraffin wax. Approximately 100 µm thick sections were cut by a sledge microtome and mounted in 1% gelatine solution on glass slides. Different staining techniques were explored (Jirkovská et al., 1994). Staining by 0.15% eosin solution was found to be suitable, ensuring good contrast between tissue compounds and sufficient penetration into the entire section. After the dehydration and clearing, the sections were mounted in entelan.

Fluorescence Microscopy and Fluorescent Probes
Edited by J. Slavík, Plenum Press, New York, 1996

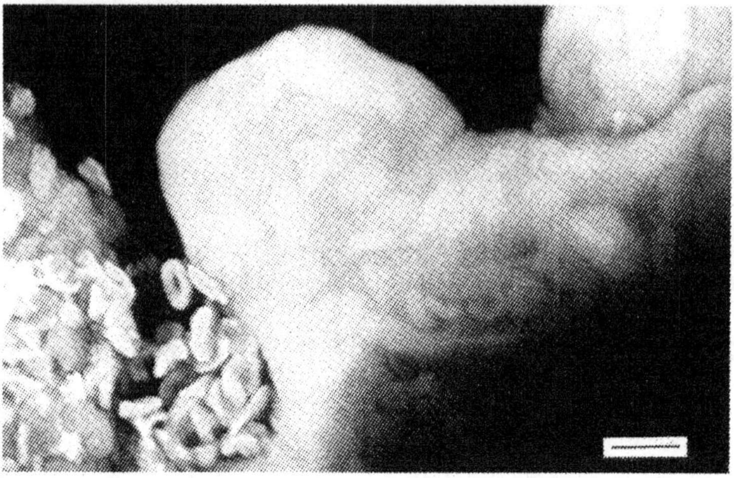

Figure 1. Terminal villus of human term placenta. Projection of 58 optical sections taken by a confocal microscope. Scale bar = 10 μm.

Terminal villi were studied, i.e. villous segments containing no vessels other than capillaries (Kaufmann et al., 1985). Individual villi (Fig.1) lying completely inside the thick sections were sampled using the disector principle (Sterio, 1984). Digitized images of thin serial optical sections (1 μm apart) were captured by a Bio-Rad MRC 600 confocal laser scanning microscope (CLSM) connected with the inverted fluorescence microscope NIKON Diaphot using a planapochromat oil immersion objective (60×, N.A.=1.4).

3-D Reconstruction

To reveal the number of capillary segments and the appearance of the capillaries in the villus, three types of 3-D reconstructions of the capillary bed were considered.

a) Topological 3-D reconstruction of the capillary bed was made and its Euler number was counted. The Euler number ($\chi(\mathrm{cap})$) of the capillary bed in the villus characterizes its topological properties, namely its connectivity ($c(\mathrm{cap})$), i.e. the number of redundant connections in the capillary bed. Taking into account that the capillary bed is connected and that it does not contain any enclosed cavities, the following equation is valid in our case:

$$\chi(\mathrm{cap}) = 1 - c(\mathrm{cap}) \,. \qquad (1)$$

Euler number was measured by the "sweeping plane" method (DeHoff, 1968) when the profiles of the capillary lumen were observed in subsequent optical sections and two types of events were counted: Firstly, the number of luminal appearances ($i(\mathrm{cap})$) and secondly, the number of luminal connections ($b(\mathrm{cap})$). The Euler number ($\chi(\mathrm{cap})$) was then evaluated by the formula:

$$\chi(\mathrm{cap}) = i(\mathrm{cap}) - b(\mathrm{cap}) \,. \qquad (2)$$

b) The wire-frame 3-D reconstruction of the capillary bed was made. A special programme for wire-frame reconstruction was designed, based on tracing the contours of villus and capillary lumen profiles by using the graphical tablet DG-1 (manufactured by the Czechoslovak Academy of Sciences) which was interfaced with an IBM PC/AT.

c) The villus and its capillary bed were rendered by using the interactive visualization programme package ANALYZE (Rob and Hanson, 1991) run on an INDY (Silicon Graphics) graphical workstation. The automatic segmentation being difficult, the contours were traced manually by a mouse.

Stereological Methods

The sampled villi were also evaluated by using stereological methods:

a) The volume density of the capillary bed in the terminal villus was estimated by the point-counting method (see, e.g. Weibel, 1979).

b) The volumes of the villus and capillary bed were estimated by Cavalieri's principle (see, e.g. Michel and Cruz-Orive, 1988).

c) The villus outer surface area and the inner surface area of the capillary bed were estimated by the method of the spatial grid (Sandau, 1987).

RESULTS

For illustration of the methods the results are shown for one of the villi (Fig.1), namely its schema (Fig.2), the topological reconstruction (Fig.3), 3-D rendering of its capillary bed (Fig.4), its stereo-pair (Fig.5), its wire-frame reconstruction (Fig.6) and estimates of its geometrical parameters (Table 1).

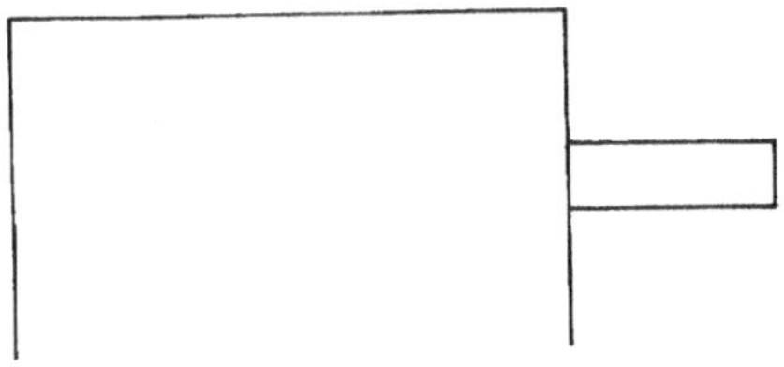

Figure 2. Schema of the terminal villus and its capillary bed

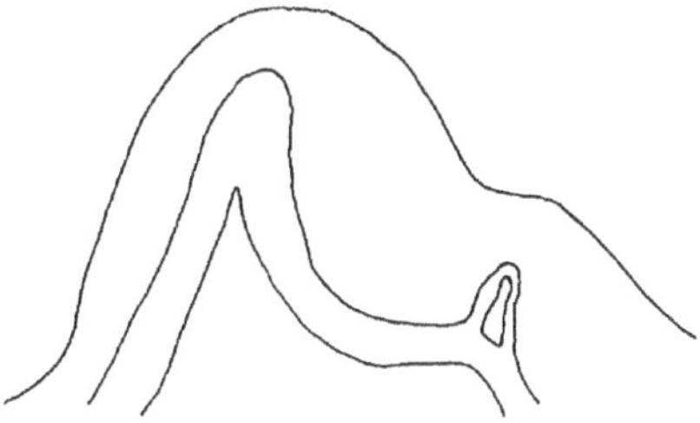

Figure 3. Topological reconstruction of the villus capillary bed.

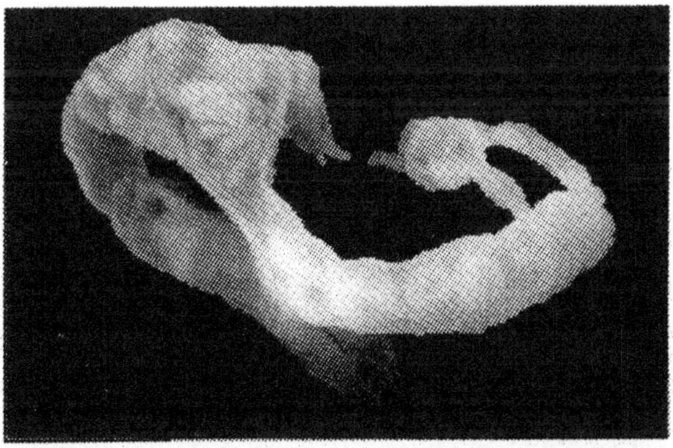

Figure 4. 3-D rendering of the villus capillary bed accomplished by ANALYZE programme package.

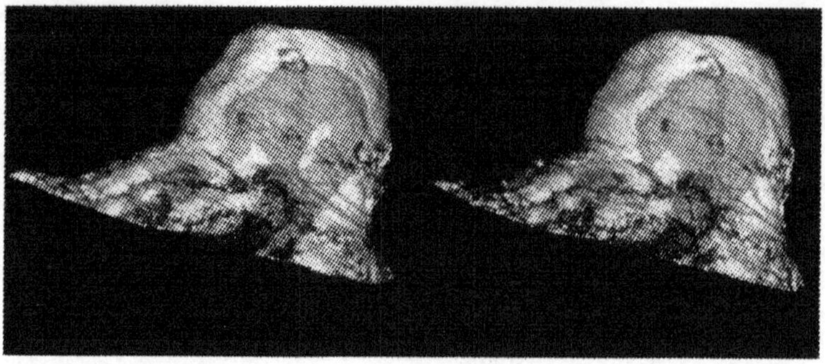

Figure 5. Stereo-pair of the terminal villus and its capillary bed accomplished by ANALYZE programme.

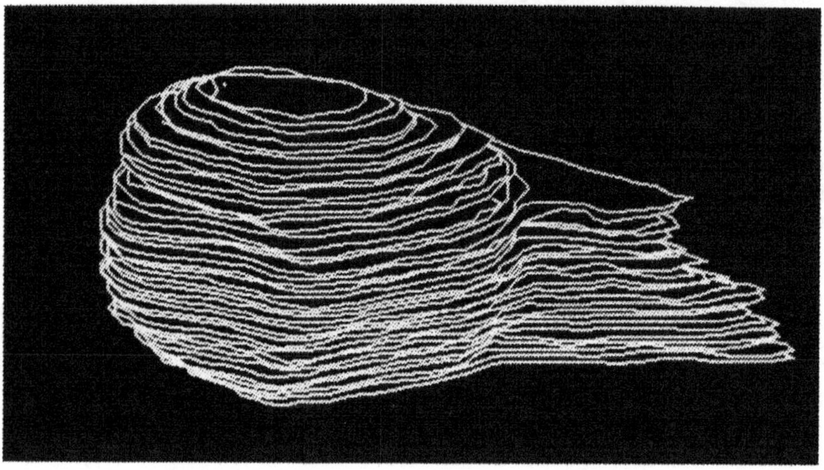

Figure 6. Wire-frame reconstruction of the terminal villus.

Table 1. The estimated villus volume (V(vil)), capillary bed volume (V(cap)), volume density of capillary lumen in villus (V_V(cap)), villus outer surface area (S(vil)), capillary inner surface area (S(cap)), ratio of capillary surface area to villus surface area (S_S(cap)), and Euler number of the capillary bed (χ(cap)).

V(vil)	V(cap)	V_V(cap)	S(vil)	S(cap)	S_S(cap)	χ(cap)
73800 μm^3	8100 μm^3	0.110	7800 μm^2	3700 μm^2	0.47	0

DISCUSSION

The application of the confocal microscope to the study of placental terminal villi allows to preserve relationships between capillaries and other components of villus. It enables to reconstruct the entire terminal placental villus and simultaneously the arrangement of its capillary bed which is not possible when using corrosive preparations (Arts, 1961) or by using SEM (Akiba et al., 1987). Moreover, the stereological methods can be easily applied and so valuable quantitative information about geometrical characteristics can be obtained.

The present study shows that confocal microscopy can be useful for 3-D study of biological objects. However, several problems and sources of errors must be taken into account.

(i) A suitable fluorescent staining of the studied specimen must be found, as contemporary confocal microscopes have an epi-illumination design. In the above example we have found the unspecific staining by eosin to be satisfactory. Eosin staining is also suitable for staining placental tissue embedded in glycol methacrylate instead of paraffin. The tissue shrinkage, occuring in our case, when we have assumed that it is uniform in all villi, thus can be minimized.

(ii) The object should be small enough to fit the limited depth of observation and field of view.

(iii) The error caused by the axial displacement due to unmatching refractive indices (see e.g. Visser et al., 1992) should be determined or minimized by matching the refractive index of the specimen to that of the medium between them as was done in our example.

(iv) Good sampling design must be followed to ensure that representative samples are examined. The unbiased sampling of particles can be guaranteed by applying the disector principle (Sterio, 1984).

ACKNOWLEDGEMENT

This study was supported by the Grant Agency of the Czech Republic (grant No.304/94/1202).

REFERENCES

Akiba, K., Kuwabara, Y., Mizuno, M., and Fukuda, S., 1987, Semiquantitative study of the vascularization of human term placenta using scanning electron microscopy, *J. Clin. Electron Microscopy* 20:5.

Arts, N.F.T., 1961, Investigations on the vascular system of the placenta, *Am. J. Obstet. Gynecol.* 82:147.

DeHoff, R.T., 1968, Curvature and topological properties of interconnected phases, in: *Quantitative Microscopy*, R.T.DeHoff and F.N.Rhines, eds. McGraw Hill, New York.

Jirkovská, M., Kubínová, L., Krekule, I., Karen, P., and Palovský, R., 1994, To the applicability of common fluorescent dyes in confocal microscopy, *Funct. Develop. Morphol.* 4:171.

Kaufmann, P., Bruns, U., Leiser, R., Luckhardt, M., and Winterhager, E., 1985, The fetal vascularisation of term human placental villi. II. Intermediate and terminal villi, *Anat. Embryol* 173:203.

Michel, R.P., and Cruz-Orive, L.M., 1988, Application of the Cavalieri principle and vertical sections method to lung: estimation of volume and pleural surface area, *J Microsc.* 150:117.

Robb, R.A., and Hanson, D.P., 1991, A software system for interactive and quantitative visualization of multidimensional biomedical images, *APESM* 14:9.

Sandau, K., 1987, How to estimate the area of a surface using a spatial grid, *Acta Stereol.* 6:31.

Sterio, D.C., 1984, The unbiased estimation of number and sizes of arbitrary particles using the disector, *J. Microsc.* 134:127.

Visser, T.D., Oud, J.L., and Brakenhoff, G.J., 1992, Refractive index and axial distance measurements in 3-D microscopy, *Optik* 90:17.

Weibel, E.R., 1979, Stereological Methods, Vol 1. Practical Methods for Biological Morphometry, *Academic Press*, London.

RETRIEVING SPATIOTEMPORAL INFORMATION FROM CONFOCAL DATA: A STUDY USING MELANOTROPE CELLS OF *XENOPUS LAEVIS*

Werner J.H. Koopman, Bruce G. Jenks, Eric W. Roubos and Wim J.J.M. Scheenen[#]

Department of Animal Physiology, Nijmegen Institute for Neurosciences, Toernooiveld 1, 6525 ED Nijmegen, The Netherlands and [#]Dept. Biomedical Sciences, University of Padova, via Trieste 75, 35121 Padova, Italy

INTRODUCTION

Confocal microscopy can be a useful tool for studies on cells in which local changes in intracellular ion concentrations need to be observed with a high spatial resolution. In time-related confocal studies high temporal resolution of changes in ion concentrations, detected as intensity changes of ion-specific fluorochromes like fura-red (Kurebayashi *et al*, 1993), can be achieved by collecting images at very short intervals (milliseconds). However, short collection periods inevitably introduce noise in the time traces, leading to difficulties in determining the starting time of changes in particular cell areas. If specimens are illuminated for prolonged periods, photobleaching becomes a very important concern. Since photobleaching is dependent on cell-specific anti-oxidant properties, it is virtually impossible to compare results obtained from different cells. In this communication we describe calculation methods carried out in experiments with isolated melanotropes of the amphibian *Xenopus laevis*, to determine changes in the intracellular concentrations of calcium using fura-red (Scheenen, 1995). Off line methods are presented to (1) correct for photobleaching and differences in loading characteristics between cells and (2) for removing high frequency noise, thus enabling the acquisition of accurate spatiotemporal data on fluorescence changes in different regions of a cell.

METHODS FOR PHOTOBLEACHING CORRECTION

For experiments conducted over a long period (e.g. 60 min), a method was developed for photobleaching correction. The method is based on the fact that a photobleaching process (as shown in Fig.1A) can be described with a decaying mono-exponential function (Wells *et al*, 1994):

$$\mathbf{F(t) = F(0) \cdot EXP\ (-\mu \cdot t)} \qquad \mathbf{(I)}$$

Fluorescence Microscopy and Fluorescent Probes
Edited by J. Slavík, Plenum Press, New York, 1996

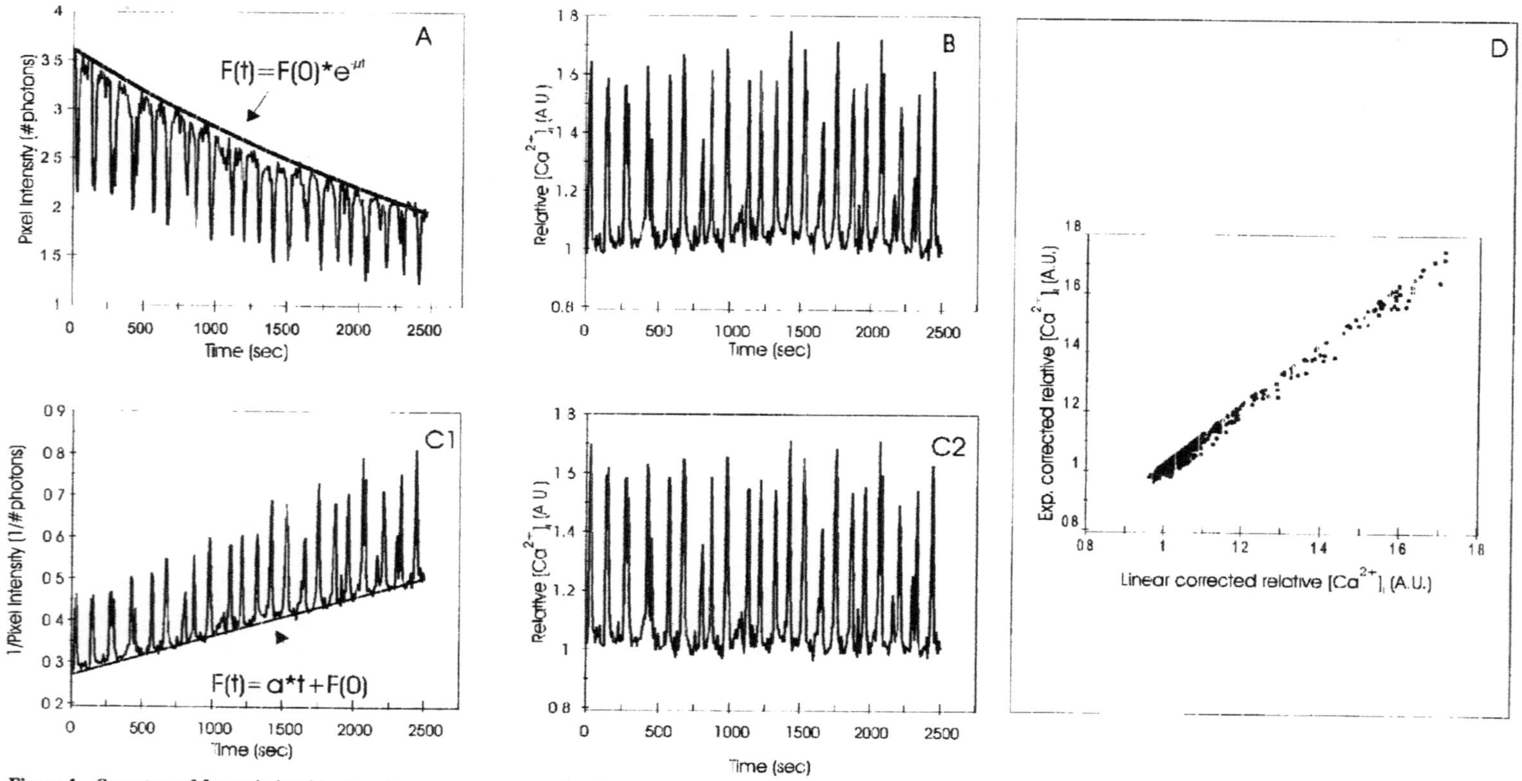

Figure 1. Correction of fura-red photobleaching by using an exponential or linear approximation. **(A)** Measured fluorescent trace showing calcium oscillations and a mono-exponential fit of the photobleaching process The fluorescence is inverse proportional to the intracellular $[Ca^{2+}]$. **(B)** The fitted exponential function divided by the experimental trace of (A) to obtain a corrected and normalized graph **(C1)** After inverting in the original signal (A) an alternative bleaching correction can be carried out by fitting a straight line to the inverted trace. **(C2)** By dividing the trace by the linear photobleaching line, the graph is normalized and corrected for photobleaching **(D)** Comparison of both methods by graphing the fluorescence intensity of the linear corrected trace against the exponential corrected one A good correlation is found (see text for details)

For effective correction of the total photobleaching, an initial value of μ is calculated from the original trace by inserting F(0) (the fluorescence level at $t = t_0$) and F_{end} (the fluorescence level at $t = t_{end}$) together with dt ($= t_{end} - t_0$) in:

$$\mu = - [LN(Fend/F(0)] / dt \qquad \textbf{(II)}$$

By manually adjusting μ and F(0) in equation (I) an optimal fit of the bleaching process can be obtained (Fig.1A). To perform the correction, the trace was divided by the calculated exponential to obtain a corrected and normalized graph (Fig.1B).

A faster method for correcting fura-red photobleaching is to calculate a linear fit to the baseline fluorescence level after first inverting the signal (Fig. 1C1). When the inverted signal is divided by this fitted line, the photobleaching effects are largely removed and the curve is normalized to 1 (Fig. 1C2). A comparison of both methods by fitting a linear equation ($y=a \cdot x+b$) shows that there is good correlation ($a=1.01$, $b=0.0042$ and $R^2=0.992$) between the exponential and linear correction methods (Fig.1D). By performing either an exponential bleaching correction, or the faster linear variant, a comparison can be made between fluorescent traces of different cells.

METHOD FOR THE REMOVAL OF HIGH FREQUENCY NOISE

When higher temporal resolution is needed, the same line is continuously scanned rather than collecting a whole image. In this linescanning mode total cell overview is lost, but the sampling interval can be very short (2 or 6 msec on the Biorad MRC-600 system) resulting in a high temporal resolution. One major drawback of using these short sampling intervals is the appearance of shot noise (Fig. 2A). Although photobleaching is present, it is of less importance when studying fast processes because the accumulated illumination time is short. An effective method for removing this noise from fluorescent traces is the discrete form of the Fast Fourier Transform (dFFT, Lam *et al* 1981). This method is a mathematical operation carried out on the complex plane, and is only effective if the frequencies of noise and signal differ considerably. Since we are studying wave speeds, processes that have a low frequency compared to that of shot noise, this criterion is met (Fig. 2A). The FFT algorithm requires the number of datapoints to be a power of 2. This is the case in our experiments because we collect 64 datapoints per frame. If the number of datapoints is not a power of 2, then a few baseline values should be added to the end of the dataseries. Adding more than 10 values should be avoided, to prevent a too strong distortion of the trace. The resulting data is then transformed from the time domain in which the experiment is carried out (Fig. 2A) to the frequency domain by inserting the measured fluorescent trace for f(x) in:

$$F(u) = 1/n \cdot SUM\ (n-1, x=0)\ \{ f(x) \cdot EXP\ (-j \cdot 2 \cdot u \cdot x/n)\} \qquad \textbf{(III)}$$

where n is the number of datapoints, u = 0,1,2...,n-1 are discrete samples of the continuous transformation at values $0, \Delta u, 2\Delta u, \ldots, (n-1)\Delta u$ and j = sqrt(-1). In our use of dFFT we insert 1/f(x) rather than f(x) in formula III, to remove the inverse proportionality of the fura red signal with the $[Ca^{2+}]_i$. In order to visualize F(u), a power spectrum (Fig.2B) is calculated by using:

$$[F(u)]^2 = [Re^2(F(u)) + Im^2(F(u))] \qquad \textbf{(IV)}$$

In this powerspectrum (shown as $LOG([F(u)]^2$ in Fig. 2B) the relative power (amplitude) of each frequency component present in the signal is plotted as a function of frequency. It is used as an aid in separating the measured signal components from noise components by setting manually all

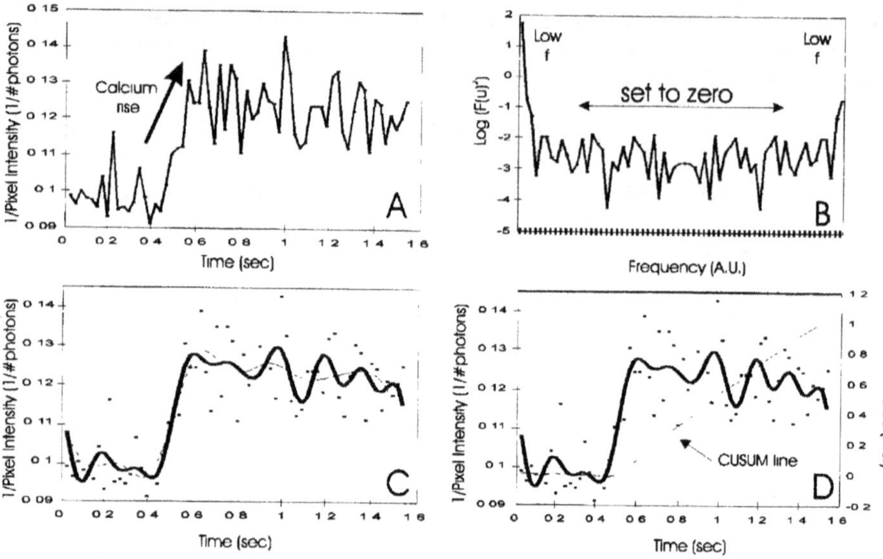

Figure 2. Analysis of fast fluorescence recordings by discrete Fast Fourier Transform (dFFT) filtering and the CUSUM algorithm **(A)** Inverted fluorescence trace of a fast linescan measurement performed in an area of a cell The introduced high frequent shot noise is disturbing the actual low frequent signal which shows a calcium rise. **(B)** The power spectrum represented as the logarithm of $[F(u)]^2$ after applying dFFT (calculated by using equation III) to the data shown in (A) The central part of the figure shows the high frequencies of the shot noise present in (A) These frequencies are set to zero to obtain $F(u)^{\#}$. **(C)** Effect of dFFT filtering by reconstruction of the signal from either 5 frequency points (stippled line) or 10 frequency points (thick line using frequency points from the shaded area in B) superimposed on the original datapoints shown in (A) Using only few points (stippled line) results in an "overfiltered" curve leading to loss of information The thick line is a better reconstruction of the original signal **(D)** Use of the CUSUM test to determine the exact timepoint of fluorescence change. Calculation of δ (see text) from the plotted CUSUM table (stippled line) shows that t=0.528 sec

amplitudes of unwanted high frequencies (in Fig.2B the central part of the curve) to zero. With the low pass filter created in this way excellent results can be obtained, even if the filter is crudely, i.e. not completely accurately, determined (Press *et al*, 1989). In this way $F(u)^{\#}$, which is identical to F(u) except for the frequencies set to zero, is obtained. Then, this dataset is transformed back again to the time domain by inserting $F(u)^{\#}$ into the inverse dFFT formula:

$$f(x) = SUM\,(n-1, u=0)\,\{F(u)^{\#} \cdot EXP\,(j \cdot 2 \cdot u \cdot x/n)\} \qquad (V)$$

This results in a f(x) of which only the real part is plotted because the complex component is very small. For estimating wavespeeds, it is important to keep the right amount of frequencies in $F(u)^{\#}$ to reconstruct the original data. Taking too few frequencies will result in a distortion of the signal (Fig. 2C, stippled line), while taking more points (shaded area in Fig. 2B) results in a better reconstruction, (Fig. 2C and D, thick line). Keeping too much frequencies deminishes the noise reduction. If the conditions mentioned above are met, the datapoints are obtained with strongly reduced noise (Fig. 2C and D, thick line).

CUSUM METHOD FOR ESTIMATING THE EXACT TIMEPOINT OF A FLUORESCENCE CHANGE

For spatiotemporal studies it is essential to know how certain changes in emission intensity correlate to the area of the cell where they are collected. In order to calculate the speed of the process studied, it is important to obtain the time point at which the change in fluorescence occurs for each trace. An objective way to do this is the CUSUM method (Wohl, 1977), based on the following formulas:

$$C_i = X_i - Avg(X) \quad \text{(VI)}$$
$$C_{i+1} = C_i + \{X_{i+1} - Avg(X)\} \quad \text{(VII)}$$

First, the average of 5 to 10 data points before the fluorescence change is calculated (= Avg(X)), then the initial C-value C_i (Eq.VI) is computed from the first measured datapoint X_i. Subsequently for each successive datapoint equation (VII) is used to calculate C-values. The only prerequisite for this method is a relatively stable baseline before the fluorescence change occurs, to enable a reliable calculation of the average signal Avg(X). The time point at which the CUSUM line starts to rise is considered as the starting point for the $[Ca^{2+}]_i$ change. This point can be estimated by calculating $\delta = C_{i+1}-C_i$ for each successive C-value. When the CUSUM line is just randomly fluctuating, the δ value will also fluctuate randomly between positive and negative sign. The δ value will permanently change to negative sign when the CUSUM line starts rising. So the first negative δ of this change correlates to the required timepoint. For example, for the trace shown in figure 2D, the starting point is 0.528 sec. This timepoint then can be used to calculate the wavespeed.

CONCLUSION

We have shown that by using an exponential curve to correct for the bleaching process for each individual cell, or a linear function as an easy alternative, time traces of long-duration experiments become more representative. As a result of this, a comparison between different cells can be made in a more reliable way. Furthermore, the combination of noise reduction by using a dFFT algorithm, and a time-point determination, using a CUSUM test, gives a standarized, easy to use method to obtain temporal information from intensity-time plots obtained through line-scanning experiments .

Acknowledgements

This study was supported by grants from the Netherlands Organization for Scientific Research (NWO), from the European Community (HCM grant ERBCHRXCT920017) and a travel grant from INSERM-NWO.

REFERENCES

Kurebayashi, N., Harkins, A.B., Baylor, S.M. (1993) Use of fura-red as an intracellular calcium indicator in frog skeletal muscle fibers, Biophys. J. **64**, 1934-1960

Lam, R.B., Wieboldt, R.C., Isenhour, T.L. (1981) Practical computation with Fourier transforms for data analysis, Anal. Chem. **53**, 889A-899A

Press, W.H., Teukolsky, S.A., Flannery, B.P., Vetterling, W.T. (1989), in Ch.12 of "Numerical recipes in Pascal: The Art of Scientific Computing", Cambridge University Press, 459-462

Scheenen, W.J.J.M. (1995) Calcium dynamics in melanotrope cells of *Xenopus laevis*, Thesis University of Nijmegen, The Netherlands, Ponsen en Looijen, 87-88

Wells, S., Johnson, I. (1994) Fluorescent labels for confocal microscopy, in Ch.5 of: "Three dimensional confocal microscopy: Volume investigation of biological specimens", Academic Press, 101-129

Wohl, H., (1977) The CUSUM plot: its utility in the analysis of clinical data. N. Eng. J. Med. **296**, 1044-1045

DYNAMICS OF ACTIN MEASURED BY FLUORESCENCE CORRELATION MICROSCOPY (FCM)

B.R. Terry[1] and E.K. Matthews[2]

[1]BioImage:S, Novo Nordisk Research Satellite, Mørkhøj Bygade 28, 3460 Søborg, Denmark
[2]Department of Pharmacology, University of Cambridge, Tennis Court Road, Cambridge, CB2 1QJ

Introduction

This article describes the use of confocal fluorescence correlation microscopy as a way to measure the translocation and dynamics of molecules on a sub-cellular scale. Using this technique it is possible with a single measurement to determine not only the size of labelled molecules in terms of their translational diffusion coefficient (D_T) but also their number density $\langle N \rangle$ within a sample[1,2]. The sample volume is defined precisely by confocal optics, and in our system is a quasi cylindrical space of approximately 1 fL, with a length to width ratio of between 6 to 8. The probe volume is therefore of subcellular dimensions. FCM has a great potential in cell biology, since most cellular events involve changes in the physical size and/or number of individual sub-cellular components e.g., the binding and release of sub-units or effectors, changes in the level of gene expression, the processes of polymerisation and depolymerisation, and changes in microviscosity or compartmentation. In such complex systems time-dependent molecular interactions can be resolved from the correlated intensity fluctuations of individual specifically-labelled, or intrinsically fluorescent, molecules in the probe volume.

We have used FCM to follow changes in $\langle N \rangle$ and D_T for both G-actin and actin polymers in free solution. Actin presents special problems which test the versatility of FCM. The theory of FCM pertains to the motion of point-like particles with simple, and uniform D_T values, through a well-defined excitation profile. Short F-actin filaments are rod-like; when labelled along their length, as for instance here with BODIPY-Phallicidin (Molecular Probes, Inc.), they become extended sources of fluorescence. Furthermore the translational diffusion coefficient of a rod-like molecule has two distinct components, relating to directions of motion orthogonal and parallel to the major axis, and it was not clear how the FCM would weight these for filaments of different length:width ratios. Nevertheless, we have been able to calibrate the FCM instrument successfully for the measurement of rod-like particles and subsequently obtain reliable and reproducible results when measuring changes in $\langle N \rangle$ and D_T of F-actin.

Experimental

The FCM was constructed from a Zeiss UPL inverted microscope (with x1.25 optavar in place), modified for epi-illumination. Attenuated light from an air cooled argon-ion laser (Omnichrome 532-AP), delivered via a single mode optical fibre (Fibercore SM450) was brought to a focus at the specimen plane of a Zeiss 100x (NA 1.3) oil immersion objective. The laser was tunable across the major emission lines, but in most experiments fluorophores were chosen for excitation at 488 nm. Maximum power delivery to the excitation volume, over the area defined by the e^{-2} diameter of the focussed beam waist, was 8 μW. For most measurements this was reduced by a factor of 30 to 100 with neutral density filters prior to launch of the beam

Fluorescence Microscopy and Fluorescent Probes
Edited by J. Slavík, Plenum Press, New York, 1996

into the fibre. The beam geometry was such that the e^{-2} beam radius (ω_0) at the specimen plane (the **r** plane, normal to the optical axis of the microscope) was 205 nm. Emitted light was brought to focus upon a 100 μm diameter aperture positioned in front of a photomultiplier tube (PMT: Thorn/EMI 9863 KB100 operated at 1.5kV) screened by a narrow band interference filter (±5 nm FWHM) appropriate for the fluorophore used. The confocal aperture, together with the beam geometry in the specimen, defined the volume from which signals were collected. This probe volume (γ) was measured using R-phycoerythrin as a calibration standard[3]. The focussed beam waist was positioned 30 μm above the lower glass/sample interface for all measurements, so that 3-D diffusion was symmetric and unimpeded within the probe volume. This spatial arrangement also ensured that signal scattering and absorption were minimised.

Signal analysis

The signal output from the PMT-discriminator unit was analysed by a Malvern K 7032 multi-8 correlator, generally configured as 16 channels in 8 blocks, with a time dilation of 2 between blocks. The minimum usable sample time of the correlator was 100 nsec, but most diffusion coefficients were accurately measured with sample times in the range 5 to 500 μsec. Correlation functions were collected for between 15 seconds to 15 minutes, the duration depending upon the size and brightness of the particles being measured. Averaging of successive functions was sometimes used to improve the signal-to-noise ratio from weakly emitting samples. Data were normalised to the measured baseline fluorescence (determined from the measured far point channels of the correlator).

Correlation functions

Normalised first order autocorrelation function for fluorescence fluctuations can be defined as

$$G_{1,1}(t)=\langle dF(t+\tau).dF(t)\rangle.\langle F(t)\rangle^{-2} \quad (1)$$

where $dF(t)=F(t)-\langle F\rangle$

If the signal is collected only from a quasi-cylindrical volume of the sample, and if the profile of the illumination intensity is Gaussian across the x-y plane (**r**) and Lorentzian along the optical axis (**z**) then the normalised function for a single mobile species approximates to

$$G_{1,1}(t)=\gamma\beta(1+t/\tau_r)^{-1}.(1+t/\tau_z)^{-0.5} \quad (2)$$

wher $\beta=\langle N\rangle^{-1}$. The time constants τ_r and τ_z apply to the apparent translational diffusion coefficient D_T of the labelled species in the **r** and **z** directions. If $z_0 \gg \omega_0$, then equation (2) simplifies to

$$G_{1,1}(t)=\gamma\beta(1+t/\tau_r)^{-1} \text{ where } \quad \tau_r=\omega_0^2(4D_T)^{-1} \quad (3)$$

Given an appropriate microscope configuration, a single species of fluorescent particle therefore generates an intensity correlation function which can be fitted by a single time constant. The time constant yields the translational diffusion coefficient in the plane orthogonal to the optical axis of the microscope, and the amplitude of the normalised function (ie., the G(0) value) will be inversely proportional to <N>, the number of mobile fluorescent particles in the sample volume.

G-actin measurements

G-actin concentration was determined by an indirect assay[4] which actually measures the activity of DNase1. Monomeric G-actin forms a 1:1 stoichiometric molecular complex with DNase1 and thereby restricts the ability of the enzyme to digest its substrate, DNA. The inhibition of DNase 1 activity is therefore dependent upon the concentration of G-actin.

Plasmid DNA was labelled with Yo-Pro 1 (Molecular Probes, Inc.), a benzoxazolium-4-quinolinium dye which becomes strongly fluorescent only when bound to DNA or RNA. The

FCM was then used to measure the rate at which known amounts of DNase1 produced DNA fragments (i.e $d\langle N\rangle/dt$) from labelled DNA in the presence or absence of varying amounts of G-actin.

In the absence of DNase 1, the correlation function of plasmid DNA (39μM base pairs), labelled with one molecule of Yo-Pro 1 per 5.5 base-pairs, was stable, reproducible and well-fitted by a regression with a single decay constant (**Fig 1a)** indicating that the DNA assumes a spherical, bundled, configuration. Addition of DNAse 1 caused a time-dependent decrease in the magnitude of G(0) corresponding to a progressive increase in <N> (see equations 2 & 3), as the labelled DNA is fragmented successively by a given amount of the enzyme over the period 2 to 20 min. To establish the concentration-dependent kinetics of the reaction it was necessary first to determine k from $\langle N_t\rangle = \langle N_0\rangle . e^{kt}$, where k=a[E], a being the activity coefficient of the enzyme and [E] the concentration of DNase 1. The concentration dependence of the change in <N> with time is illustrated in **Fig 1b** for different concentrations of DNase I.

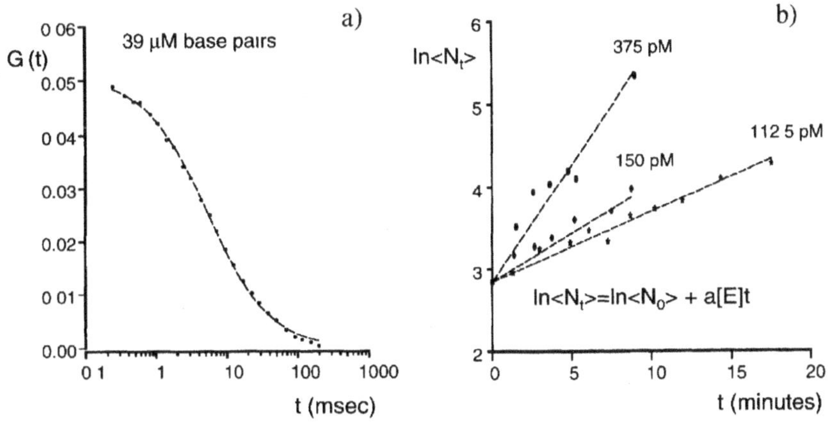

Fig. 1. a) Correlation function of plasmid DNA with a labelling ratio of one Yo-Pro 1 molecule for each 5.5 base pairs. b) Time-dependent increase in <N> (i.e., decrease in G(0)), as fluorophore-labelled DNA is fragmented progressively by DNase I at three different concentrations of enzyme.

From similar experiments in the presence of known amounts of enzyme and G-actin we measured the K_a of DNase1 for G-actin as 1.03 x 10^9 M^{-1} (SEM ± 10%), in good agreement with values obtained by other means. With this method we were able to assay G-actin concentrations down to 100 pM.

F-actin measurements

Values for the measured translational diffusion coefficient (M_T) of F-actin were systematically less than those predicted by Broersma's empirical equations[5,6] for the diffusion coefficients of rigid rods. However, a simple linear relationship was found by calibration which allowed values of D_T for rod-like molecules to be determined for any measured value of M_T, the required correction factor f being expressed in terms of $1/M_T$ (**Fig 2a**). The FCM was calibrated to measure D_T in the range 1 x 10^{-7} to 1.5 x 10^{-9} $cm^2.s^{-1}$ corresponding approximately to filaments from 10 to 500 actin units in length.

The FCM also required calibration for the determination of <N> when filament length exceeded certain limits. Measurement of <N> for short filaments was straightforward, with values of $\gamma/G(0)$ accurately reflecting the expected number density, $\langle N\rangle_{exp}$. As filament length increases beyond 200 nm (equivalent approximately 70 actin units), *i.e.*, of the order of ω_0, <N> tends to be overestimated. Once calibrated, however, the likely overestimate in $\langle N\rangle_m$, the measured number density, can be calculated directly from the value of M_T determined for each correlation function. Fitted data are illustrated in **Fig 2b**. The relationship between M_T and the ratio of

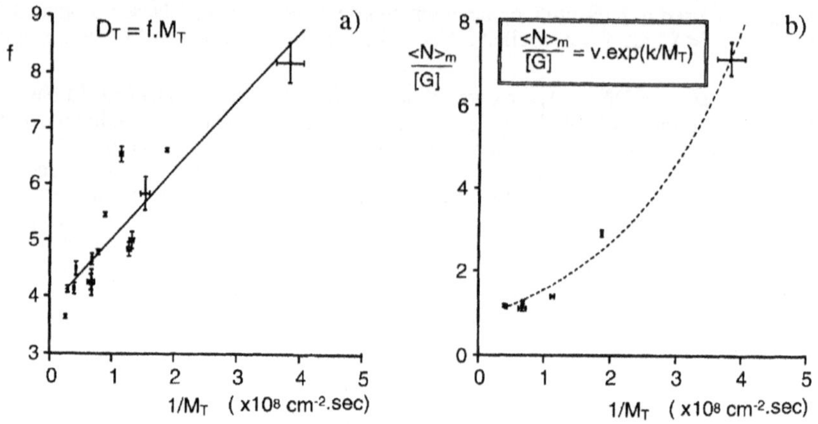

Fig. 2. a) Determination of the translational diffusion coefficient, D_T, from the measured value of M_T and the correction factor 'f'. b) Calibration of $<N>_m$ as an inverse function of M_T.

$<N>_m/[G]$ (where [G] in this instance is the concentration of gelsolin used for nucleation, and hence the maximum possible filament number) provides a further, separate, means of confirming v, the nucleating activity of the gelsolin samples used. Fits to the calibration data gave values of v equal to 0.92, in agreement with that determined by independent methods.

Once calibrated the FCM was used to measure changing mobilities of actin filaments to quantify both the rate and extent of polymerisation upon the pointed ends of pre-formed actin filaments. Filaments 27 actin units in length were pre-polymerised from gelsolin-actin nuclei (GA_2 nuclei) and labelled with BODIPY-phallicidin. These were then incubated with additional G-actin (non-labelled) under polymerising conditions. The reaction volume contained 70 filaments.fL^{-1} and sufficient monomeric G-actin was added to allow the attachment of a further 60 actin units per filament **(Fig 3)**.

In a polymerising system, $(N_{act})_t$, the number of actin units per filament at any time t after initiating polymerisation should be defined by :

$$(N_{act})_t = (N_{act})_0 + ([A]/[F]).[1-\exp(-qt)] \quad (4)$$

where [F] is the concentration of filaments and [A] that of G-actin monomers. N_{act} values were calculated from the Broersma equation following correction of M_T to D_T. The experimental data

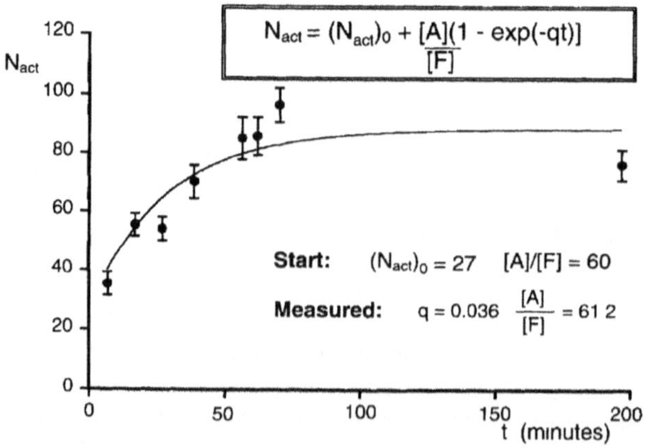

Fig. 3. Polymerisation of G-actin and determination of the rate constant for F-actin formation

displayed a very good fit to equation 4, filaments increasing from an $(N_{act})_0$ of 27 units to a maximum of 88 actin units per filament after 200 min. The time constant of the assembly process was 27.8 minutes at 298°K and polymerisation was almost complete by 60 min.

In summary, we have demonstrated that FCM can be used to make reliable and non-invasive measurements of systems which involve, separately or together, changes in particle number and particle size with time, and that such measurements can be made within microscopic volumes of subcellular dimensions. We have also shown that such measurements can accomodate "difficult" particles whose diffusional behaviour departs from that of ideal spheres. Elsewhere[7] we report on the successful use of the FCM to detect and characterise green fluorescent protein (GFP). The availability of the cDNA for this unique fluorophore[8] will allow *in situ* expression of protein components such as actin, each molecule labelled specifically (and covalently) with an intrinsic fluorescent marker. FCM has the potential to quantify the dynamics of microscopic events which involve changes in $\langle N \rangle$ or D_T. GFP now makes it possible to realise this potential within intact living cells.

This work was supported by the Wellcome Trust.

References

1. D.Magde, E.L. Elson, and W.W. Webb, Fluorescence correlation spectroscopy. II. An experimental realisation, *Biopolymers* 13:29 (1974)
2. R.Rigler, Ü. Mets, J. Widgren, and P. Kask, Fluorescence correlation spectroscopy with high count rate and low background: Analysis of translational diffusion, *Eur. Biophys. J.* 22:169 (1993)
3. B.R. Terry and E.K. Matthews, G-actin concentrations and F-actin dynamics measured within femtolitre volumes by fluorescence correlation microscopy, *Biophys J.*, submitted (1995)
4. Z. Huang, S. Yue, W. You and R.P. Haugland, *Anal. Biochem.*, 214:272 (1993)
5. S. Broersma, Viscous force constant for a closed cylinder, *J. Chem. Phys.*, 32:1632 (1960)
6. S. Broersma, Viscous force and torque constants for a cylinder, *J. Chem. Phys.*, 74:6989 (1981)
7. B.R. Terry, E.K. Matthews, and J. Haseloff, Molecular characterisation of recombinant green fluorescent protein by fluorescence correlation microscopy, *Biophys. Biochem. Res. Comms.* submitted (1995)
8. D.C. Prasher, V.K. Eckenrode, W.W. Ward, F.G. Prendergast, and M.J. Cormier, Primary structure of the *Aequorea victoria* green-fluorescent protein, *Gene* 111:229 (1992)

INDEX

Acceptor, 9
Actin, 296
AFIC, 7
Agarose electrophoresis gel, 181
Aggregates, 160
Aldehyde
  dehydrogenase, 218
  oxidase, 218
Ammonium chloride, 98
Amylase, 253
ANEPPS, 59
Anisotropy, 9, 16, 57
9,10 Anthracene dicarbonitrile, 82
Anthroyl, 59
Antibody, 38, 60
  fibronectin, 257
  HLA-DR, 236
  incubation, 179
  MAC-1, 236
  MnSOD, 236
  RAM-11, 235, 236
  W 6/32, 236
APTA, 58
APTRA, 58
Arterial hypertension, 217
Arteriosclerosis, 235, 247
Artifacts, 119-123
Autofluorescence, 120, 127, 213

Ba^{2+}, 58
Bacteria, 261
Bacterial β-galactosidase, 211
BCECF, 58, 91-93, 96, 97, 107-111, 113-118, 273-276
Bead loading, 60
Beer's law, 1
Bifidobacterium longum, 263
Binding site, 146
Biotin, 186
BKEz-7 endothelial cells, 71
Blood preasure, 134
Bone marrow cells, 186
Bovine
  serum albumin, 104, 113-118, 140, 212
  prothrombin, 227
Brightness, 3

Ca^{2+}, 5, 11, 39, 58, 101-105, 111, 224-228, 275, 293
  oscilation, 85
Calcein, 241-244, 275
Calcium crimson, 5
Calcium green, 5, 39, 58, 87
Calcium phosphate, 279
cAMP, 192, 214
Cancer, 203
Capilary bed, 286
Carbocyanines, 58-60, 157
Carboxyfluorescein, 58
Cardiac rat myocyte, 40
CCD-detector, 4, 7, 11, 29, 69, 91, 96, 198, 206, 263
CCCP, 151
Cd^{2+}, 58
Cell
  73/73, 248
  nuclei, 7, 169, 186, 191, 207, 248, 266
  population, 65
  surface area, 267
  volume, 267-271
  wall, 153, 253
Chenopodium rubrum, 126
Chlorophyll, 38, 49, 76, 127
Chloroquine, 98
Cholesterol, 247
Chromosome, 169, 175, 176, 182
Chromosomal abnormalities, 185
CMFDG, 212-215
Collagen, 72
Commelina communis, 128
Computer, 7, 29, 63, 80, 175, 181, 205, 269
Confocal microscopy, 7, 35-46, 47-56, 62, 79-82, 87, 91, 101-105, 120, 126, 169, 182, 248, 267-271, 273, 279-283, 285-290, 291, 297
Convolution, 48
Coumarin, 180, 205
CPM, 231, 237
C.SNAFL-1, 41, 58
CUSUM method, 295
Cyanines, 204
Cytochrome, 72
Cytofluorogram, 280
Cytoplasm, 145

Cytosol, 113

D12 10x80 cells, 244
DACM, 231
DAPI, 60
Deconvolution, 54
Depolarization, 10, 154
Detector, 4
Dictyostelium discoideum, 97
Digoxigenin, 186
DiS-C_3(3), 139-143, 145, 151-154, 157-162
DiS-C_3(5), 157-162
DMEM, 212, 248
DNA, 169, 181, 185-189, 197-200, 263, 298
    replication, 191, 213
    topology, 200
DNase 1, 298
Donor, 9
DPH, 59
Dual wavelength, 85, 101, 114, 126, 192
Dual labelling, 250
Dye leakage, 274

Egg yolk lecithin, 140
Elastin, 72
Endogenous marker, 64
Energy transfer, 23
Epi-fluorescence microscope, 7, 85, 91, 198, 236
Epi-illumination, 2, 29, 297
Escherichia coli, 81
Esterase, 193
Excimer, 59
Exogenous marker, 64

Fe^{3+}, 58
Fibroblast cell, 212, 279
Filter, 2, 5, 38, 67, 80, 87, 140, 170, 175, 198
    inner effect, 114
FISH, 8, 60, 69, 170, 175, 179, 185-189, 263
FITC, 38, 60, 96, 169, 192, 236, 248, 257, 263, 280
Flavin, 72
FLIM, 9-33, 71-78
Flow cytometry, 61-66, 182, 192, 229, 241
Fluo-3, 5, 11, 58, 87
Fluorescein, 58, 91, 180, 269
Fluorescence
    anisotropy microscopy, 9
    contrast, 203
    decay, 17, 23, 35-46, 72-77, 80-82, 141, 224
    image, 4, 23, 40, 65, 91, 207, 236, 242
        3-D, 50, 122, 135, 263, 267-271, 279
    imaging, 16, 67-70, 107, 170, 206, 242
    intensity, 8, 25, 57, 105, 114, 157, 198
    lifetime, 1, 57, 76, 81, 141, 203
    microscope, 15, 22, 47-56, 71, 126, 182, 206
    microscopy, 1, 47, 61-66, 164, 169, 175, 231
    resonance energy transfer Microscopy, 8
Fluorescent marker, 63
Fluorescent probes, 57-60
Fourier transform, 30, 293

FRAP, 6, 59
Froskolin, 212
Fura-2, 58
Fura red, 58, 87, 101-105, 291

Gain, 121
Gene
    mapping of, 182
    sequences, 182
Geometric distortion, 4
Glioma cells (U 118 MG), 108, 212
Glucose-induced oscillations, 85
Glutathione, 229
GOSMH, 163
Growth medium YEPG, 151, 163
Growth phase, 153
Gymnodinium nagasakience, 38

Halothane, 133
Hematocrit, 136
Hematopoetic disorders, 188
Hepatocyte, 273
Heterodyne detection, 27
Histogram, 280
Hg^{2+}, 58
Hoechst
    33258, 60
    33342, 197, 266
    33842, 60
Homodyne detection, 27
HPV-16 genome, 8
Human melanoma cells, 103
Hybridization, 180, 185
Hypersensitive rat, 217
Hypotonic stress, 275

IBMX, 212
Image
    analysis, 62, 127, 170, 177, 231, 236, 261
    detector (camera), 4, 69, 72, 122, 134, 176, 179, 231, 236
    intensifier, 29, 91, 96, 205
    processing, 120, 135, 175, 182, 261, 280
        computer, 4
        operations, 4
    segmentation techniques, 8, 261-266
    steady-state, 29
Indo-1, 58
Ion concentration, 5, 107
Iron-binding proteins, 95
Isosbestic point, 102, 135

K^{+}, 58, 116, 143
Killer toxin, 163-166
Kidney cells, 273

La^{3+}, 58
Lactoferin, 95, 97
LacZ, 211-215

Laser, 29, 37, 63, 72, 81, 140, 152, 192, 205, 229, 273, 279, 297
Leukemia, 185
Light scatter, 63
Light
  background, 120
    corrections, 198
  out-of-focus, 119
Liposome, 140-143
Lookup tables, 177
Low density lipoprotein, 247-250
Lymphocyte, 169

MA-DFO, 58
Macrophage J774, 42
Macrophages, 235
Magnesium green, 58
Malignant cell, 203
Malpighian tubules, 113
MAPB, 136
Medical diagnostics, 71
Me_2CF, 117
Melanotrope cell, 291
Membrane
  fluidity, 11, 59
  potential, 58, 139, 145-149, 151-154, 157-162
Merocyanines, 58
Messenger RNA, 182
Mg^{2+}, 58
Microfluoreimetri, 85
Microinjection, 60, 212
Mild sonication, 60
Mitochondria, 146
Modulation of the light, 21
MONAL
  -62, 218-222
  -71, 218-222
  -41, 218-222
Monensin, 98
Monobromobimane, 237
Multidrug-resistant cell, 241
Murine mastocytoma cells, 81
Muscle cell, 247

Na^+, 58
NAD^+, 218
NADH, 72
NADPH, 72
NBD-DFO, 58
NERF, 58
Nernst equation, 160
Nigericin, 107
Noise 29, 69, 178, 262, 293, 298
Nuclear protein, 200
Nucleid acid, 179, 279
Nucleotide, 180

O_2, 42
Objective, 3, 170, 206
  immersion, 268, 297
  multi immersion, 38

Offset, 121
Optical distortion, 47-56
Optical fibre, 205, 297
Optical multichannel analyzer, 71, 140, 152, 205
Orthophthaldialdehyde, 237
Osmotic shock, 60
Osmotic effect, 140
Oxonols, 58
Oxysterols, 248

Pancreatic cells, 85
Patman, 223
Pb^{2+}, 58
PCR, 8
PETG, 212
P-glycoprotein, 241-244
pH, 39, 41, 58, 91-93, 95-99, 107-111, 113-118, 125-130, 134-138
  apoplastic, 128
Phallacidin, 60
Phalloidin, 60
Phase
  fluorimetry, 36
  contrast, 261-266
Phosphatidylcholine, 223
Photobleaching, 6, 59, 274, 291
  correction, 293
Photometric errors, 49, 50
Photon detector (photomultiplier), 4, 37, 80, 122, 298
Photosynthesis, 76
Picture element (pixel), 4, 23, 30, 107, 111, 135, 178, 261, 280
pK, 58, 108, 114, 135
Placenta, 285
Plasma membrane, 9, 247
  lipid, 10, 11
  protein, 11
Plasmid, 212
Pluronic, 103
Point spread function, 50, 170
Polarization, 10, 57, 59
Polarized light, 9
Porphyrins, 72
Primulin, 127
PRINS, 179
Prodan, 223
Propidium iodide, 169, 197, 248
Prothrombin, 223-228
Protoplast, 152-154
Pyranine, 58

Quantum yield, 1, 16, 18, 39, 57, 203, 229
Quenching, 42, 157, 275
Quin2, 58

Rabbit, 133, 236
Rat organs, 220
Ratio imaging, 38, 58, 69, 96, 97, 107, 119-123, 126, 135
Ratio of intensities, 36, 114, 143, 146, 198

Reconstruction, 69
  3-D, 285-290
Redistribution of dye, 151
Refractive index, 269
Reh-6 cell, 197
Relaxation time, 224
Respiratory
  activity, 152
  function, 74
RET, 8
Rhodamine, 58-60, 82, 140, 163, 180, 229
Ringer solution, 113
RNA, 185
R-Phycoerythrin, 241
RTDP, 42
Rye cell, 253

Saccharomyces, 74
  cerevisiae, 91, 151, 163, 191
  diastaticus, 165
  kluyveri, 165
  oviformis, 165
Scrape loading, 60
Scratch loading, 60
Second messenger, 5
Shading, 4, 7
Single cell, 65, 241, 275
Single photon counting, 140, 224
SNARF, 58, 91-93, 107-111, 126-129
Spatial resolution, 1, 169
Spatiqal distortions, 48, 49
Spectral shift, 115-118, 146, 152, 205
Stern-Volmer equation, 42
Stokes shift, 223-228
Styryl dyes, 59

Temperature, 103
Thiol, 229
Thresholding, 261-266
Time-gating, 36, 79-82
Time-resolved microscope, 79-82
TIRF, 7
TMA-DPH, 59
TMB-8, 275
TOTO-1, 207
Toxicity of the probes, 232
Transferrin, 95
Translational diffusion, 299
TRITC (tetramethylrhodamine isothiocyanate), 60
Tritrichomonas foetus, 95
Tritrichomonas vaginalis, 98
Tryptophan, 72
TSQ, 58
Tumor cell, 205, 241
TUNEL technique, 181
Two-photon technique, 48-56, 88
Tyrosine, 72

Umbelliferone, 135

V79 cell, 232
Valinomycin, 59, 139
Videomicroscopy, 4, 63

Xenopus laevis, 291
X-gal cytochemistry, 211

Yeast cell, 74, 91, 151, 163, 191
YOPRO-1, 197, 298-301

Zn^{2+}, 58

GPSR Compliance

The European Union's (EU) General Product Safety Regulation (GPSR) is a set of rules that requires consumer products to be safe and our obligations to ensure this.

If you have any concerns about our products, you can contact us on ProductSafety@springernature.com

In case Publisher is established outside the EU, the EU authorized representative is:

Springer Nature Customer Service Center GmbH
Europaplatz 3
69115 Heidelberg, Germany

Batch number: 10761599

Printed by Printforce, the Netherlands